VLSI Design Handbook

Volume I

VLSI Design Handbook
Volume I

Edited by **Martin Limestone**

CLANRYE
INTERNATIONAL

New Jersey

Published by Clanrye International,
55 Van Reypen Street,
Jersey City, NJ 07306, USA
www.clanryeinternational.com

VLSI Design Handbook: Volume I
Edited by Martin Limestone

International Standard Book Number: 978-1-63240-519-7 (Hardback)

Printed in the United States of America.

Contents

Preface

The abbreviation VLSI stands for Very Large Scale Integration. Integrated circuit technology allows billions of transistors to be fabricated into a single chip. The development of this technology only occurred in the twentieth century, somewhere in the mid-1920s, when numerous people tried to create devices which intended to convert solid-state diodes into triodes by controlling current. However, it was only in 1947, with the creation of transistors at Bell Labs that vacuum tubes were replaced by solid-state devices.

Factually the Moore's Law was always validated for prediction of exponential complexity growth and advancement in the performance of integrated circuits. Most semiconductor based industries, face extreme problems in maintaining all aspects of production process during designing of the chip. These issues range from scientific research in discovering novel materials and devices to advanced technology developments and finding new killer applications. This book has been compiled in order to emphasize the latest developments in the vast field of VLSI design.

The contributors have made no attempt to be comprehensive on the topics. Instead, they tried to provide some promising concepts, such as problems and challenges for the introduction of new–generation electronic design automation tools, optimization, modeling and simulation methodologies, thermal and power reduction and management, parasitic interconnects, etc.

I would like to thank all the authors for their excellent contributions in different applications of VLSI. Despite the rapid advances in the field, I believe that the examples provided here will allow us to look through some main researches. I hope that this book will prove to be a worthy contribution in the field of VLSI. I also wish to thank the publisher and the publishing team for their outstanding support at every level of the editing process. Lastly, I wish to convey my regards to my friends and family for supporting me in every endeavor of my life.

Editor

Design of Low Power Multiplier with Energy Efficient Full Adder Using DPTAAL

A. Kishore Kumar,[1] **D. Somasundareswari,**[2] **V. Duraisamy,**[3] **and T. Shunbaga Pradeepa**[4]

[1] ECE Department, Hindusthan College of Engineering & Technology, Coimbatore 641 032, India
[2] Department of Electrical Sciences, Adithya Institute of Technology, Coimbatore 641 107, India
[3] Maharaja Institute of Technology, Coimbatore 641 407, India
[4] ECE Department, Coimbatore Institute of Technology, Coimbatore 641 014, India

Correspondence should be addressed to A. Kishore Kumar; kishore_hindusthan@yahoo.in

Academic Editor: Wieslaw Kuzmicz

Asynchronous adiabatic logic (AAL) is a novel lowpower design technique which combines the energy saving benefits of asynchronous systems with adiabatic benefits. In this paper, energy efficient full adder using double pass transistor with asynchronous adiabatic logic (DPTAAL) is used to design a low power multiplier. Asynchronous adiabatic circuits are very low power circuits to preserve energy for reuse, which reduces the amount of energy drawn directly from the power supply. In this work, an 8×8 multiplier using DPTAAL is designed and simulated, which exhibits low power and reliable logical operations. To improve the circuit performance at reduced voltage level, double pass transistor logic (DPL) is introduced. The power results of the proposed multiplier design are compared with the conventional CMOS implementation. Simulation results show significant improvement in power for clock rates ranging from 100 MHz to 300 MHz.

1. Introduction

Over the past few decades, low power design solution has steadily geared up the list of researcher's design concerns for low power and low noise digital circuits to introduce new methods to the design of low power VLSI circuits. Moore's law describes the requirement of the transistors for VLSI design which gives the experimental observation of component density and performance of integrated circuits, which doubles every two years. Transistor count is a primary concern which largely affects the design complexity of many function units such as multiplier and arithmetic logic unit (ALU). The significance of the digital computing lies in the multiplier design. The multipliers play a significant role in arithmetic operations in DSP applications. Recent developments in processor designs also focus on low power multiplier architecture usage in their circuits. Two significant yet often conflicting design criteria are power consumption and speed. Taking into consideration these constraints, the design of low power multiplier is of great interest. As reported

in [1], to get the best power and area requirements of the computational complexities in the VLSI circuits, the length and width of transistors are shrunk into the deep submicron region, handled by process engineering.

In recent years, the literatures have identified several types and designs of adiabatic circuits. For instance, 2N2N2P, PFAL, pass transistor adiabatic logic, clocked adiabatic lLogic, improved pass-gate adiabatic logic, and adiabatic differential switch Logic were designed and achieved considerable energy savings, compared with conventional CMOS design [3–9]. In [10], complementary pass transistor adiabatic logic circuit was discussed, in which the nonadiabatic energy loss of output loads has been completely eliminated. In [11], adiabatic CPL circuits using two-phase power clocks were presented. In [12], energy saving design technique achieved by latched pass transistor with adiabatic logic was presented. Many research methods in the adiabatic logic have been attempted to reduce the power dissipation of VLSI circuits, reported in [9–16]. Many research efforts in the multiplier design have been introduced to obtain energy efficiency in VLSI circuits. In [17], a 1.5 ns 32-b CMOS ALU in double pass-transistor logic was proposed to improve the circuit performance at reduced supply voltage ranges. In [18], a low power multiplier using 4-2 compressor based on adiabatic CPL circuit is described. By the scaling rules set by Dennard, smart optimization can be achieved by means of timely introduction of new processing techniques in device structures and materials [19]. In [1], low power multiplier design using complementary pass transistor asynchronous adiabatic logic is investigated, which exhibits low power and reliable logical operations.

In this paper, design of low power multiplier with energy efficient full adder using double pass transistor asynchronous adiabatic logic (DPTAAL) is proposed and discussed in further sections.

2. Adiabatic Logic Design

"Adiabatic" is a term of Greek origin which spent most of its history related to classical thermodynamics. It refers to a system in which a transition occurs without energy (usually in the form of heat) being either lost to or gained from the system. In the context of use of electronic systems, electronic charge is preserved rather than heat. Adiabatic logic is viewed on issues related to the thermodynamics of computation. By considering this branch of physics that usually looks at mechanical engines and applying it to computing engines, research areas such as reversible computation as well as adiabatic logic have been developed. By moving to a computing paradigm that is reversible, energy can be reprocessed from a computing engine and reused to perform further calculations.

This style of logical approach differs from CMOS circuits, which dissipate energy during switching. To reduce the dynamic power, there are some conventional approaches such as reducing supply voltage, decreasing physical capacitance, and reducing switching activity. These approaches are not conforming enough to meet today's power requirement. On the other hand, most research has focused on building adiabatic logic, which is a hopeful design for low power applications. Adiabatic technique works with the concept of switching activities which reduces the power by giving stored energy back to the supply. Thus, the term adiabatic logic is applied in low power VLSI circuits which execute reversible logic. In the adiabatic techniques, the main design changes are focused on power clock which plays the essential role in the principle of operation. The following major design rules for the adiabatic circuit design are achieved in each phase of the power clock.

(1) Never turn on a transistor if voltage exists across it ($V_{DS} > 0$).

(2) Never turn off a transistor if current exists across it ($I_{DS} \neq 0$).

(3) Never pass current through a diode.

In all the four phases of power clock, if these conditions are satisfied, recovery phase will restore the energy to the power clock, resulting in considerable energy saving. Even some complexities in adiabatic logic design perpetuate. Two such complexities are circuit implementation for time-varying power sources that needs to be done and computational implementation by low overhead circuit structures that needs to be followed [1].

3. Asynchronous Adiabatic Logic (AAL)

Asynchronous adiabatic logic is a unique design technique which combines the energy saving benefits of asynchronous logic and adiabatic logic. Like adiabatic circuits, asynchronous circuits are also a promising technology to focus on low power, highly modular digital circuits. One of the properties of asynchronous systems which make them useful in these applications is that circuits include a built-in insensitivity to variations in power supply voltage, with a lower voltage resulting in slower operation rather than the functional failures that would be seen if traditional synchronous systems were used. Another benefit is the fact that when an asynchronous system is idle, it will not utilize clock signals, whereas in synchronous systems, these clock signals are propagated throughout the entire system and convert energy to heat, often without performing any useful computations.

In contrast to the synchronous circuits, asynchronous circuits perform handshaking between their components to perform all necessary synchronization, communication, and sequencing of operations. Asynchronous circuits fall into different classes, each offering different advantages. The main privilege of this circuit is its low power consumption, stemming from its elimination of clock drivers and the fact that no transistor ever transitions unless it is performing a useful computation.

4. Proposed Design

The main objective of this paper is to design low power multiplier with energy efficient full adder cell using double pass transistor with asynchronous adiabatic logic. The logic scheme for full adder cell is illustrated in Figure 1. In this, entire system consists of two main blocks, such as logical block and control and regeneration (C&R) block.

As in Figure 1, data output signal of any logical block is not only going into next logical block as data input, but at the same time, it is used to generate a control signal for the next logical block using C&R block 1 as reported in [15]. This technique helps to save the required power clock generator with less power.

4.1. Power Clock. In adiabatic circuits, the supply voltage behaves as the clock of the circuit by providing the power, to the circuit and for this reason, it is called power clock. One of the main concerns in the adiabatic logic circuits is the power clock generation. In these circuits, the supply voltage is desired to be a ramping voltage. In the conventional synchronous adiabatic circuits, rather driving each adiabatic logic unit with an externally supplied clock phase, each block is controlled and powered by control signal generated by the

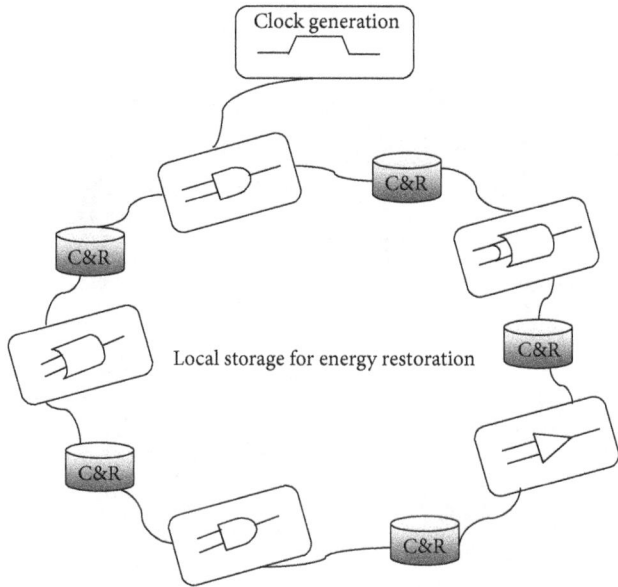

FIGURE 1: Logic scheme for fulladder cell.

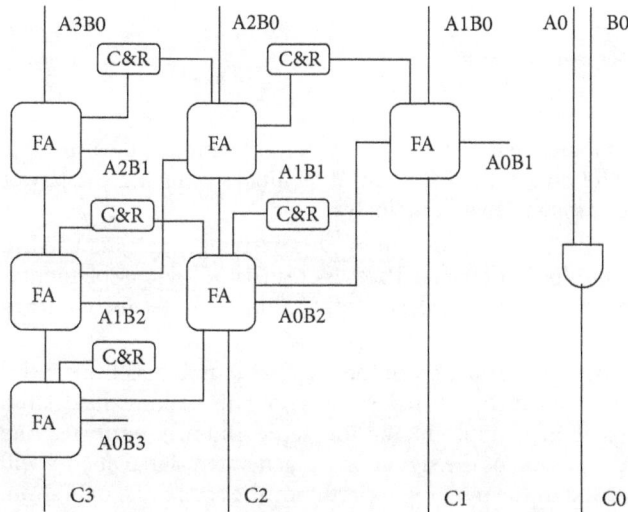

FIGURE 2: Proposed multiplier design scheme.

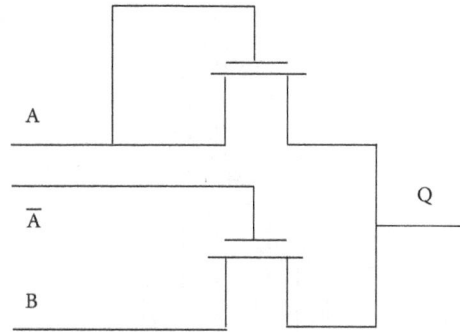

FIGURE 3: Control and regeneration (C&R) block.

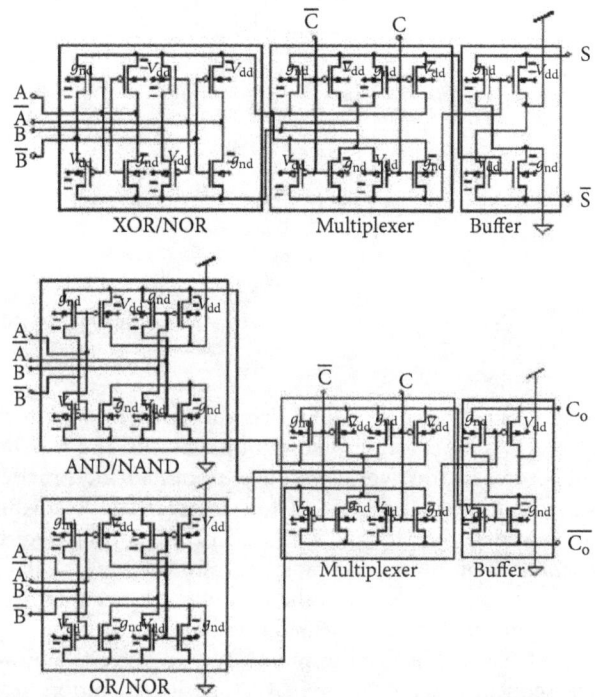

FIGURE 4: DPL full adder cell [2].

C&R block with the help of the logical output of the previous stage. In the design of VLSI circuits, power clock design is a major issue, because the whole transistor logic system shares the power clock. The power clock switching circuit will also dissipate the most power in the logic. Nowadays multiple phase clocks and clock pipelining are the most followed techniques to reduce power dissipation in the power clocks.

The synchronous clock system utilizes the clock source globally; that is, single clock is shared and restored by the large number of logical gates in parallel. Here switching loss of the power clock generator is more as in the CMOS circuit operation.

The simple construction of the pass transistor logic makes it easy to adjust the sizing of transistors to get the desired charging and discharging time; hence the slope of the output

control signal minimizes the power. The clock energy in the asynchronous clock system is locally stored in the C&R block, and it has been used for later gates; the loss of energy of each operation will be taken from its clock source. The local regeneration stores the intermediate energy. This energy is provided to the required operations for the next level of logic. However, the initial requirement of power from the clock generator remains the same; after powering up the logical sequence, power taken from the power clock is reduced drastically.

The proposed multiplier design scheme is illustrated in Figure 2. In this, data out signal of any full adder is not only going into next full adder as data input. But at the same time, it is used to generate a control signal for the next full adder using C&R block 1. This technique helps to save the required power clock generator with less power [15].

This approach gives the feasibility of using the adiabatic logic in real-time implementations. Also to reduce the initial

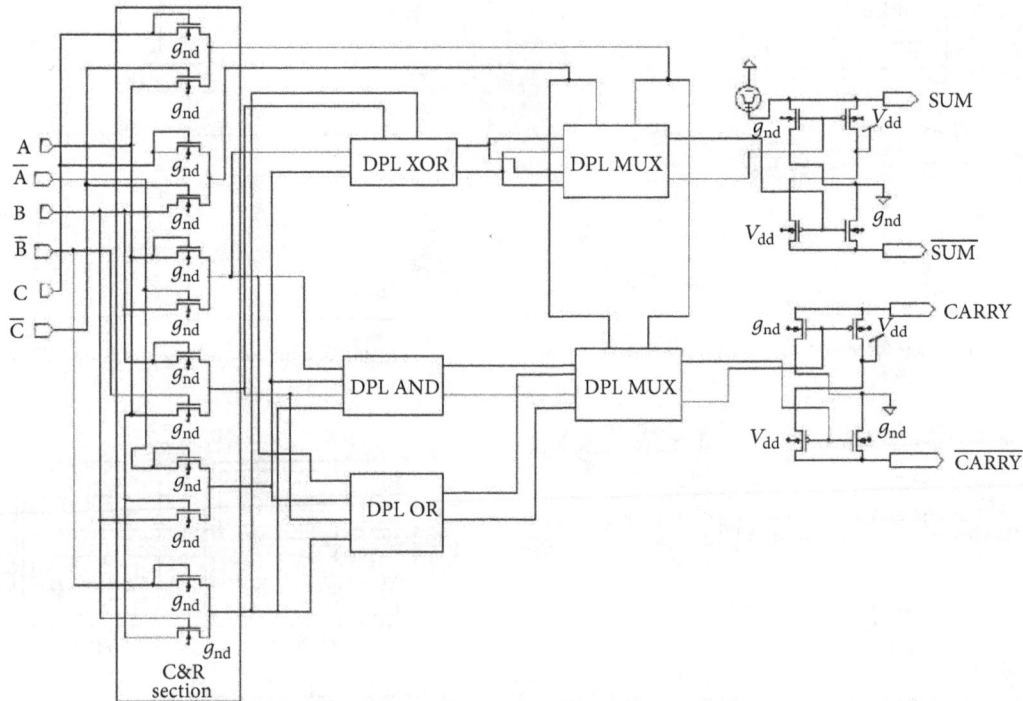

FIGURE 5: DPTAAL full adder logic diagram.

power dissipation, we can utilize the conventional techniques for compensation, like multiple clocks and pipeline architecture. In this work, we have examined the practical approach of adiabatic logic in full adiabaticity. According to the Landauer's principle method to charge/discharge the capacitances of input nodes adiabatically, the input voltages must be reconstructed from the outputs. It is accomplished by using the control and regeneration block.

Control block is used to follow and preserve the power clock sequences with the input vectors. Regeneration gives power saving strategy. All logic gates or logic sequences are connected through C&R structure. The throughput of the logical systems is reduced by the intermediate C&R blocks due to the asynchronous mode of operation. The speed of operations can be compensated for the higher input frequency due to the improvement of speed grade of proposed asynchronous adiabatic logic, as discussed in [1].

5. Control and Regeneration (C&R) Block

The control and regeneration (C&R) block is given in Figure 3. C&R block generates the control signal for the next logical stage with the help of the previous stage output signal. The regeneration technique makes the control signal strong enough to drive the next logical block. In the proposed design, asynchronous operation has been achieved by the control and regeneration part. This C&R controls and regenerates the energy, required for the next operation to the next logical block. The system energy will be circulating among the logical circuits and the minimum power is required from the power clock generator for the operation. Generally, the regenerated signal is stored and circulated between the C&R

and logical part. Thus, there will not be much power reverse to the power clock system. It facilitates reducing the power clock system switching losses.

The proposed design of DPTAAL logic gates is used to design logical blocks. The pass transistor logic implementation is used for the design of C&R block in terms of energy efficiency and functionality.

The NOR portion of the OR gate is acting as the control part whereas the NOT portion is not only making the desired logical inversion. But at the same time, it performs the regeneration of the signal. The regenerated signal energy will be used in the next logic circuit for the sequential operation. The NOT portion will again regenerate the signal whereas the operation gets completed. The construction of the C&R promotes the local storage of the energy and switching circuit for the recovery. The power reduction is not achieved in C&R block. However 60% to 70% of power saving and 1/3 of the speed improvement are achieved, compared to the adiabatic logic with the power clock generator, as discussed in [15].

6. Double Pass Transistor (DPL)

Double pass transistor (DPL) is a modified version of complementary pass transistor logic (CPL) that meets the requirement of reduced supply voltage designs. In DPL circuits full swing operation is achieved by simply adding PMOS transistors in parallel with the NMOS transistors. Thus, the problems of noise margin and speed degradation at reduced supply voltages associated in CPL circuits are avoided. The circuit diagram of the DPL full adder cell is given in Figure 4.

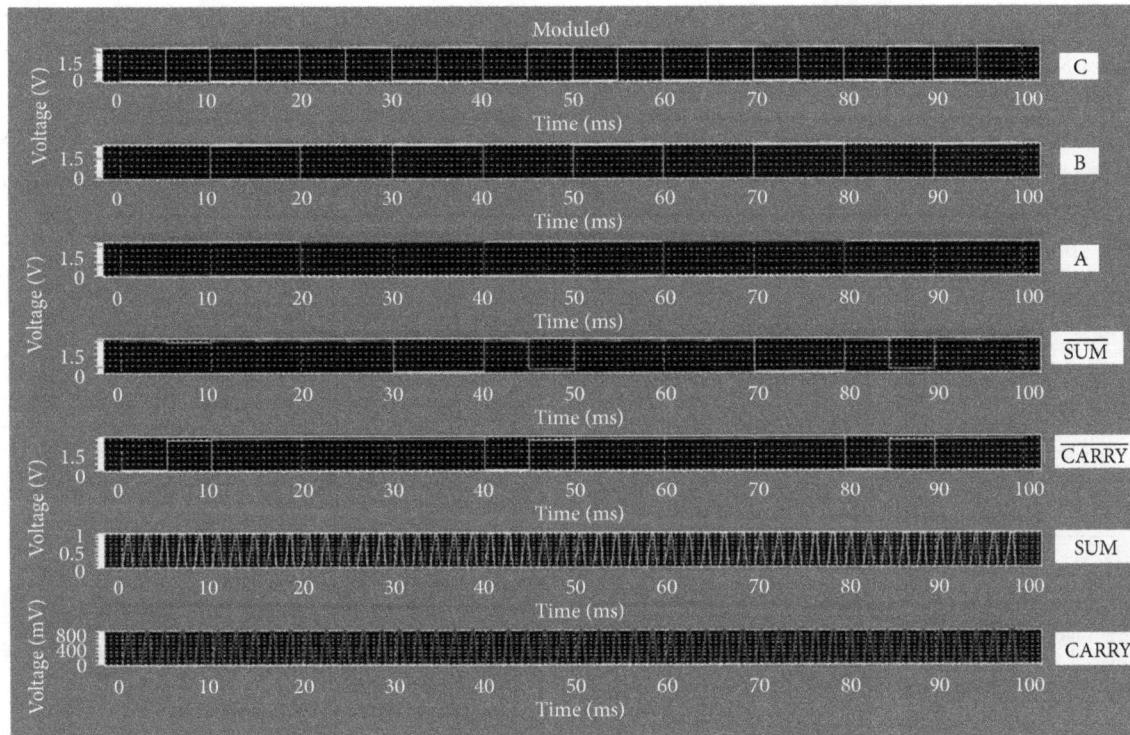

FIGURE 6: Simulation results of DPTAAL full adder cell.

TABLE 1: Transistors sizes used in each design block of the DPTAAL full adder.

Design blocks	PMOS		NMOS	
	Minimum length (μm)	Width (μm)	Minimum length (μm)	Width (μm)
Adiabatic DPL gates, MUX, and buffer	0.18	5.0	0.18	5.0
C&R section	0.18	5.0	0.18	2.0

In this, sum output consists of XOR/XNOR gates, a multiplexer, and a CMOS output buffer. The carry output consists of AND/NAND gates, OR/NOR gates, a multiplexer, and a CMOS output buffer. These DPL gates consist of both NMOS and PMOS pass transistors, in contrast to CPL gates, where only NMOS pass transistors are used. The outputs S bar and C_o bar are acting as the current paths, where inputs A, B, and C are all low. These current paths include two pass transistors, and there are two current paths for each output. In the double pass transistor logic gates, the inputs to the gates of the PMOS transistors are changed from A to B. This arrangement compensates for the speed degradation of CMOS pass transistors in two ways. First, it is a symmetrical arrangement whereby any input is connected to the gate of one MOSFET and the source of another. In the case of XOR/XNOR, it is perfectly symmetrical. Any of the inputs A, A bar, B, and B bar is connected to the gates of NMOS and to the sources of the NMOS and PMOS. This results in balanced input capacitance and reduces the dependence of the delay time on data. Secondly, it has double transmission characteristics.

In the DPL gate, both A and B are passed when A&B are low. In both the CPL and CMOS implementations, the gate input A or A bar controls the pass transistors. When A is low, B is passed, and B is passed when A is high. In the DPL gate, on the other hand, there are two types of pass transistors: one is controlled by A and the other by B. The A controlled pass transistors operate in the same way as CPL and CMOS. For the B controlled pass transistors, when B is low, A is passed, and A bar is passed when B is high. As a result, there are always two current paths driving the buffer stage [17].

7. Double Pass Transistor with Asynchronous Adiabatic Logic (DPTAAL)

In the DPL design, the widths of the NMOS and PMOS pass transistors are one-third and two-thirds, respectively, of the NMOS pass transistor in the CPL gate, so the input capacitance and the gate area are nearly the same for all these architectures. The resistance including that of the CMOS buffer of the previous stage is smallest for the DPL gate due to its double-transmission property. In multiplier circuits, the DPL full adder is as fast as CPL, 18% faster than the conventional pass transistor logic, and 37% faster than CMOS, reported in [17].

TABLE 2: Qualitative comparison of logic designs [2].

Logic designs	No. of MOS logic networks	Output driving capability	Input/output decoupling	Signal rails	Robustness
CMOS	n + p	Medium-good	Yes	Single	High
CPL	2n	Good	Yes	Dual	Medium
DPL	2n + 2p	Good	Yes	Dual	High

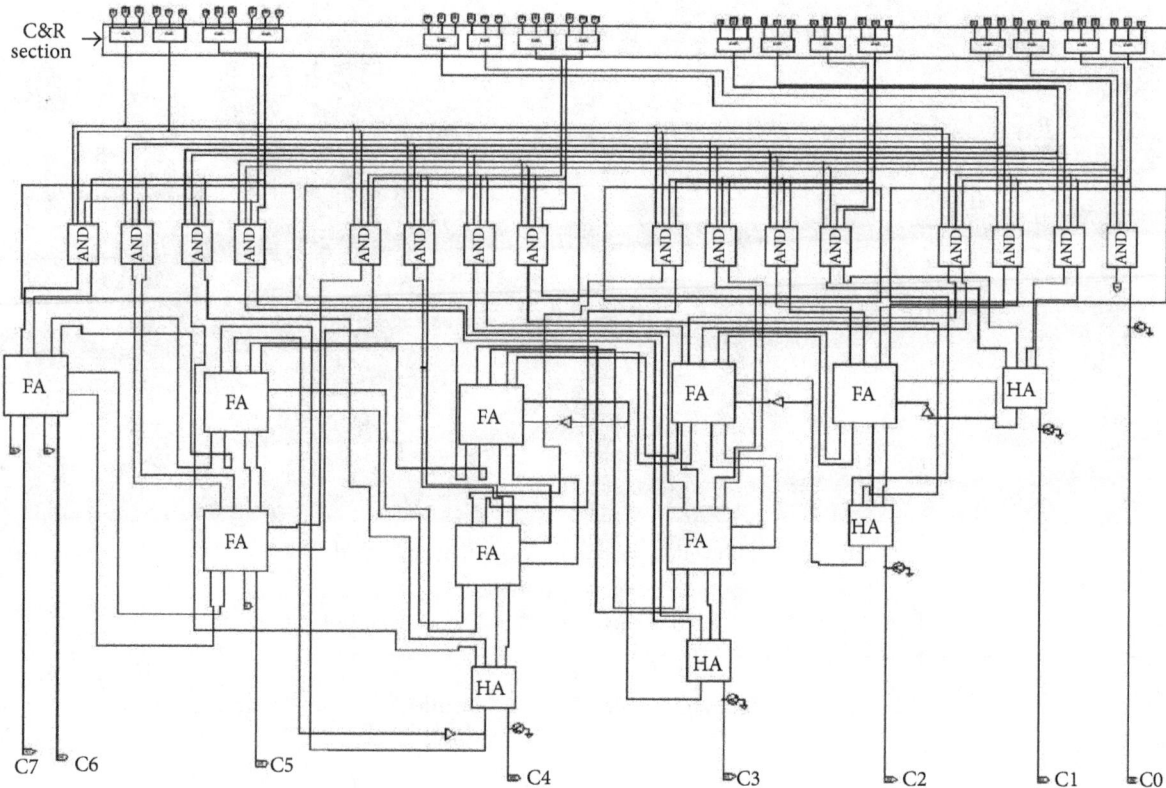

FIGURE 7: Schematic of DPTAAL multiplier.

In the proposed design, double pass transistor technique is combined with asynchronous adiabatic logic (AAL) design technique to obtain the significant power benefits in the digital circuits. Asynchronous adiabatic full adder uses double pass transistor logical block with C&R structures. A simple implementation of this system is depicted. It is a full adder cell, with the logical part designed using DPTAAL, whereas the control part of the C&R block and regeneration part is made of pass transistor logic. This pass transistor logic is functioning as transmission gate in the output logic of each gate structure. The DPTAAL design of full adder cell is presented in Figure 5, which consists of C&R section, adiabatic DPL full adder circuit, multiplexer section, and an output buffer. In this DPTAAL full adder, sum circuit section includes DPL XOR gate, a DPL multiplexer, C&R section, and an output buffer. The carry output section consists of DPL AND gate, DPL OR gate, a DPL multiplexer, C&R section, and an output buffer. These adiabatic DPL gates consist of both NMOS and PMOS pass transistors to achieve full swing operation. When the inputs A, B, and C are all low, the outputs SUM bar and CARRY bar will be acting as the current paths.

Thus, two current paths for each output can be achieved. In this DPTAAL design, power and clock lines are mixed into a single power clock line which has both functions of powering and timing the circuit. C&R section is the main concept of this DPTAAL design, which generates the control signal for the next logical gate using the output signal of the previous gate. The regeneration technique makes the control signal strong enough to drive the next logical gate. Thus, power consumption from the power clock is reduced drastically. A multiplexer chooses the output to be one of several inputs based on a select signal. In this full adder design, multiplexer is used to select the required outputs of DPL-XOR, DPL-AND, and DPL-OR, based on the inputs C and C bar.

All these full adder blocks have been designed with PMOS/NMOS transistors, focusing on low power consumption and high efficient operation. The dimensions of all gate lengths (L) of these transistors have been taken as 0.18 μm. The width (W_p & W_n) of PMOS/NMOS transistors has been taken as 5.0 μm. For C&R section, the width (W_n) of NMOS transistors has been taken as 2.0 μm and the width (W_p) of PMOS transistors has been taken as 5.0 μm, with gate length

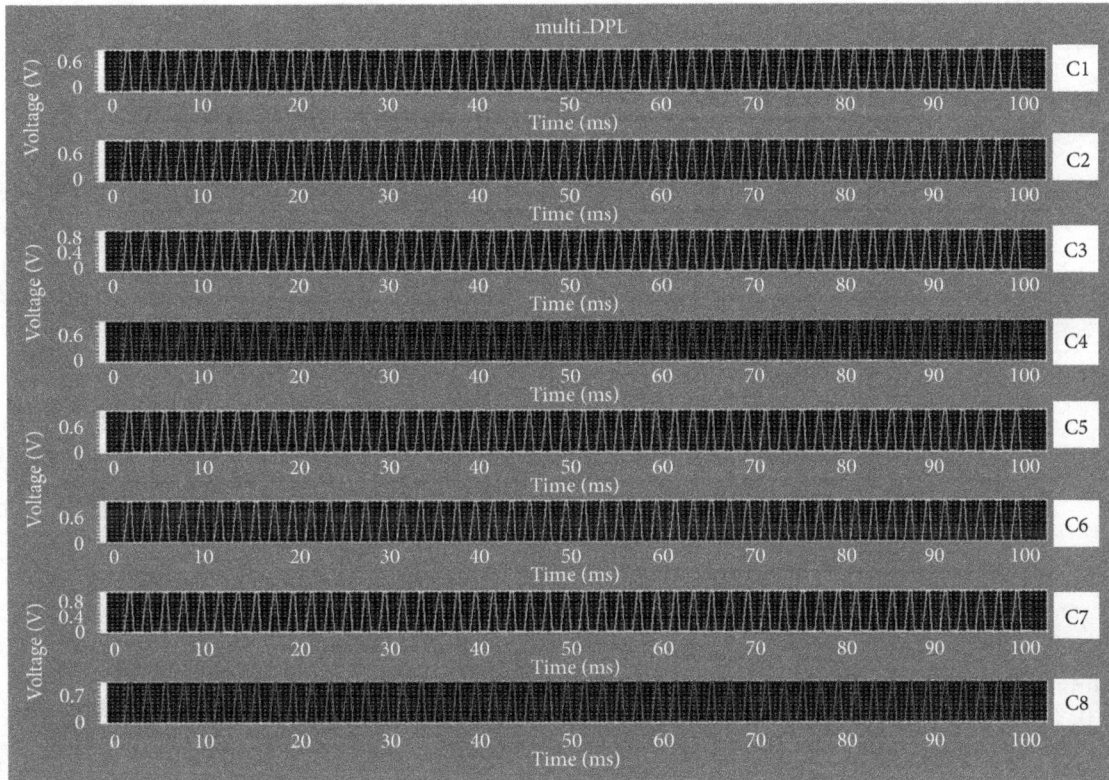

FIGURE 8: Simulation results of DPTAAL multiplier.

of 0.18 μm. Table 1 illustrates the final sizes of the transistors used in each design block of the DPTAAL full adder.

8. Simulation Results and Performance Analysis

The qualitative comparison of three logic designs CMOS, complementary pass transistor logic (CPL), and DPL is given in Table 2, which influences circuit performance and energy consumption. In particular, the number of MOS logic networks, the output driving capabilities, the presence of input/output decoupling, the number of signal rails, and the robustness with respect to voltage scaling are given for the logic styles discussed [2]. In DPL, the robustness with respect to voltage scaling is high, which improves circuit performance at reduced supply ranges. Its symmetrical arrangement and double transmission characteristics compensate for the speed degradation arising from the use of PMOS and NMOS pass transistors.

Energy efficient full adder cell design using asynchronous adiabatic logic with double pass transistor has been implemented. The simulation results of the DPTAAL full adder are presented in Figure 6, for the various combinations of the inputs.

The power clock sequence for this full adder structure is shown in these simulation results and it is based on the conventional structure. These simulation results are obtained for a periodic sequence as represented in the figure, propagated through the buffer chain.

TABLE 3: Transistor count comparison of full adders.

Logic style	No. of transistors
Conventional CMOS	28
Double pass transistor logic (DPL)	48
DPTAAL full adder design	65*

*DPTAAL is 35% larger than DPL.

4-bit and 8-bit multipliers with energy efficient full adder using DPTAAL have been implemented and compared with conventional CMOS logic. The proposed design of DPTAAL multiplier is presented in Figure 7 and its simulation results are presented in Figure 8.

The transistor count comparison of CMOS, DPL, and DPTAAL full adder cells is given in Table 3. As presented in Table 3, the transistor count of the DPTAAL full adder is increased as compared with existing designs; hence a large on-chip area overhead is associated with AAL design. Taking into consideration the gain in energy efficiency, the area overhead is acceptable. The area of AAL can be reduced by using more area efficient logical blocks, as discussed in [15].

The energy performance of the DPTAAL full adder is compared with the conventional CMOS full adder and given in Table 4. The obtained energy of these full adders is specified in fJ (femto Joules), for operating frequencies from 1 MHz to 300 MHz. The symmetrical arrangement and double transmission characteristics of the adiabatic DPL gates improve the circuit performance in the DPTAAL full

TABLE 4: Energy consumption of DPTAAL full adder.

	Logic designs	Frequency				
		1 MHZ	10 MHZ	100 MHZ	200 MHZ	300 MHZ
Energy (fJ)	CMOS	75	75	75	75	75
	DPTAAL	5	5.4	7.5	12	17.2

TABLE 5: Power comparison of conventional CMOS versus DPTAAL multiplier.

No. of bits	Frequency				
	1 MHZ	10 MHZ	100 MHZ	200 MHZ	300 MHZ
Conventional CMOS (μW)					
4 bit	0.24	0.49	2.74	3.34	5.01
8 bit	0.35	0.67	3.44	6.81	11.84
DPTAAL (μW)					
4 bit	0.19	0.39	1.75	2.42	3.21
8 bit	0.25	0.51	2.42	5.02	8.09

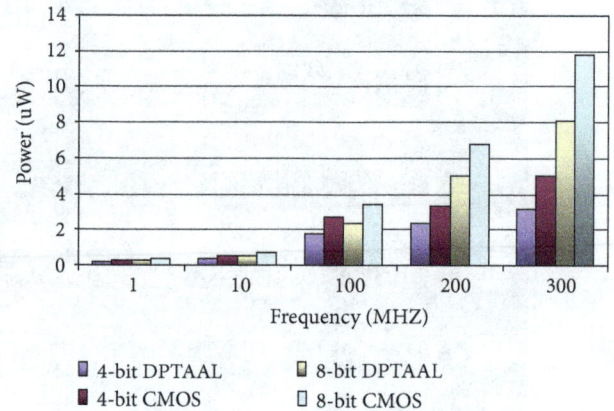

■ 4-bit DPTAAL   ■ 8-bit DPTAAL
■ 4-bit CMOS   □ 8-bit CMOS

FIGURE 9: Power comparison graph.

TABLE 6: Process parameters used for this simulation.

Parameters	Value
Process technology	0.18 μm
Supply voltage (V_{DD})	1.8 V
Ambient temperature	0–70°C
Gate oxide thickness (T_{ox})	4 nm
Gate capacitance (C_g)	2 fF/μm^2
Minimum gate length (L_{min})	0.18 μm
NFET threshold voltage (V_{tn})	0.39 V
PFET threshold voltage (V_{tp})	−0.42 V
NFET drain current (I_{Dsat})	600 mA
PFET drain current (I_{Dsat})	260 mA

adder. Asynchronous operation is achieved by the control and regeneration (C&R) block to reduce the power clock system switching losses.

By combining these techniques, DPTAAL full adder design features the lowest energy consumption per addition as compared with conventional CMOS design. HSPICE simulations showed energy savings up to 84% in this full adder design, maintaining proper functionality.

The power consumption of the simulated 4×4 and 8×8 multipliers is reported in Table 5, for operating frequencies as low as 1 MHz and as high as 300 MHz. It can be observed by comparing the data presented in Table 5 that DPTAAL design achieves significant power savings for clock rates ranging from 200 MHz to 300 MHz. The power comparison graph for 4×4 and 8×8 multipliers is shown in Figure 9. The multiplier design is studied on TSMC 0.18 μm CMOS process models in Tanner EDA tools with SPICE support, at 1.8 V supply voltage. The standard values of gate capacitances and other MOSFET model parameters were included in this simulation. The simulation parameters of this process technology are summarized in Table 6.

9. Conclusion

In this paper, we have proposed a framework for designing a low power multiplier using energy efficient full adder. A unique approach, double pass transistor with asynchronous adiabatic logic (DPTAAL), has been followed in the full adder design. Double pass transistor logic (DPL) is a modified version of complementary pass transistor logic, which is used to improve the circuit performance at reduced voltage level. This technique is combined with asynchronous adiabatic logic (AAL) to obtain the energy saving benefits with improved circuit performance in full adder design.

In this DPTAAL design, asynchronous operation has been achieved by the control and regeneration (C&R) section, which generates the control signal for the next logical gate

using the output signal of the previous gate. The regeneration technique makes the control signal strong enough to drive the next logical gate. It facilitates reducing the power clock system switching losses.

The energy performance of the DPTAAL full adder is compared with the conventional CMOS full adder for the various frequency ranges and achieved significant energy savings up to 84%. This energy efficient full adder cell is used in the multiplier design.

The performance of this design is analyzed with 4-bit and 8-bit multipliers for operating frequencies as low as 1 MHz and as high as 300 MHz. The power results of the proposed multiplier design are compared with the conventional logic designs. It is observed that for frequencies between 200 MHz and 300 MHz, DPTAAL multiplier circuits consume less power than the conventional designs.

The DPTAAL multiplier circuits have been implemented and studied using 0.18 μm TSMC technology file with 1.8 V supply voltage and have shown great prospect for the development of power aware systems. This approach confirms the feasibility of asynchronous adiabatic multiplier in low power computing applications.

Acknowledgments

The authors would like to thank all the researchers who have contributed to this field of research. The comments of anonymous reviewers to improve the quality of this paper are also acknowledged.

References

[1] A. Kishore Kumar, D. Somasundareswari, V. Duraisamy, and M. Pradeepkumar, "Low power multiplier design using complementary pass-transistor asynchronous adiabatic logic," *International Journal on Computer Science and Engineering*, vol. 2, no. 7, pp. 2291–2297, 2010.

[2] R. Zimmermann and W. Fichtner, "Low-power logic styles: CMOS versus pass-transistor logic," *IEEE Journal of Solid-State Circuits*, vol. 32, no. 7, pp. 1079–1090, 1997.

[3] V. S. K. Bhaaskaran, S. Salivahanan, and D. S. Emmanuel, "Semi-custom design of adiabatic adder circuits," in *Proceedings of the 19th International Conference on VLSI Design Held Jointly with 5th International Conference on Embedded Systems Design*, pp. 745–748, January 2006.

[4] A. Blotti and R. Saletti, "Ultralow-power adiabatic circuit semi-custom design," *IEEE Transactions on Very Large Scale Integration (VLSI) Systems*, vol. 12, no. 11, pp. 1248–1253, 2004.

[5] V. G. Oklobdzija, D. Maksimovic, and F. Lin, "Pass-transistor adiabatic logic using single power-clock supply," *IEEE Transactions on Circuits and Systems II*, vol. 44, no. 10, pp. 842–846, 1997.

[6] D. Maksimović, V. G. Oklobdzija, B. Nikolic, and K. W. Current, "Clocked CMOS adiabatic logic with integrated single-phase power-clock supply," *IEEE Transactions on Very Large Scale Integration (VLSI) Systems*, vol. 8, no. 4, pp. 460–463, 2000.

[7] L. Verga, F. Kovacs, and G. Hosszu, "An improved pass gate adiabatic logic," in *Proceedings of the 14th Annual International ASIC/SOC Conference*, pp. 208–211, 2001.

[8] Y. Zhang, H. H. Chen, and J. B. Kuo, "0.8 V CMOS adiabatic differential switch logic circuit using bootstrap technique for low-voltage low-power VLSI," *Electronics Letters*, vol. 38, no. 24, pp. 1497–1499, 2002.

[9] J. Hu, W. Zhang, and Y. Xia, "Complementary pass-transistor adiabatic logic and sequential circuits using three-phase power supply," in *Proceedings of the 47th Midwest Symposium on Circuits and Systems*, vol. 2, pp. 201–204, July 2004.

[10] J. Hu, T. Xu, and H. Li, "A lower-power register file based on complementary pass-transistor adiabatic logic," *IEICE Transactions on Information and Systems*, vol. E88-D, no. 7, pp. 1479–1485, 2005.

[11] J. Dai, J. Hu, W. Zhang, and L. Wang, "Adiabatic CPL circuits for sequential logic systems," in *Proceedings of the 49th Midwest Symposium on Circuits and Systems (MWSCAS '06)*, pp. 713–717, August 2007.

[12] J. Parkand, S. Je Hong, and J. Kim, "Energy-saving design technique achieved by latched pass-transistor adiabatic logic," in *Proceedings of IEEE International Symposium on Circuits and Systems (ISCAS '05)*, vol. 5, pp. 4693–4696, May 2005.

[13] A. G. Dickinson and J. S. Denker, "Adiabatic dynamic logic," *IEEE Journal of Solid-State Circuits*, vol. 30, no. 3, pp. 311–315, 1995.

[14] M. Arsalan and M. Shams, "Charge-recovery power clock generators for adiabatic logic circuits," in *Proceedings of the 18th International Conference on VLSI Design: Power Aware Design of VLSI Systems*, pp. 171–174, January 2005.

[15] M. Arsalan and M. Shams, "Asynchronous adiabatic logic," in *Proceedings of IEEE International Symposium on Circuits and Systems (ISCAS '07)*, pp. 3720–3723, May 2007.

[16] J. Hu, W. Zhang, X. Ye, and Y. Xia, "Low power adiabatic logic circuits with feedback structure using three-phase power supply," in *Proceedings of the International Conference on Communications, Circuits and Systems*, vol. 2, pp. 1375–1379, May 2005.

[17] M. Suzuki, N. Ohkubo, T. Shinbo et al., "1.5-ns 32-b CMOS ALU in double pass-transistor logic," *IEEE Journal of Solid-State Circuits*, vol. 28, no. 11, pp. 1145–1151, 1993.

[18] L. Wang, J. Hu, and J. Dai, "A low-power multiplier using adiabatic CPL circuits," in *Proceedings of the International Symposium on Integrated Circuits (ISIC '07)*, pp. 21–24, September 2007.

[19] R. H. Dennard, F. H. Gaensslen, H. N. Yu, V. L. Rideout, E. Bassous, and A. R. LeBlanc, "Design of ion-implanted MOSFET's with very small physical dimensions," *IEEE Journal of Solid-State Circuits*, vol. 9, no. 5, pp. 256–268, 1974.

Performance Analysis of High Speed Hybrid CMOS Full Adder Circuits for Low Voltage VLSI Design

Subodh Wairya,[1] Rajendra Kumar Nagaria,[2] and Sudarshan Tiwari[2]

[1] Department of Electronics Engineering, Institute of Engineering & Technology (IET), Lucknow 226021, India
[2] Department of Electronics and Communication Engineering, Motilal Nehru National Institute of Technology (MNNIT), Allahabad 211004, India

Correspondence should be addressed to Subodh Wairya, swairya@gmail.com

Academic Editor: Jose Carlos Monteiro

This paper presents a comparative study of high-speed and low-voltage full adder circuits. Our approach is based on hybrid design full adder circuits combined in a single unit. A high performance adder cell using an XOR-XNOR (3T) design style is discussed. This paper also discusses a high-speed conventional full adder design combined with MOSCAP Majority function circuit in one unit to implement a hybrid full adder circuit. Moreover, it presents low-power Majority-function-based 1-bit full addersthat use MOS capacitors (MOSCAP) in its structure. This technique helps in reducing power consumption, propagation delay, and area of digital circuits while maintaining low complexity of logic design. Simulation results illustrate the superiority of the designed adder circuits over the conventional CMOS, TG, and hybrid adder circuits in terms of power, delay, power delay product (PDP), and energy delay product (EDP). Postlayout simulation results illustrate the superiority of the newly designed majority adder circuits against the reported conventional adder circuits. The design is implemented on UMC 0.18 μm process models in Cadence Virtuoso Schematic Composer at 1.8 V single-ended supply voltage, and simulations are carried out on Spectre S.

1. Introduction

It is time we explore the well-engineered deep submicron CMOS technologies to address the challenging criteria of these emerging low-power and high-speed communication digital signal processing chips. The performance of many applications as digital signal processing depends upon the performance of the arithmetic circuits to execute complex algorithms such as convolution, correlation, and digital filtering. Fast arithmetic computation cells including adders and multipliers are the most frequently and widely used circuits in very-large-scale integration (VLSI) systems. The semiconductor industry has witnessed an explosive growth of integration of sophisticated multimedia-based applications into mobile electronics gadgetry since the last decade. However, the critical concern in this arena is to reduce the increase in power consumption beyond a certain range of operating frequency. Moreover, with the explosive growth, the demand, and the popularity of portable electronic products, the designers are driven to strive for smaller silicon area,

higher speed, longer battery life, and enhanced reliability. The XOR-XNOR circuits are basic building blocks in various circuits especially arithmetic circuits (adders & multipliers), compressors, comparators, parity checkers, code converters, error-detecting or error-correcting codes and phase detector.

Adder is the core element of complex arithmetic circuits like addition, multiplication, division, exponentiation, and so forth. There are standard implementations with various logic styles that have been used in the past to design full-adder cells [1–4] and the same are used for comparison in this paper. Although they all have similar function, the way of producing the intermediate nodes and the transistor count is varied. Different logic styles tend to favor one performance aspect at the expense of the others. The logic style used in logic gates basically influences the speed, size, power dissipation, and the wiring complexity of a circuit. The circuit *delay* is determined by the number of inversion levels, the number of transistors in series, transistor sizes (i.e., channel widths), and the intracell wiring capacitances. Circuit *size* depends upon the number of transistors, their sizes and on the wiring complexity. Some of them use one logic style for the whole full adder while the other use more than one logic style for their implementation.

Power is one of the vital resources, hence the designers try to save it while designing a system. Power dissipation depends upon the switching activity, node capacitances (made up of gate, diffusion, and wire capacitances), and control circuit size. At the device level, reducing the supply voltage V_{DD} and reducing the threshold voltage accordingly would reduce the power consumption. Scaling the supply voltage appears to be the well-known means to reduce power consumption. However, lower-supply voltage increases circuit delay and degrades the drivability of the cells designed with a certain logic style. One of the most significant obstacle in decreasing the supply voltage is the large transistor count and V_{th} loss problem. By selecting proper (W/L) ratio we can minimize the power dissipation without decreasing the supply voltage.

To summarize, some of the performance criteria are considered in the design and evaluation of adder cells and some are utilized for the ease of design, robustness, silicon area, delay, and power consumption. The paper is organized section wise. Section 2 describes the review of full adder circuit topologies. Section 3 illustrates the concept of SUM function-based hybrid full adders topologies and highlights some 1-bit adder cells, which is based on XOR-XNOR (3T) circuits. A review of Majority function, MOS capacitor characteristics, and three-input and five-input Majority function (MOSCAPs) based full adder topologies has been discussed in Section 4. In Section 5, implementations of Hybrid XOR-XNOR (3T) and Majority-function-based full adder methodologies are discussed. The simulation results are analyzed and compared in Section 6. Finally, Section 7 concludes the paper.

2. Review of Full Adder Topologies

In recent years, several variants of different logic styles have been proposed to implement 1-bit adder cells [5–28]. There are two types of full adders in case of logic structure. One is static and the other is dynamic style. Static full adders are commonly more reliable, simpler and are lower power consuming than dynamic ones. Dynamic is an alternative logic style to design a logic function. It has some advantages over the static mode such as faster switching speeds, no static power consumption, nonratioed logic, full swing voltage levels, and lesser number of transistors. An N input logic function requires N+2 transistors versus 2N transistors in the standard CMOS logic. The area advantage comes from the fact that the pMOS network of a dynamic CMOS gate consists of only one transistor. This also results in a reduction in the capacitive load at the output node, which is the basis for the delay advantage. There are various issues related to the full adder like power consumption, performance, area, noise immunity, regularity and good driving ability. Many researchers have combined these two structures and have proposed hybrid dynamic-static full adders. They have investigated different approaches realizing adders using CMOS technology each having its own pros and cons.

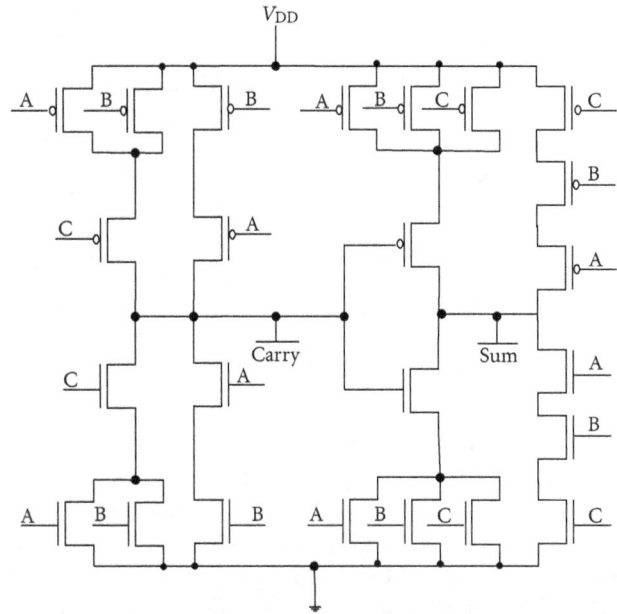

FIGURE 1: C-CMOS adder cell.

Full adder circuits can be divided into two groups on the basis of output. The first group of full adders have full swing output. C-CMOS, CPL, TGA, TFA, Hybrid, 14T, and 16T belong to the first group [5–20, 29–31]. The second group comprises of full adders (10T, 9T and 8T) without full swing outputs [21–28]. These full adders usually have low number of transistors- (3T-) based XOR-XNOR circuit, less power consumption, and less area occupation. The nonfull swing full adders are useful in building up larger circuits as multiple bit input adder and multipliers. One such application is the Manchester Carry-Look Ahead chain. The full adders of first group have good driving ability, high number of transistors, large area, and usually higher power consumption in comparison to the second group.

There are standard implementations for the full-adder cells which are used as the basis of comparison in this paper. Some of the standard implementations are as follows.

CMOS logic styles have been used to implement the low-power 1-bit adder cells. In general, they can be broadly divided into two major categories: the Complementary CMOS and the Pass-Transistor logic circuits. The complementary CMOS (C-CMOS) full adder (Figure 1) is based on the regular CMOS structure [3, 4, 29]. The advantage of complementary CMOS style is its robustness against voltage scaling and transistor sizing, which are essential to provide reliable operation at low voltage with arbitrary transistor sizes.

The pass-transistor logic (PTL) is a better way to implement circuits designed for low power applications. The low power pass-transistor logic and its design analysis procedures were reported in [12, 13]. Its advantage is that one pass-transistor network (either pMOS or nMOS) is sufficient to implement the logic function, which results in lower number of transistors and smaller input load. Moreover, direct V_{DD}-to-ground paths, which may lead to short-circuit energy dissipation, are eliminated.

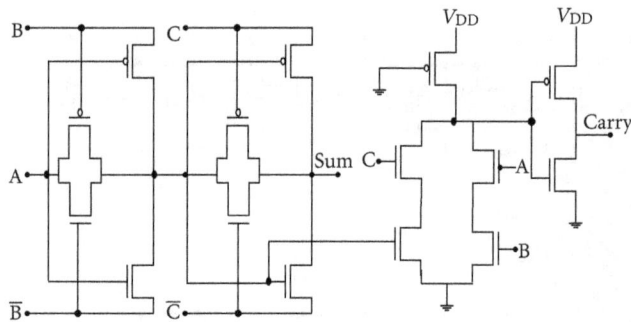

FIGURE 2: TG-Pseudo adder cell.

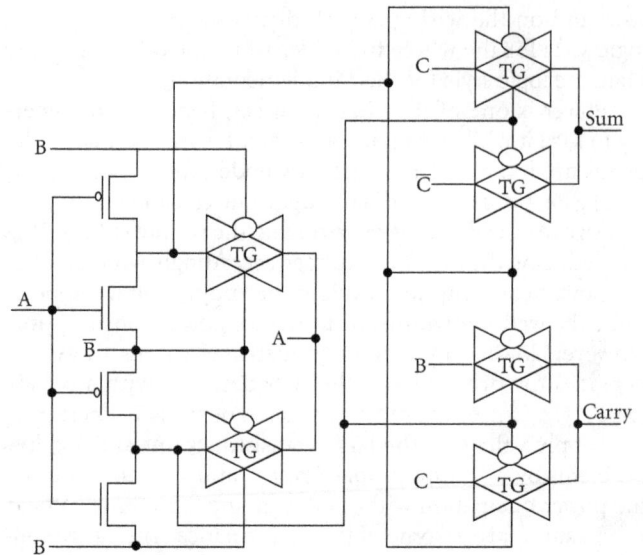

FIGURE 3: TG adder cell.

Pseudo nMOS full adder cell operates on pseudo logic, which is referred to as ratioed style. This full adder cell uses 14 transistors to realize the negative addition function. The advantage of pseudo nMOS adder cell is its higher speed (compared to conventional full adder) and less transistor count. The disadvantage of pseudo nMOS cell is the static power consumption of the pull-up transistor as well as the reduced output voltage swing, which makes this adder cell more susceptible to noise. To increase the output swing, CMOS inverter is added to this circuit.

Newly designed full adder [20] is a combination of low power transmission gates and pseudo nMOS gates as depicted in Figure 2. Transmission gate consists of a pMOS transistor and an nMOS transistor that are connected in parallel, which is a particular type of pass-transistor logic circuit. There is no voltage drop at output node, but it requires twice the number of transistors to design similar function.

Another full adder is the Complementary Pass Transistor Logic (CPL) with swing restoration, which uses 32 transistors [5, 6, 30, 31]. CPL adder produces many intermediate nodes and their complement to give the outputs. The most important features of CPL include the small stack height and low output voltage swing at the internal node which contribute to reduction in power consumption. The CPL suffers from static power consumption due to the low swing at the gates of the output inverters. *Double pass-transistor logic* (DPL) [8] and *swing restored pass-transistor logic* (SRPL) [9, 10] are related to CPL.

Some designs of the full adder circuit based on transmission gates are shown in Figure 3. Transmission gate logic circuit is a special kind of pass-transistor logic circuit [4, 5, 25]. The main disadvantage of transmission gate logic is that it requires twice the number of transistors than pass-transistor logic or more to implement the same circuit. TG gate full adder cell has 20 transistors. Similarly, transmission function full adder (TFA) cell has 16 transistors [4, 29]. It exhibits better speed and less power dissipation than the conventional CMOS adder due to the small transistor stack height.

3. Sum Function-Based Hybrid Full Adder Topologies

More than one logic style is used for implementation of the hybrid full adders. The hybrid adder cells may be classified into various categories depending upon their structure and logical expression of the Sum and Carry output signals. All hybrid designs use the best available modules implemented using different logic styles or enhance the available modules in an attempt to build a low power consuming full-adder cell [17–19]. Most full adder topologies are based on two XOR circuits: one to generate H (XOR) with $\overline{H}$ (XNOR), and the other to generate the Sum output. The carry signal is obtained by using one MUX (multiplexer):

$$\text{Sum} = A \oplus B \oplus C, \qquad \text{Carry} = AB + C(A \oplus B),$$

$$H = A \oplus B, \qquad \text{Sum} = H \oplus C, \qquad (1)$$

$$\text{Carry} = A \cdot \overline{H} + C \cdot H.$$

3.1. XOR-XNOR Topologies. In [28, 32–35], the XOR-XNOR circuit designed with static CMOS logic with complementary pull-up pMOS and pull-down nMOS networks is the conventional one, but it requires more number of CMOS transistors. This circuit may operate with full output voltage swing. Different XOR/XNOR topologies are illustrated in Figure 4. A PTL based 6-transistor XOR-XNOR circuit presented in [34] has full output voltage swing and better driving capability.

A new set of low power four transistor (4T) XOR and XNOR circuits called powerless P-XOR and Groundless G-XNOR, respectively, is proposed in [25–28, 32]. The P-XOR and G-XNOR circuits consume less power than other designs because they have no direct supply voltage (V_{DD}) or ground connection. The performance of the complex logic circuits is affected by the individual performance of the XOR-XNOR circuits that are included in them. An XOR and XNOR function with low circuit complexity can be achieved with only three transistors (3T) in PTL. Despite the saving in transistor count, the output voltage level is degraded at certain input signal combinations.

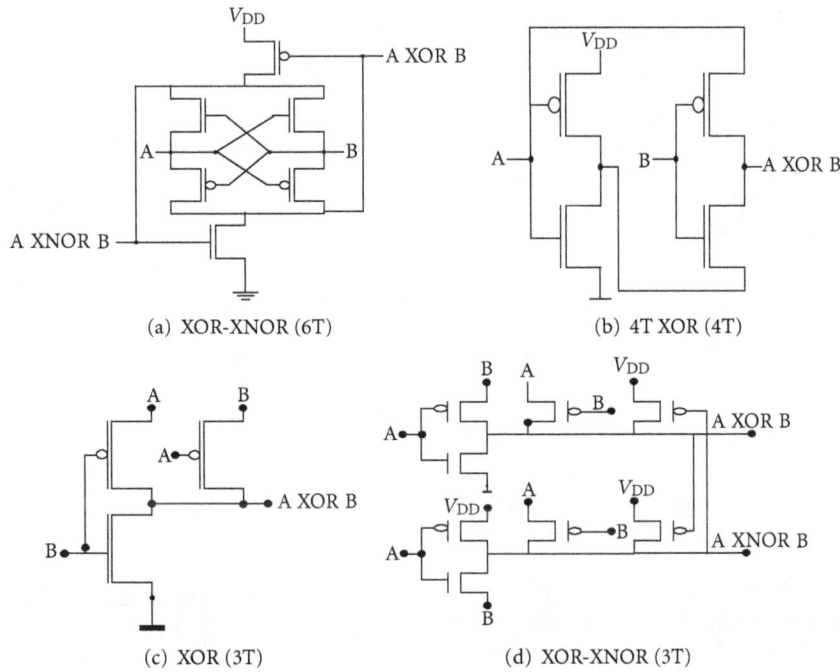

(a) XOR-XNOR (6T)

(b) 4T XOR (4T)

(c) XOR (3T)

(d) XOR-XNOR (3T)

FIGURE 4: Basic designs of XOR-XNOR gate found in literature.

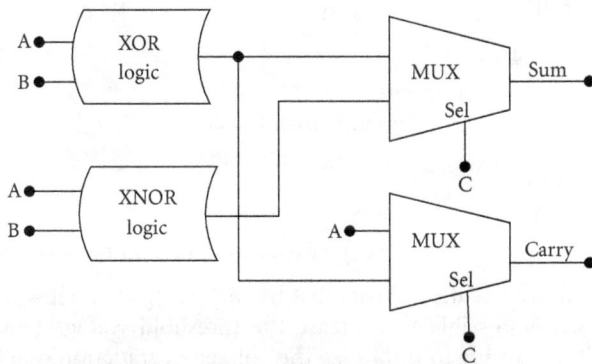

FIGURE 5: *Cascaded* XOR-XNOR based-*adder*.

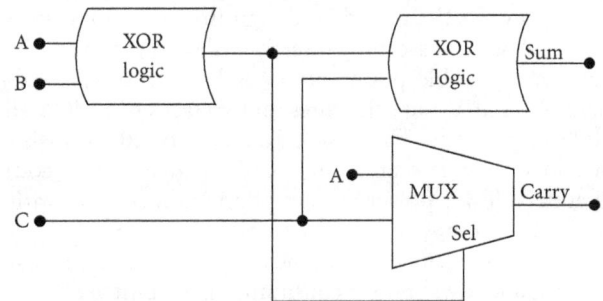

FIGURE 6: *Centralized* XOR-XNOR based-*adder*.

Generally, the main aim is to reduce the number of transistors in the adder cell and consequently to reduce the number of power dissipating nodes. This is achieved by utilizing intrinsically low power consuming logic styles like TFA, TGA or simply passing transistors. There are three main components to design a hybrid full adder circuit [19]. These are XOR or XNOR, Carry generator and Sum generator. Hybrid adders may be classified into two groups which are as follows.

3.2. Cascaded Output Based Adders (Group 1). In this category, signal Sum is generated using, either two cascaded XOR or two cascaded XNOR modules. Figure 5 shows the basic blocks of this category. Almost all the circuits in this category suffer from high delay in generating Sum and Carry signals. The Static Energy Recovery full adder (SERF) falls under this category [23].

3.3. Centralized Output Based Adders (Group 2). In this category, Sum and Carry are generated using intermediate signals XOR and XNOR. In this group, output Sum and Carry are generated faster than the outputs in cascaded output full adders. The key point here is to produce intermediate signals simultaneously. Otherwise, there may be glitches, unnecessary power consumption, and longer delay. Figure 6 shows the basic blocks of this category. TGA and TFA are in this category. Some of the hybrid full adders do not belong to any of these two groups, such as Complementary and Level Restoring Carry Logic (CLRCL) full adder [26] and Multiplexer based (MBF 12T) full adder [18].

3.4. 10T Full Adder. In [24] different components have been combined to make 41 new 10T full adder full adders. Some 10T full adders can be designed by interchanging the inputs of the module having lowest propagation delay amongst all the 10T full adder circuits. The design of the 10T adder cell is based on an optimized design for the XOR function

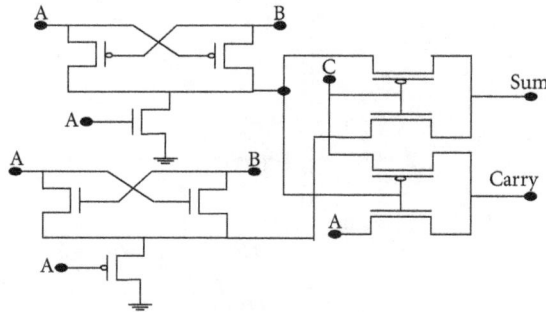

FIGURE 7: XOR-XNOR- (3T-) based 10T full adder.

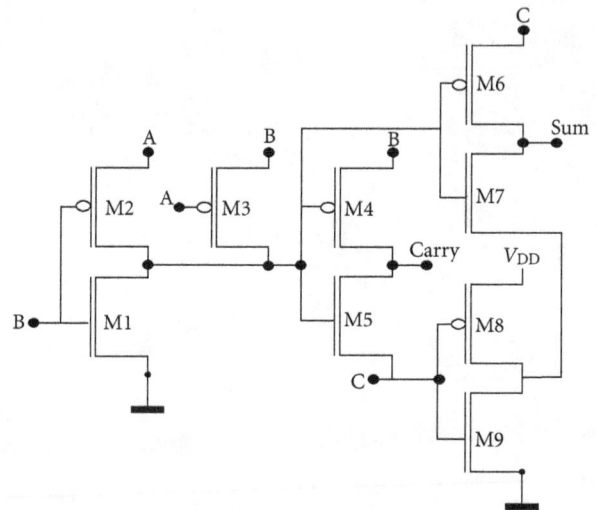

FIGURE 8: 9T full adder.

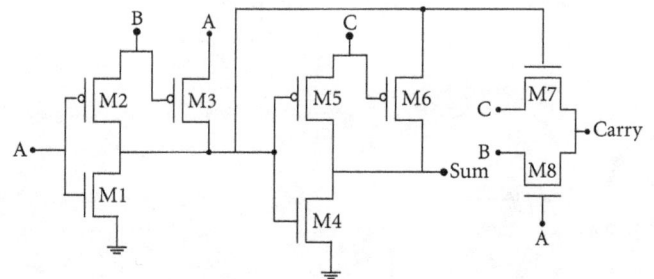

FIGURE 9: 8T full adder.

and pass transistor logic to implement the addition logic function. Two XOR operations are required to calculate the Sum function. Each XOR operation requires four transistors (4T). 2X1 MUX is used for Carry function implemented using two transistors.

Another 10T full adder based on centralized structure is shown in Figure 7. Intermediate XOR and XNOR are generated using three transistor (3T) XOR and XNOR gate. Sum and Carry are generated using two double transistors multiplexers. 3T XOR and XNOR consume high energy due to short circuit current in ratio logic. They all have double threshold losses in full adder output terminals. This problem usually prevents the full adder design from operating at low supply voltage or cascading directly without extra buffering. The lowest possible power supply is limited to $2V_{tn} + V_{tp}$ where V_{tn} and V_{tp} are the threshold voltages of nMOS and pMOS respectively. The basic advantages of 10T transistor full adders are: less area compared to higher gate count full adders, lower power consumption and lower operating voltage. It becomes very difficult and even obsolete to keep full voltage swing operation as the designs with fewer transistor count and lower power consumption are pursued.

3.5. 9T Full Adder. In nine transistor (9T) full adder circuit, we have only one 3T XOR gate as is shown in the Figure 8 [36]. The design of 3T (M1–M3) XOR circuit is based on a modified version of a CMOS inverter and a pMOS pass transistor. When $A = 1$ and $B = 0$, voltage degradation due to threshold drop occurs across transistor M3 and consequently the output (M3) is degraded with respect to the input. The voltage degradation due to threshold drop can be minimized by increasing the W/L ratio of transistor M3. An equation relating threshold voltage of a MOS transistor to the channel length and width is given as

$$V_T = V_{T0} + \gamma\left(\sqrt{V_{SB} + \phi_0} - \sqrt{\phi_0}\right) - \alpha_l \frac{t_{ox}}{L}(V_{SB} + \phi_0)$$
$$- \alpha_v \frac{t_{ox}}{L}(V_{DS}) + \alpha_W \frac{t_{ox}}{L}(V_{SB} + \phi_0), \tag{2}$$

where

V_{T0} is the zero bias threshold voltage,

γ is bulk threshold coefficient,

ϕ_0 is $2\phi_F$, where ϕ_F is the Fermi potential,

t_{ox} is the thickness of the oxide layer,

$\alpha_l, \alpha_v,$ and α_w are the process dependent parameters.

The above equation shows that by increasing channel width (W) it is possible to decrease the threshold voltage (V_{th}). So it is possible to minimize the voltage degradation due to threshold voltage by increasing the width of M3 transistor & keeping the length constant. In 9T full adder circuit pass transistor M4, M5 and M6, M7 are used for Carry and Sum function respectively.

3.6. 8T Full Adder. The design of an eight transistor (8T) full adder using 3T XOR gates is shown in Figure 9 [37]. The Boolean equations for the design of the eight transistor full adder are as follows:

$$Sum = A \oplus B \oplus C,$$
$$Carry = BC + CA + AB = C(A \oplus B) + AB. \tag{3}$$

The Sum output function is obtained by a cascade of 3T XOR gates. Carry can be realized using a wired OR logic in accordance with the above equation.

Another 8T full adder using centralizer output condition contains three modules—two 3T XOR gates and one multiplexer (2T). It can work at high speed with low power dissipation due to minimum number of transistors and small transistor delay.

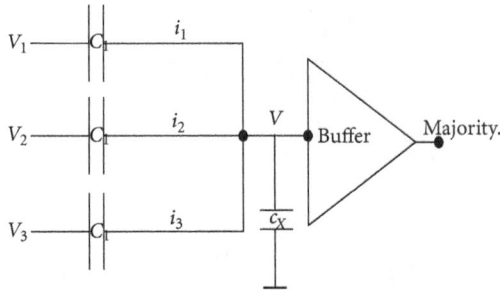

FIGURE 10: Implementation of Majority functions (MOSCAP).

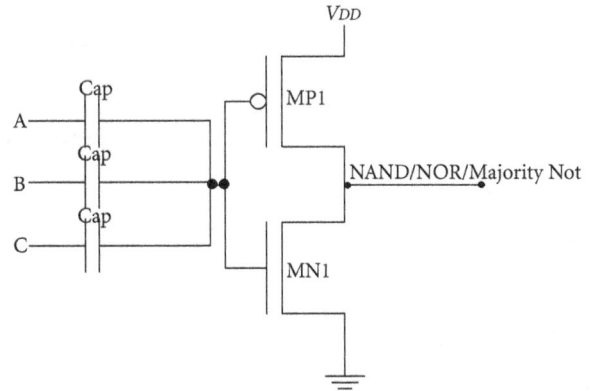

FIGURE 11: Majority function- (MOSCAP-) based logic gates.

4. Carry (Majority) Function-Based Hybrid Full Adder Topologies

The Majority function is a logic circuit that functions as a majority vote to determine the output of the circuits [38]. This function has only odd number of inputs. Its output is equal to "1" when the number of input logic "1" is more than logic "0". Comparing to the XOR implementations of full adder cells, Majority-based full adders are more reliable and robust [38]. Moreover, the bridge style full adder circuits [39] by sharing transistors can operate faster and are smaller than the conventional CMOS full adder circuits.

4.1. Literature Review of Majority Functions. Boolean algebra with three variables is used to facilitate the conversion of a sum-of-products expression to minimize majority logic as shown in Table 1 [38]. Three binary variables can only produce eight unique minterms. Any three-variable Boolean function can be represented by the combinations of up to eight of these minterms. The three-variable Boolean function of 5–7 minterms can be represented using the complement form of 3–1 minterms. Based on DeMorgan's theorem, a Boolean function, expressed as the sum of several minterms, can also be expressed as the complement of the sum of the remaining minterms. The simplified majority expressions for 13 standard functions are given in Table 1.

4.2. Circuit-Interpretation-of-MOS Capacitor- (MOSCAP-) Based Majority Not Function. The majority structure is implemented by three input capacitors. These three input capacitors prepare an input voltage that is applied for driving static CMOS buffer. The majority gates may be designed with more inputs by this method by increasing the number of input capacitors. The capacitor network is used to provide voltage division for implementing majority logic as explained below.

Total current I at node $V = I_1 + I_2 + I_3$,

$$(V)c_X s = (V_1 - V)c_1 s + (V_2 - V)c_1 s + (V_3 - V)c_1 s$$

$$= (c_X + 3c_1)V = (V_1 + V_2 + V_3)c_1, \qquad (4)$$

$$V = (V_1 + V_2 + V_3)\left(\frac{c_1}{3c_1 + c_X}\right).$$

The input capacitors shown in Figure 10 are used to prepare an input voltage that is applied for driving static

inverter. When the majority of inputs are "0", the output of capacitor network is considered as logic "0" by the CMOS buffer and consequently the output of buffer is 0 V. When the majority of inputs are logic "1", the output of capacitor network is considered logic "1" by the CMOS buffer and consequently the output of buffer is V_{DD}. The input capacitance of the CMOS buffer is negligible and has no effect on operation of the circuit. Three capacitors perform voltage summation to implement scaled-linear sum. Through superposition of input capacitors, increased input voltage is scaled at point V as shown in Figure 10 and given in Table 2 [40].

4.3. MOS Capacitor (MOSCAP) Structure. In this section hardware implementation and construction of MOSCAP are discussed. Tying the drain and source of a MOSFET together results in a MOSCAP. Many realizable alternatives such as Poly-Insulator-Poly capacitors (PIPCAP), Metal-Insulator-Metal capacitors (MIMCAP), or Metal-Oxide-Semiconductor capacitors (MOSCAP) can be utilized for realizing the capacitor network. However, MOSCAP has an advantage of more capacitance; less chip area. The nMOSCAP usually has lesser capacitance in comparison to pMOSCAP for the same area, so pMOSCAP is used for implementing the capacitor network. Table 3 shows that the variation of MOS capacitor with respect to channel width of MOS transistor.

4.4. Implementation of (NAND, NOR and Majority Not) Gates Using MOSCAP Majority Function. Figure 11 shows the circuit used to implement Majority Not function with inverter utilizing high-V_{th} for both nMOS and pMOS. This circuit can be used to implement NAND gate using high-V_{th} nMOS and low-V_{th} pMOS, and NOR gate using low-V_{th} nMOS and high-V_{th} pMOS. The Majority gates may be designed with more inputs by this method by increasing the number of input capacitors. The capacitor network is used to provide voltage division for implementing majority logic.

There are two methods to design the NAND and NOR logic circuits. First method is the transistor sizing that shifts the voltage transfer curve (VTC) to the left and right by changing the ratio of (W/L)n to (W/L)p. Raising this ratio moves VTC to the left; therefore, this circuit will operate as

TABLE 1: Majority expression of standard logic functions.

Standard Boolean function	Majority expression	Function implementation diagram
$F = A$	$M(A, 0, 1)$	
$F = A \cdot B$	$M(A, B, 0)$	
$F = A \cdot B \cdot C$	$M(M(A, B, 0), C, 0)$	
$F = A \cdot B \cdot C + \overline{A} \cdot \overline{B} \cdot \overline{C}$	$M\{M(M(A, 0, B), C, 0),\ M(M(\overline{A}, \overline{B}, 0), \overline{C}, 0), 1\}$	
$F = A \cdot B + \overline{A} \cdot \overline{B} \cdot C$	$M\{M(A, 0, B), M(M(\overline{A}, \overline{B}, 0), C, 0), 1\}$	
$F = A \cdot B + \overline{B} \cdot C$	$M\{M(A, 0, B), M(\overline{B}, 0, C), 1\}$	
$F = A \cdot B + \overline{A} \cdot \overline{B}$	$M\{M(A, 0, B), M(\overline{A}, 0, \overline{B}), 1\}$	
$F = A \cdot B + B \cdot C$	$M(B, M(A, 1, C), 0)$	

TABLE 1: Continued.

Standard Boolean function	Majority expression	Function implementation diagram
$F = A \cdot B + B \cdot C + \overline{A} \cdot \overline{B} \cdot \overline{C}$	$M\{M(B, M(A, C, 1), 0), M(\overline{A}, M(\overline{B}, \overline{C}, 0), 0), 1\}$	
$F = A \cdot B \cdot C + A \cdot \overline{B} \cdot \overline{C}$	$M\{M(A, B, \overline{C}), M(A, \overline{B}, C), 0\}$	
$F = A \cdot B \cdot C + \overline{A} \cdot B \cdot \overline{C} + A \cdot \overline{B} \cdot \overline{C}$	$M\{M(A, C, 0), M(A, B, \overline{C}), M(\overline{A}, \overline{B}, \overline{C})\}$	
$F = A \cdot B + B \cdot C + A \cdot C$	$M(A, B, C)$	
$F = A \cdot B \cdot C + \overline{A} \cdot \overline{B} \cdot C + A \cdot \overline{B} \cdot \overline{C} + \overline{A} \cdot B \cdot \overline{C}$	$M\{M(\overline{A}, B, C), M(A, \overline{B}, C), \overline{C}\}$	

TABLE 2: Switching voltage at output node V of the capacitance network.

Inputs			Voltage at V node	Majority Not
A	B	C	V_{DD}	$\overline{\text{Carry}}$
0	0	0	0V	1
0	0	1	$V_{DD}/3$	1
0	1	0	$V_{DD}/3$	1
0	1	1	$2V_{DD}/3$	0
1	0	0	$V_{DD}/3$	1
1	0	1	$2V_{DD}/3$	0
1	1	0	$2V_{DD}/3$	0
1	1	1	V_{DD}	0

TABLE 3: Channel width v/s MOS capacitor in 0.18 μm Tech.

Cap	2.89 fF	4.89 fF	6.89 fF	8.89 fF	10.91 fF
Width (W) μm	1.59	2.71	3.83	4.95	6.07

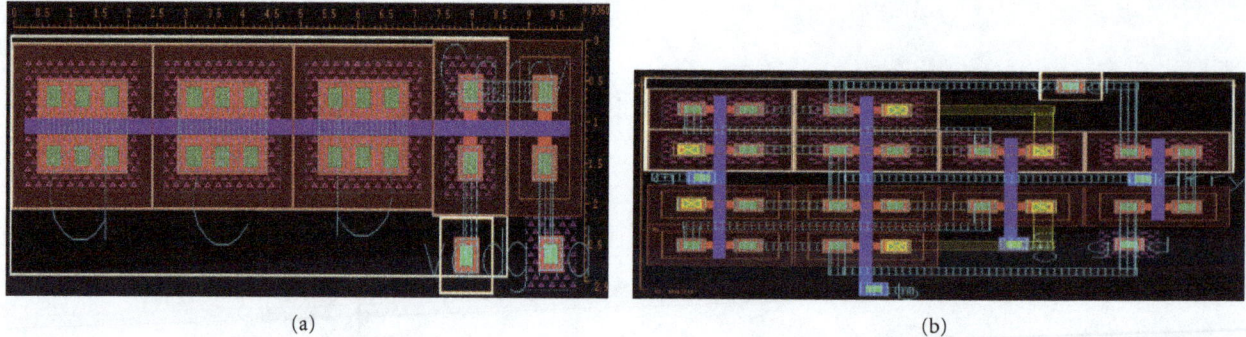

(a)   (b)

FIGURE 12: (a) MOSCAP Majority Not function layout. (b) Static CMOS bridge (Majority function) layout.

NOR function. Contrary to this, decreasing the ratio makes the NAND function. The second method uses high-threshold voltage (V_{th}) transistors (MP1 & MN1) as shown in Figure 4.

Simulation results in Table 4 illustrate the comparison of static logic gates with MOSCAP-based majority function, static and dynamic logic style.

4.5. Layout and Area Analysis of Majority Circuits. The layout of Majority Not function (MOSCAP) and static CMOS bridge-type Majority function circuits are shown in Figures 12(a) and 12(b), respectively, and the area is given in Table 5. The area of the MOSCAP Majority function (MOSCAP) circuit is 50% less than that of the bridge type Majority function circuit. At low voltages (say 1 V) delay and power consumption is much more improved in comparison to the static one, and hence MOSCAP Majority function is more reliable, power efficient with less occupation of chip area in VLSI circuit designing. By a perfect layout design, even more reduction in the area is possible and thus a more compact design can be implemented.

4.6. A Review of Majority-Function-Based Full Adder Topologies. As Table 6 exhibits, Sum is different at merely two places with Majority Not function when inputs are 000 or 111. The values of these two functions are not equal at $A = B = C = $ "0" and $A = B = C = $ "1". Therefore, we correct these two states by using a pMOS and an nMOS transistor. These transistors must be arranged in a way that ensures the correctness of the circuit [39].

The basic logic design of a full adder includes two 3-input NAND and NOR gates with Majority Not function inputs as shown in Figure 13. The MajFA1 adder is designed using pass-transistor logic as shown in Figure 13 similar to the [39]. The logic (NAND and NOR) gates designed with pass transistor logic styles have less power dissipation and delay than in standard CMOS.

In six mid-states of Table 6, the Sum output is equal to $\overline{\text{Carry}}$ (Majority Not Function) and the MP1 and MN1 transistors are off. But, in all one input state and all zero

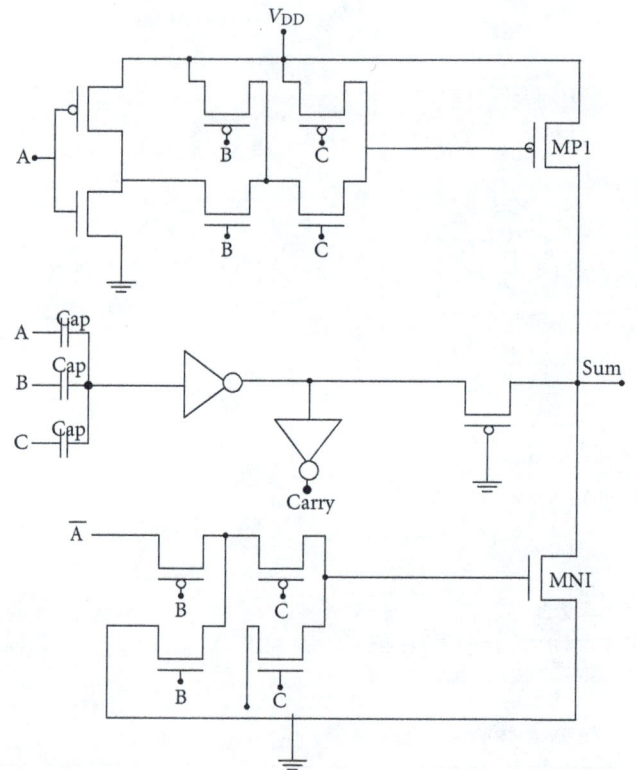

FIGURE 13: Design methodologies for Majority-function-based full adder (MajFA1).

input state the Sum is obtained by the NAND and NOR gates, respectively. In order to design circuit operations in the given state one nMOS and one pMOS pass transistor are added to the circuit. These transistors are used to disconnect the path between $\overline{\text{Carry}}$ and Sum in all "0" and "1" input state.

4.7. Majority Full Adder Using 3-Input Majority Not Function (MOSCAP). In this section full adder based on low power design of 3-input Majority Not function (MOSCAP) with

TABLE 4: Simulation results of NAND, NOR, and majority Not logic gates at 1 V.

Design	Static Majority function			MOSCAP Majority function		
	Delay (ps)	Power (μw)	PDP (10^{-18} j)	Delay (ps)	Power (μw)	PDP (10^{-18} j)
NAND	36	0.041	1.47	23	0.038	0.87
NOR	40	0.042	1.68	27	0.039	1.05
Maj. Not	43	0.048	2.06	18	0.038	0.68

TABLE 5: Simulation layout comparisons of Majority function logic.

μm	Bridge Majority function			MOSCAP Majority function		
Layout	Length (μm)	Width (μm)	Area (μm^2)	Length (μm)	Width (μm)	Area (μm^2)
Dimen.	8.8	6.9	60.7	9.9	2.95	29.2

standard logic gates is discussed. The Boolean expression may be expressed as

$$\text{Sum} = \overline{\text{Carry}} \cdot (A + B + C) + A \cdot B \cdot C. \quad (5)$$

$\overline{\text{Carry}}$ logic output will be generated by 3-inputs MOSCAP Majority Not function.

The MajFA2 full adder uses 12 transistors, and 3 capacitors are based on pseudo CMOS structure with MOSCAP Majority function. Full adder output $\overline{\text{Carry}}$ function is designed with 3 input Majority Not function logic. In this design, "a" and "b" inverters implement NOR and NAND functions, respectively.

The full adder (MajFA3) is based on MOSCAP Majority Not function with only static CMOS inverter as shown in Figure 14(b). Simulation results illustrate that the reported adder circuits having low PDP works efficiently at low voltages [41]. Outputs of the circuit will be connected to power supply or ground and therewith, the circuit has good driving capability. These inverter-based full adders are a suitable structure for the construction of low-power and high-performance VLSI systems.

4.8. Majority Full Adder Using 5-Input Majority Not Function (MOSCAP). Here if we exert a Majority function of five inputs out of which two are $\overline{\text{Carry}}$ and the other three are logic inputs (A, B, C), we will get Sum of the output as explained in the given equation. Consequently, according to this fact $\overline{\text{Sum}}$ is generated by means of two Majority Not functions. The first one is a three input Majority Not function which results in the $\overline{\text{Carry}}$ function and the second one is a five-input Majority Not function which creates $\overline{\text{Sum}}$:

$$\begin{aligned}
\text{Sum} &= ABC + \overline{A}\,\overline{B}C + \overline{A}B\overline{C} + A\overline{B}\,\overline{C} \\
&= ABC + \left(\overline{AB} \cdot \overline{AC} \cdot \overline{BC}\right) \cdot (A + B + C) \\
&= ABC + \overline{\text{Carry}} \cdot \text{Carry} + \overline{\text{Carry}}\,(A + B + C) \\
&= ABC + \overline{\text{Carry}}\,(AB + AC + BC) + \overline{\text{Carry}}\,(A + B + C) \\
&= \text{Majority}\left(A, B, C, \overline{\text{Carry}}, \overline{\text{Carry}}\right).
\end{aligned} \quad (6)$$

Reference [42]. MajFA4 full adder design has two stages. Carry is implemented by means of a Majority Not function in the first stage and in the second stage a five-input Majority Not function is used for implementing Sum function.

In the full adder circuit shown in Figure 15, first Majority Not gate is made of 3-input MOSCAP with a CMOS inverter. Three Cap1 capacitors with input signal and CMOS inverter are used to generate $\overline{\text{Carry}}$ signal. These three input capacitors prepare an input voltage that is applied for driving CMOS inverter. If more than two inputs become high then the M1 transistor will turn-on and in this case the $\overline{\text{Carry}}$ will fall to "0" logic. Therefore, Carry will be "1" logic. Otherwise, M1 and M3 will turn-off and turn-on, respectively, and output Carry will fall to "0" logic. Second Majority Not function is based on five-input capacitors and CMOS inverter (M2 & M4 transistors). It has two capacitors Cap2 and three inputs Cap2. Based on function, Sum = Maj$(A, B, C, \overline{\text{Carry}}, \overline{\text{Carry}})$, the value of Cap2 is two times the value of Cap1, because we are providing two $\overline{\text{Carry}}$ as inputs with two parallel capacitors, and these two capacitors are added. One $2 \times$ Cap2 capacitance is attached between $\overline{\text{Carry}}$ output and input of transistor M2. The basic scheme of this full adder circuit utilizes only 7 capacitors and 8 transistors. The main advantage of this design is its simplicity, modularity, and lesser number of transistors being used.

As reported in MajFA5, hybrid full adder circuit in Figure 16 uses 16 transistors. Its output Sum function is based on 5-input Majority Not gates. In this design, the first Majority Not gate is implemented with a high-performance CMOS bridge circuit [43]. This design uses more transistors, called bridge transistors, sharing transistors of different paths to generate new paths from supply lines to circuit outputs. The bridge design offers more regularity and higher performance than the other CMOS design styles and is completely symmetric in structure. Using the bridge circuit leads to reduction in delay and power consumption of the full adder cell and it also increases the robustness of the circuit.

5. Proposed Hybrid Full Adder Topologies

5.1. XOR-XNOR- (3T-) Based Full Adders. The general structure of a XOR-based full adder consists of one exclusive

TABLE 6: Truth table for Majority-function-based full adder.

Inputs				Full adder logic outputs	
A	B	C	Carry	$\overline{\text{Carry}}$	Sum = Maj$(A, B, C, \overline{\text{Carry}}, \overline{\text{Carry}})$
0	0	0	0	1	0
0	0	1	0	1	1
0	1	0	0	1	1
0	1	1	1	0	0
1	0	0	0	1	1
1	0	1	1	0	0
1	1	0	1	0	0
1	1	1	1	0	1

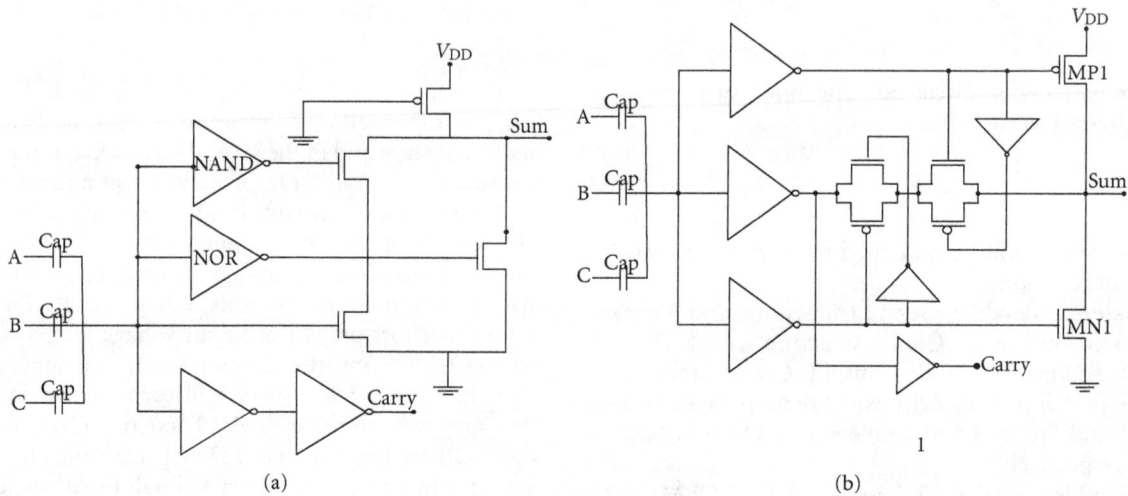

FIGURE 14: (a) Majority-function-based full adder (MajFA2). (b) Inverter-based Majority full adder (MajFA3).

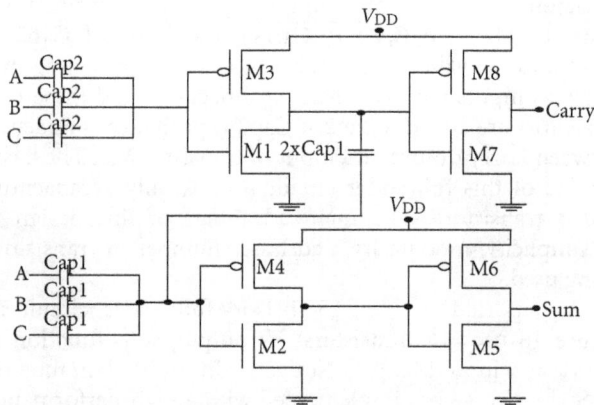

FIGURE 15: 3-input MOSCAP Majority full adder (MajFA4).

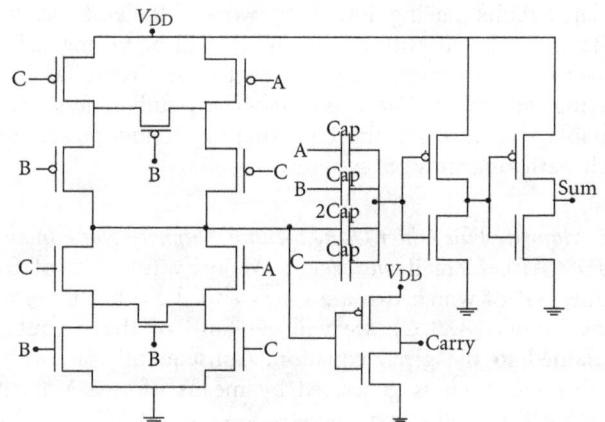

FIGURE 16: 5-input MOSCAP Majority full adder (MajFA5).

OR/NOR function (XOR/XNOR), two transmission gates in the middle, and one XOR gate to the right as shown in Figure 17. The complementary outputs of the XOR/XNOR gate are used to control the transmission gate which together realizes a multiplexer circuit producing the carry.

The circuit is a combination of two logic styles and offers high-speed, low-power consumption and energy efficiency. Lowering the supply voltage appears to be a well-known means of reducing power consumption. However, lowering the supply voltage also increases the circuit delay and

degrades the drivability of cells designed with certain logic styles. By selecting proper (W/L) ratio, we can optimize the circuit performance parameters without decreasing the power supply. The 3T XOR/XNOR gates are used in a designed full adder circuits as shown in Figures 18 and 19.

In design1 full adder circuit, XOR circuit comprises M1, M2 and M3 transistors and the output of M4 and M5 transistor is XNOR circuit. TG (M6, M7) and TG (M8, M9) give the carry and restored output swing. TG (M10, M11) and pass transistor M12, M13 are used for Sum output

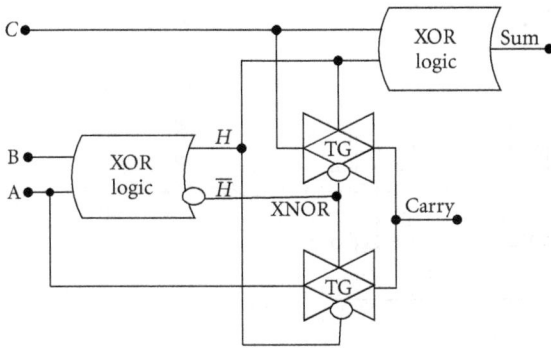

FIGURE 17: General structure of proposed XOR-XNOR-based adder.

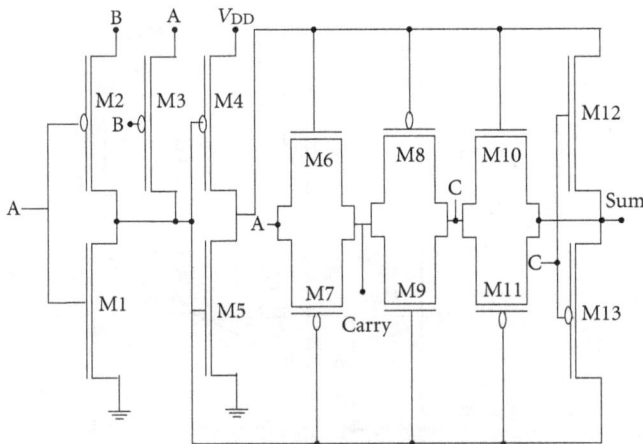

FIGURE 18: XOR- (3T-) based design 1 full adder.

and to restore the output swing as shown in Figure 18. It implements the complementary pass-transistor logic to drive the load.

A novel 16-transistor full adder circuit that generates XOR-XNOR outputs simultaneously is shown in Figure 19. Similarly in design 2 full adder circuits M1, M2 and M3 are used as XOR and the output of M4, M5, M6 is XNOR circuit. The cross-coupled PMOS transistors are connected between XOR and XNOR output to alleviate threshold problem for all possible input combination at low voltage ($0.8V_{DD}$) and reduce short-circuit power dissipation. The cross-coupled two pMOS transistors (M7, M8) are connected between XOR and XNOR outputs to eliminate the nonswing operation at low voltage.

5.2. Majority-Function-Based Full Adder. In the proposed methodology, we have designed two full adder topologies, one is based on static bridge logic style and other is based on dynamic bridge logic style. The proposed adder modules enjoy advantages of the bridge style including low-power consumption and the simplicity of the design. The proposed full adder structure design (PMajFA1) is based on capacitor network and Majority Not function as shown in Figure 20.

The proposed Majority-function-based adder design has some advantages which improves the metrics of the proposed

design significantly. In the reported previous full adder design [43], the CMOS bridge circuit does not have high driving power to drive the capacitor (2Cap) and an inverter. This increases the delay at low voltages in nanotechnology. However, in the proposed design, an inverter with high driving power drives four transistor gates (bridge circuit) and an inverter. Besides, the more driving power of the inverter in comparison to the bridge circuit and the sum of the gate capacitances of four transistors being less than the capacitance of the capacitor (2Cap) of the reported design (MajFA5) illustrate the superiority of the proposed full adder design (PMajFA1).

Furthermore, as in the proposed design three capacitors perform voltage summation to implement scaled-linear sum instead of five capacitors. It has larger noise margins than the previous design. Moreover, the proposed design have no threshold loss problem at its nodes and has higher noise margin compared to MajFA3 (minimum no of transistor) because its inverters has normal VTC curve, which works on inverters with shifted VTC and its operation is highly dependent on the proper operation of these inverters.

The Majority-function-based proposed design 2 (PMajFA2) adder uses 15 transistors and is based on regular dynamic CMOS bridge transistors. Full adder output $\overline{Carry}$ function is designed with 3-input Majority Not function logic and output Sum function is generated using dynamic CMOS bridge logic style as shown in Figure 21. The advantage of these adder cells are higher speed, lower transistor count and it compromises noise margin. This type of circuit is preferred in smaller area requirement with lesser delay at low voltage. It has larger noise margins in comparison to the previous designs and reported full adder circuits.

6. Simulation Results

The simulation has been performed for different supply voltage ranging from 0.8 V to 1.8 V, which allows us to compare the speed degradation and average power dissipation of the reported and newly designed adder topologies. The results of the designed circuits in this paper are compared with a reported standard CMOS full adder circuit. To compare one-bit full adder's performance, we have evaluated delay and power dissipation by performing simulation runs on a Cadence environment using 0.18-μm CMOS technology at room temperature.

The simulation test bench used for load analysis is shown in Figure 22. Output loads have been added according to the test bench. The two inverters with same *W/L* have been used to make output buffers. Output load was added at the input of the output buffers to evaluate driving capability of the circuits without output buffers. We used buffers to check the output logic levels. Power and delay of inverters have been included in power and delay calculation of the whole circuit. The transistor size for buffers is two for pMOS and one for nMOS.

The transistors that are used in XOR-XNOR- (3T-) based full adder designed circuits (13T & 16T) are using 3T

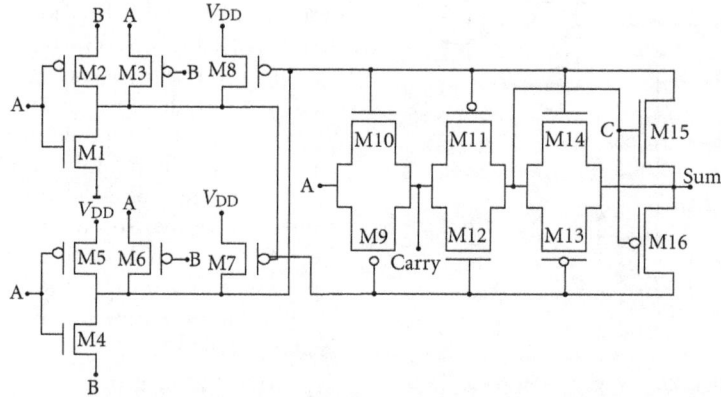

FIGURE 19: XOR-XNOR- (3T-) based design 2 full adder.

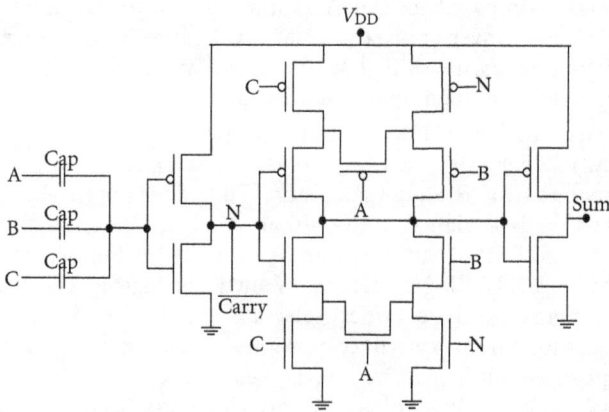

FIGURE 20: Majority-function-based adder design 1 (PMajFA1).

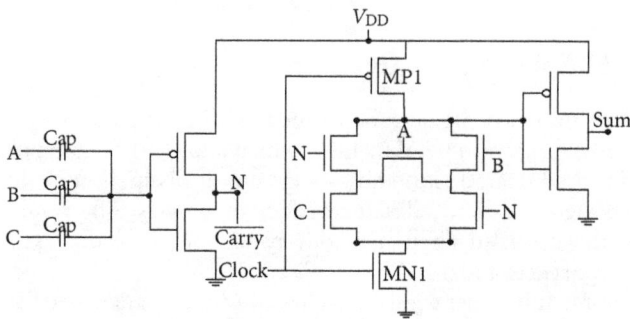

FIGURE 21: Majority function-based adder design 2 (PMajFA2).

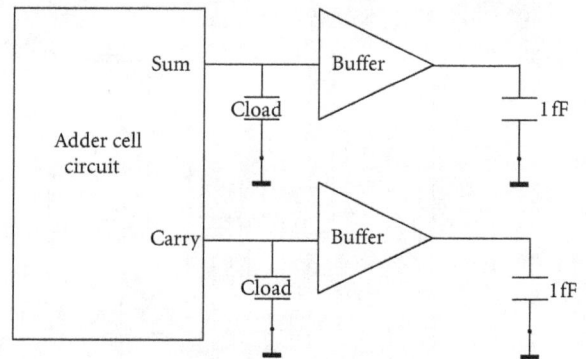

FIGURE 22: Simulation test bench for load Analysis.

transistors XOR logic. Thus the area overhead of the designed circuits is lower than that of the reported conventional adders and also some other adder circuits. By optimizing the transistor size of full adders considered, it is possible to reduce the delay of all the adders without significantly increasing the power consumption, and transistor sizes can be set to achieve minimum *power delay product* (PDP) and energy delay product (EDP). All adders were designed with minimum transistor sizes initially and then simulated. The PDP (10^{-18} j) and EDP (10^{-30} sj) are a quantitative measure of the efficiency and a compromise between power dissipation and speed. PDP and EDP are particularly important

when low power and high speed operation are needed. At low voltages, design 1 is better than 9T and design 2. From the simulation results, it is perceptible that design 1 is superior in PDP to all the other designs at all simulation conditions.

Each one-bit full adder has been analyzed in terms of propagation delay, average power dissipation, and their products. By the value of delay, power, power-delay product and energy delay product of C-CMOS, hybrid and newly designed full adders are measured. The smallest voltage that could work on 10T is 1.4 V. The lowest supply voltage for simulation comparison for conventional CMOS, and newly designed full adder circuits, is 0.8 V (V_{DD}). For each transition, the delay is measured from 50% of the input voltage swing to 50% of the output voltage swing. The maximum delay is taken as the cell delay.

High speed of the designed full adders is due to the short path between input and output logic circuit. Simulation results (Figure 23(a)) show that design 2 is the best circuit in terms of speed at all voltages since XOR and XNOR logic is generated separately in a single circuit. It has high delay and high sensitivity against voltage scaling. Design 2 is miles ahead than design 1 and shows better performance even than 9T full adder. At low voltages, design 2 shows better delay than 9T. 9T has minimum number of transistors but high delay because XNOR logic is generated using XOR with

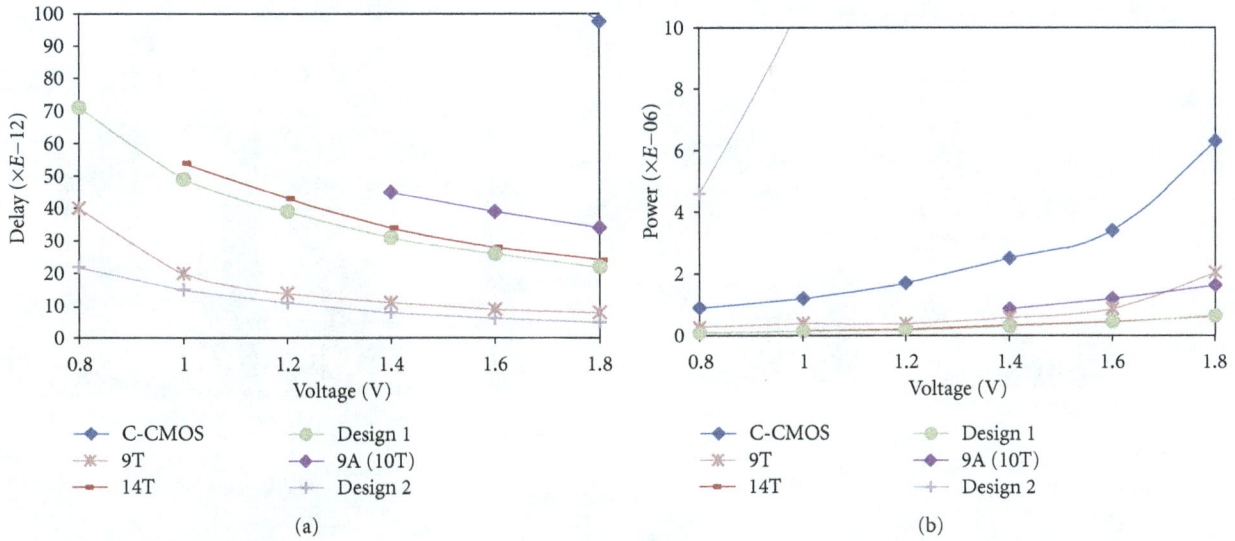

FIGURE 23: (a) Delay (ps) of XOR-XNOR-based adders. (b) Power (μW) XOR-XNOR-based adders.

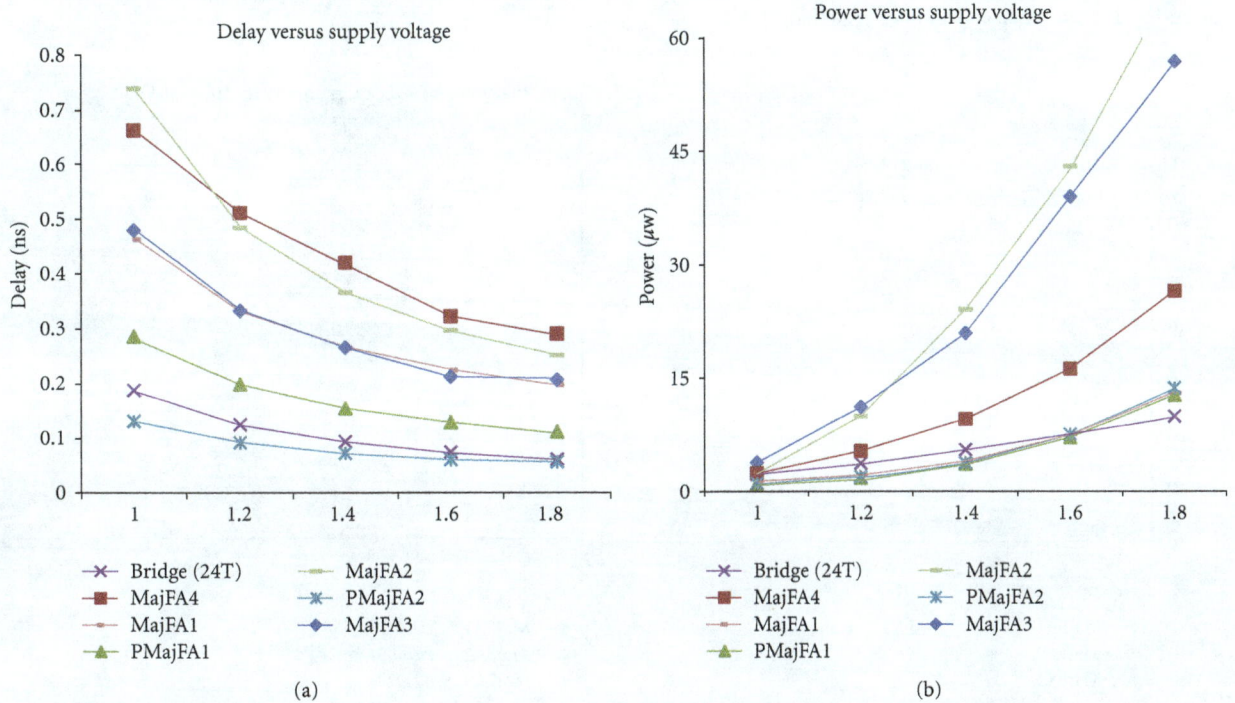

FIGURE 24: (a) Delay of Majority-function-based full adder circuits. (b) Power of Majority-function-based full adder circuits.

CMOS inverter. However, at all supply voltage variations Design 2 is faster than 9T full adder.

Figure 23(b) shows that proposed design 2 full adder is the most power consuming circuit at 1.8 V. The power consumption worsens as we increase the supply voltage. Design 1 has the least power consumption in comparison to the other simulated adder circuits. It worked successfully even at low voltage. Design 2 full adder consumes higher power due to the use of high power consuming 3T XOR and a 3T XNOR gate in a single unit.

Simulation results (Figure 24) show that Majority function based design 2 full adder (PMajFA2) is the best circuit

in terms of speed at all voltages. It has low delay and high sensitivity against voltage scaling. Design 2 is miles ahead than the reported design and shows better performance.

6.1. Load Analysis. Output load is one of the important parameters that affects power and performance of the circuits. Here we changed the output loads from 2 fF to 500 fF. A fixed value 1 fF capacitance has been added at the output of the buffer circuit. Minimum output load for all the simulation is 2 fF, except for the case in which we study the effect of output load on full adder. The effect of output load is shown in Figures 25 and 26. All the circuits have been

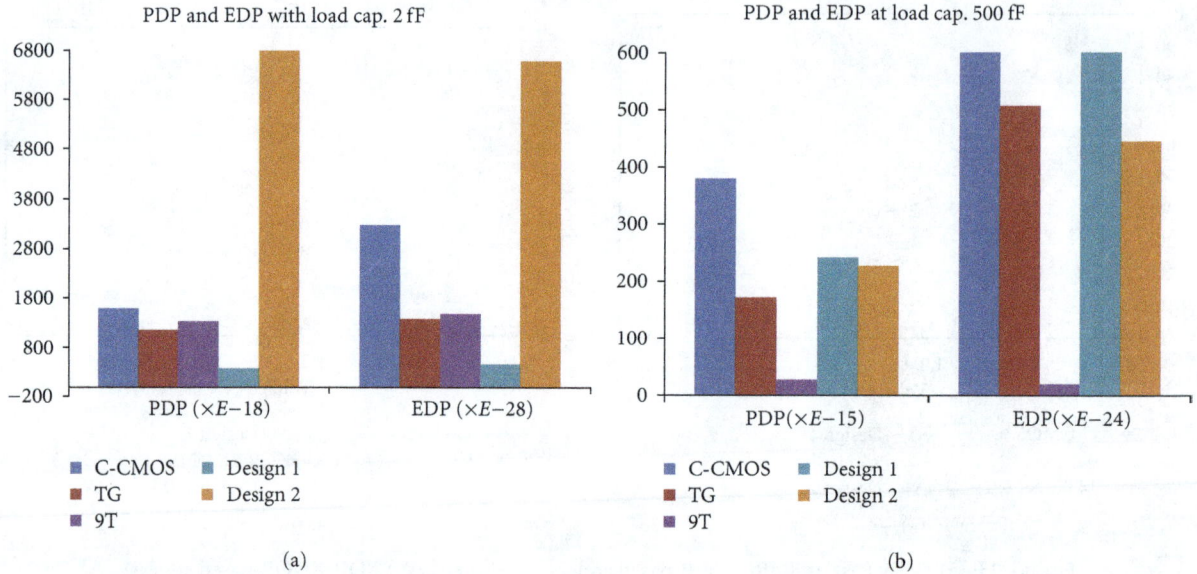

FIGURE 25: (a) PDP and EDP of XOR-XNOR based full adder cells with load capacitance (2 fF) at 1.8 V. (b) PDP and EDP of XOR-XNOR based full adder cells with load capacitance (500 fF) at 1.8 V.

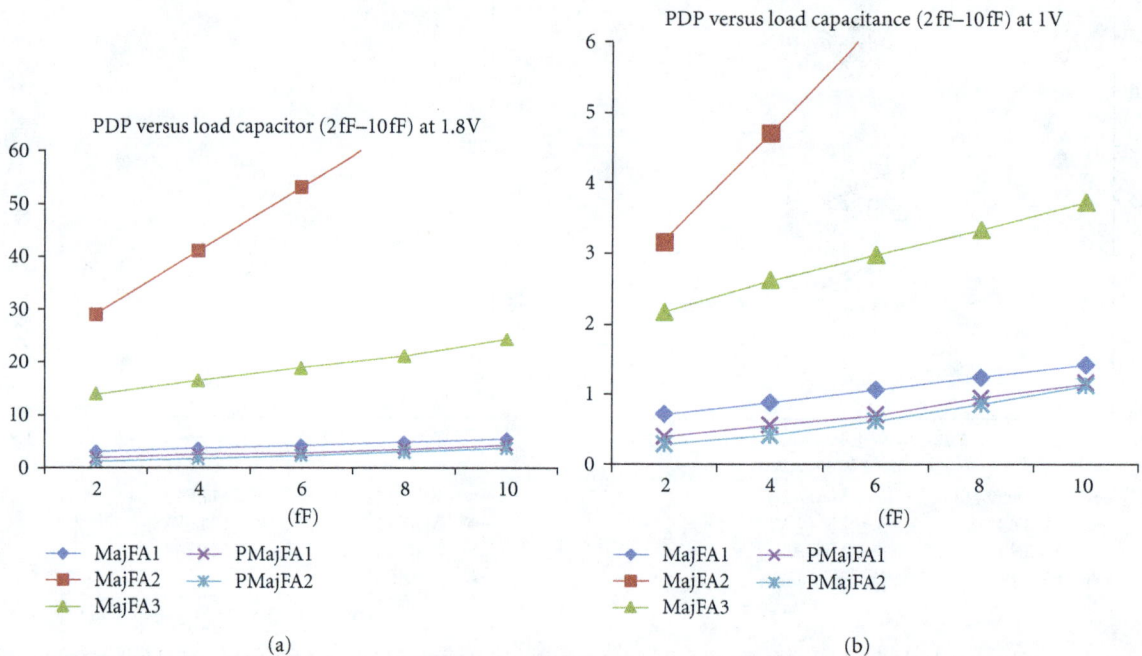

FIGURE 26: (a) PDP comparison of Majority-function-based full adder cells with capacitance load variation at 1.8 V. (b) PDP comparison of Majority-function-based full adder cells with capacitance load variation at 1 V.

optimized at 1.8 V supply voltage with 2 fF output load. For fair comparisons, the conditions were kept unchanged for all circuits.

9T is the best circuit in terms of power consumption since it has the least power consuming for all values of output load. The power of the designed circuits changes sharply by increasing the output load capacitance value as shown in Table 3 at 1.8 V. At 2 fF load, design 2 is the fastest circuit. Design 2 full adder is, however, placed second after 9T in terms of delay in high output load capacitance 500 fF. As

shown in Figure 25, design 1 has the lowest PDP for all output loads below 500 fF. In the case of 500 fF output load, 9T shows huge improvement in terms of PDP in comparison to the other designed circuits. At 2 fF, 9T has better EDP than all other designed circuits. As shown in Figures 25 and 26, design 1 has lowest EDP in all output loads below 500 fF. In case of 500 fF output load, 9T has the lowest EDP. Design 2 shows improvement in terms of EDP in comparison to the other circuits at maximum load condition. At all output load values, 9T is better than design 1 in terms of EDP.

TABLE 7: Area comparisons of the XOR-XNOR-based adders.

Designs	CMOS	TGA	10T	9T	Design 1	Design 2
Length (μm)	17.5	14	11.2	10.1	15.5	15.2
Width (μm)	7.1	9.6	6.3	8.2	5.15	6.6
Area (μm^2)	124.2	135	71	82.8	80	100.3

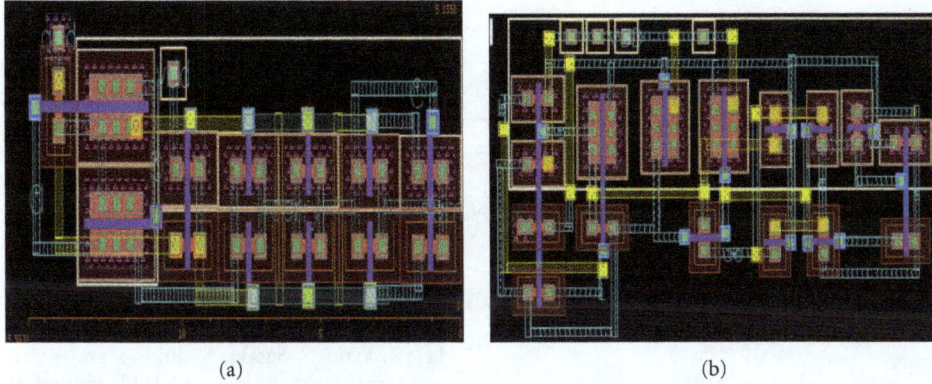

(a)   (b)

FIGURE 27: (a) Layout of design 1 (13T) full adder cell. (b) Layout of design 2 (16T) full adder cell.

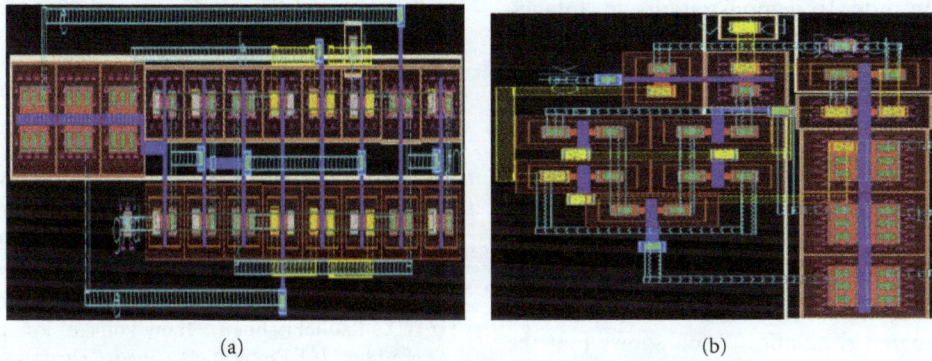

(a)   (b)

FIGURE 28: (a) Layout of design 1 (PMajFA1) full adder cell. (b) Layout of design 2 (PMajFA2) full adder cell.

Majority-function-based design 1 full adder (PMajFA1) is the best circuit in terms of power consumption for all values of output loads. The power of the designed circuits changes sharply by increasing the output load capacitance value at 1 V. At 2 fF load, Design 2 full adder (PMajFA2) is the fastest circuit. According to the simulation results, design 1 (PMajFA1) and design 2 (PMajFA2) has the lowest PDP among the other circuits for all output load capacitors as shown in Figure 26.

6.2. Layout and Area Analysis. With regard to the implementation area obtained from the layouts, it can be seen that the proposed full adders require the smallest area, which can also be considered as one of the factors for the lower delay and power consumption, as it implies smaller parasitic capacitances being driven inside the full adder. Table 7 illustrates that the layout of TGA full adders occupies the maximum silicon area. TGA adder is composed of transmission gates, which has more area due to the inefficient usage of the n-type wells. CPL adder needs the most number of metal lines to connect the complementary inputs. 10T adder has the lowest area because of the number of transistors, but the overall performance is inferior at low supply voltage (less than 1.4 V). The compact designed layout of the newly design full adders using 0.18 μm technology is all shown in Figures 27 and 28. The layout of the design 1 circuit occupies the least silicon chips area amongst all the simulated full adder cells that are performed well below 1 V. The schematic and layout editors are Cadence Virtuoso and Cadence Virtuoso XL, respectively, which are used for layout designing.

The values of layout circuit length, width, and overall area are listed in Table 7. Simulation layout results show that design 1 has the minimum power consumption due to the lowest area. 9T has minimum number of transistors but its area is much more due to the optimization of transistor parameter (W/L) which works at low voltage. Power consumption is lower than the 10T full adder and it can work up to 0.8 V satisfactorily. Design 2 has highest power dissipation when compared to the other designed full adder circuits. By

TABLE 8: Area comparisons of the Majority-function-based full adder cells.

Designs	MajFA1	MajFA3	MajFA4	PMajFA1	PMajFA2
Area (μm^2)	104.5	96	97	128	64

a perfect layouts design, more reduction in area is possible and more compact design will be implemented.

The compact designed layout of the newly design full adders using $0.18\,\mu$m technology are all shown in Figure 28. The layout of the design-2-Majority-function-based full adder circuit occupies less silicon area amongst all the simulated full adder cells that are performed well below 1 V. The value of layout circuit overall area of the conventional and newly designed full adder cells is listed in Table 8. Majority-function based Design 2 full adder (PMajFA2) has the lowest layout area.

7. Conclusion

An alternative internal logic structure for designing full adder cells is introduced. In order to demonstrate its advantages, four full adders were built in combination with pass-transistor powerless/groundless logic styles. Different adder logic styles have been implemented, simulated, analyzed, and compared. Using the adder categorization and hybrid-CMOS design style, many full adders can be conceived. As an example, new full adders designed using hybrid-CMOS design style with pass transistor are presented in this paper that targets low PDP. The hybrid-CMOS full adder shows better performance than most of the other standard full-adder cells owing to the new design modules proposed in this paper. The compared simulation result shows that the performance of the new designs is far superior to the other reference design of full adder circuits under different load conditions and for other simulation parameters.

Acknowledgments

The authors wish to thank Professor Jose Carlos Monteiro and the anonymous reviewers for their constructive comments and suggestions.

References

[1] N. Weste and K. Eshraghian, *Principles of CMOS VLSI Design: A System Perspective*, Addison-Wesley, Reading, Mass, USA, 1993.

[2] J. P. Uyemura, *Introduction to VLSI Circuits and Systems*, John Wiley & Sons, New York, NY, USA, 2002.

[3] S.-M. Kang and Y. Leblebici, *CMOS Digital Integrated Circuits: Analysis and Design*, Tata McGraw-Hill, New York, NY, USA, 2003.

[4] N. Weste and D. Harris, *CMOS VLSI Design*, Pearson Wesley, 2005.

[5] M. M. Vai, *VLSI Design*, CRC & Taylor & Francis, Boca Raton, Fla, USA, 2001.

[6] I. S. Abu-Khater, A. Bellaouar, and M. I. Elmasry, "Circuit techniques for CMOS low-power high-performance multipliers," *IEEE Journal of Solid-State Circuits*, vol. 31, no. 10, pp. 1535–1546, 1996.

[7] U. Ko, P. T. Balsara, and W. Lee, "Low-power design techniques for high-performance CMOS adders," *IEEE Transactions On Very Large Scale Integration (VLSI) Systems*, vol. 3, no. 2, pp. 327–333, 1995.

[8] A. Bellaouar and M. I. Elmasry, *Low-Power Digital VLSI Design: Circuits and Systems*, Kluwer Academic, 1995.

[9] A. Parameswar, H. Hara, and T. Sakurai, "A high speed, low power, swing restored pass-transistor logic based multiply and accumulate circuit for multimedia applications," in *Proceedings of the IEEE Custom Integrated Circuits Conference*, pp. 278–281, San Diego, Calif, USA, May 1994.

[10] A. Parameswar, H. Hara, and T. Sakurai, "A swing restored pass-transistor logic-based multiply and accumulate circuit for multimedia applications," *IEEE Journal of Solid-State Circuits*, vol. 31, no. 6, pp. 804–809, 1996.

[11] K. Yano, Y. Sasaki, K. Rikino, and K. Seki, "Top-down pass-transistor logic design," *IEEE Journal of Solid-State Circuits*, vol. 31, no. 6, pp. 792–803, 1996.

[12] D. Radhakrishnan, S. R. Whitaker, and G. K. Maki, "Formal design procedures for pass-transistor switching circuits," *IEEE Journal of Solid-State Circuits*, vol. 20, no. 2, pp. 531–536, 1984.

[13] R. Zimmermann and W. Fichtner, "Low-power logic styles: CMOS versus pass-transistor logic," *IEEE Journal of Solid-State Circuits*, vol. 32, no. 7, pp. 1079–1090, 1997.

[14] A. M. Shams and M. A. Bayoumi, "Structured approach for designing low power adders," in *Proceedings of the 31st Asilomar Conference on Signals, Systems & Computers*, vol. 1, pp. 757–761, November 1997.

[15] A. M. Shams and M. A. Bayoumi, "A novel high-performance CMOS 1-bit full-adder cell," *IEEE Transactions on Circuits and Systems II*, vol. 47, no. 5, pp. 478–481, 2000.

[16] D. Radhakrishnan, "Low-voltage low-power CMOS Full Adder," *IEE Proceedings: Circuits, Devices and Systems*, vol. 148, no. 1, pp. 19–24, 2001.

[17] S. Goel, S. Gollamudi, A. Kumar, and M. Bayoumi, "On the design of low-energy hybrid CMOS 1-bit full adder cells," in *Proceedings of the 47th IEEE International Midwest Symposium on Circuits and Systems*, pp. 209–212, July 2004.

[18] Y. Jiang, A. Al-Sheraidah, Y. Wang, E. shah, and J. Chung, "A novel multiplexer-based low power full adder," *IEEE Transaction on Circuits and Systems*, vol. 51, no. 7, pp. 345–348, 2004.

[19] S. Goel, A. Kumar, and M. A. Bayoumi, "Design of robust, energy-efficient full adders for deep-submicrometer design using hybrid-CMOS logic style," *IEEE Transactions on Very Large Scale Integration (VLSI) Systems*, vol. 14, no. 12, pp. 1309–1321, 2006.

[20] S. Wairya, R. K. Nagaria, and S. Tiwari, "A novel CMOS Full Adder topology for low voltage VLSI applications," in *Proceedings of the International Conference on Emerging Trends in Signal Processing & VLSI Design (SPVL '10)*, pp. 1142–1146, Hyderabad, India, June 2010.

[21] A. M. Shams, T. K. Darwish, and M. A. Bayoumi, "Performance analysis of low-power 1-bit CMOS Full Adder cells," *IEEE Transactions on Very Large Scale Integration (VLSI) Systems*, vol. 10, no. 1, pp. 20–29, 2002.

[22] M. Vesterbacka, "14-Transistor CMOS Full Adder with full voltage-swing nodes," in *Proceedings of the IEEE Workshop Signal Processing Systems*, pp. 713–722, October 1999.

[23] R. Shalem, E. John, and L. K. John, "Novel low power energy recovery Full Adder cell," in *Proceedings of the 9th Great Lakes Symposium on VLSI (GLSVLSI '99)*, pp. 380–383, March 1999.

[24] H. T. Bui, Y. Wang, and Y. Jiang, "Design and analysis of low-power 10-transistor Full Adders using novel XOR-XNOR gates," *IEEE Transactions on Circuits and Systems II*, vol. 49, no. 1, pp. 25–30, 2002.

[25] C. H. Chang, J. Gu, and M. Zhang, "A review of 0.18-μm Full Adder performances for tree structured arithmetic circuits," *IEEE Transactions on Very Large Scale Integration (VLSI) Systems*, vol. 13, no. 6, pp. 686–694, 2005.

[26] J. F. Lin, Y. T. Hwang, M. H. Sheu, and C. C. Ho, "A novel high-speed and energy efficient 10-transistor full adder design," *IEEE Transactions on Circuits and Systems I*, vol. 54, no. 5, pp. 1050–1059, 2007.

[27] S. Veeramachaneni and M. B. Sirinivas, *New Improved 1-Bit Full AdderCells*, CCECE/CGEI, Ontario, Canada, 2008.

[28] J. M. Wang, S. C. Fang, and W. S. Feng, "New efficient designs for XOR and XNOR functions on the transistor level," *IEEE Journal of Solid-State Circuits*, vol. 29, no. 7, pp. 780–786, 1994.

[29] N. Zhuang and H. Wu, "A new design of the CMOS full adder," *IEEE Journal of Solid-State Circuits*, vol. 27, no. 5, pp. 840–844, 1992.

[30] A. P. Chandrakasan, S. Sheng, and R. W. Brodersen, "Low-power CMOS digital design," *IEEE Journal of Solid-State Circuits*, vol. 27, no. 4, pp. 473–484, 1992.

[31] A. P. Chandrakasan and R. W. Brodersen, *Low Power Digital CMOS Design*, Kluwer Academic Publishers, 1995.

[32] H. Lee and G. E. Sobelman, "New XOR/XNOR and Full Adder circuits for low voltage, low power applications," *Microelectronics Journal*, vol. 29, no. 8, pp. 509–517, 1998.

[33] D. Radhakrishnan, "A new low power CMOS Full Adder," in *Proceedings of the the International Conference on Software Engineering (ISCE '99)*, pp. 154–157, Melaka Malaysia, 1999.

[34] M. Vesterbacka, "New six-transistor CMOS XOR circuit with complementary output," in *Proceedings of the 42nd IEEE Midwest Symposium on Circuits and Systems (MWSCAS '99)*, pp. 796–799, Las Cruces, NM, USA, August 1999.

[35] S. S. Mishra, S. Wairya, R. K. Nagaria, and S. Tiwari, "New design methodologies for high speed low power XOR-XNOR circuits," *Journal of World Academy Science, Engineering and Technology*, vol. 55, no. 35, pp. 200–206, 2009.

[36] S. Wairya, R. K. Nagaria, and S. Tiwari, "New design methodologies for high-speed low-voltage 1 bit CMOS Full Adder circuits," *International Journal of Computer Technology and Application*, vol. 2, no. 3, pp. 190–198, 2011.

[37] S. R. Chowdhury, A. Banerjee, A. Roy, and H. Saha, "A high speed 8 transistor Full Adder design using novel 3 transistor XOR gates," *International Journal of Electronics, Circuits and Systems, WASET Fall*, pp. 217–223, 2008.

[38] W. Ibrahim, V. Beiu, and M. H. Sulieman, "On the reliability of majority gates Full Adders," *IEEE Transactions on Nanotechnology*, vol. 7, no. 1, pp. 56–67, 2008.

[39] K. Navi, O. Kavehei, M. Ruholamimi, A. Sahafi, S. Mehrabi, and N. Dadkhahi, "Low-power and high-performance 1-bit CMOS Full Adder cell," *Journal of Computers*, vol. 3, no. 2, pp. 48–54, 2008.

[40] S. Wairya, R. K. Nagaria, and S. Tiwari, "New design methodologies for high speed mixed mode Full Adder circuits," *International Journal of VLSI and Communication Systems*, vol. 2, no. 2, pp. 78–98, 2011.

[41] K. Navi, V. Foroutan, M. Rahimi Azghadi et al., "A novel low-power Full-Adder cell with new technique in designing logical gates based on static CMOS inverter," *Microelectronics Journal*, vol. 40, no. 10, pp. 1441–1448, 2009.

[42] K. Navi, M. Maeen, V. Foroutan, S. Timarchi, and O. Kavehei, "A novel low-power Full-Adder cell for low voltage," *Integration, the VLSI Journal*, vol. 42, no. 4, pp. 457–467, 2009.

[43] K. Navi, M. H. Moaiyeri, R. F. Mirzaee, O. Hashemipour, and B. M. Nezhad, "Two new low-power Full Adders based on majority-not gates," *Microelectronics Journal*, vol. 40, no. 1, pp. 126–130, 2009.

3

A Signature-Based Power Model for MPSoC on FPGA

Roberta Piscitelli and Andy D. Pimentel

Computer Systems Architecture Group, Informatics Institute, University of Amsterdam, 1098 XH Amsterdam, The Netherlands

Correspondence should be addressed to Roberta Piscitelli, r.piscitelli@uva.nl

Academic Editor: Luigi Raffo

This paper presents a framework for high-level power estimation of multiprocessor systems-on-chip (MPSoC) architectures on FPGA. The technique is based on abstract execution profiles, called event signatures, and it operates at a higher level of abstraction than, for example, commonly used instruction-set simulator (ISS)-based power estimation methods and should thus be capable of achieving good evaluation performance. As a consequence, the technique can be very useful in the context of early system-level design space exploration. We integrated the power estimation technique in a system-level MPSoC synthesis framework. Subsequently, using this framework, we designed a range of different candidate architectures which contain different numbers of MicroBlaze processors and compared our power estimation results to those from real measurements on a Virtex-6 FPGA board.

1. Introduction

The complexity of modern embedded systems, which are increasingly based on multiprocessor SoC (MPSoC) architectures, has led to the emergence of system-level design. System-level design tries to cope with the design complexity by raising the abstraction level of the design process. Here, a key ingredient is the notion of high-level modeling and simulation in which the models allow for capturing the behavior of system components and their interactions at a high level of abstraction. These high-level models minimize the modeling effort and are optimized for execution speed. Consequently, they facilitate early architectural design space exploration (DSE).

An important element of system-level design is the high-level modeling for architectural power estimation. This allows to verify that power budgets are approximately met by the different parts of the design and the entire design and evaluate the effect of various high-level optimizations, which have been shown to have much more significant impact on power than low-level optimizations [1].

The traditional practice for embedded systems evaluation often combines two types of simulators, one for simulating the programmable components running the software and one for the dedicated hardware parts. However, using such a hardware/software cosimulation environment during the early design stages has major drawbacks: (i) it requires too much effort to build them, (ii) they are often too slow for exhaustive explorations, and (iii) they are inflexible in quickly evaluating different hardware/software partitionings. To overcome these shortcomings, a number of high-level modeling and simulation environments have been proposed in recent years. An example is our Sesame system-level modeling and simulation environment [2], which aims at efficient design space exploration of embedded multimedia system architectures.

Until now, the Sesame framework has mainly been focused on the system-level performance analysis of multimedia MPSoC architectures. So, it did not include system-level power modeling and estimation capabilities. In [3], we initiated a first step towards this end; however, by introducing the concept of *computational event signatures*, allowing for high-level power modeling of microprocessors (and their local memory hierarchy). This signature-based power modeling operates at a higher level of abstraction than commonly used instruction-set simulator (ISS)-based power models and is capable of achieving good evaluation performance. This is important since ISS-based power estimation generally is not suited for early DSE as it is too slow for evaluating a large design space: the evaluation of a single-design point via ISS-based simulation with a realistic benchmark program may take in the order of seconds to hundreds of seconds. Moreover, unlike many other high-level power estimation techniques, the signature-based power modeling technique still incorporates an explicit microarchitecture model of a processor, and thus is able to perform micro-architectural DSE as well.

In this paper, we extend the aforementioned signature-based power modeling work, and we present a full system-level MPSoC power estimation framework based on the Sesame framework, in which the power consumption of all the system components is modeled using signature-based models. The MPSoC power model has been incorporated into Daedalus, which is a system-level design flow for the design of MPSoC-based embedded multimedia systems [4, 5]. Daedalus offers a fully integrated tool flow in which system-level synthesis and FPGA-based system prototyping of MPSoCs are highly automated. This allows us to quickly validate our high-level power models against real MPSoC implementations on FPGA.

In the next section, we briefly describe the Sesame framework. Section 3 introduces the concept of *event signatures* and explains how they are used in the power modeling of architectures. Section 4 gives an overview of our MPSoC power modeling framework and the different components used for modeling processors, memories, and communication channels. Section 5 presents a number of experiments in which we compare the results from our models against real measurements of real MPSoC implementations on a Virtex-6 FPGA board. In Section 6, we describe related work, after which Section 7 concludes the paper.

2. The Sesame Environment

Sesame is a modeling and simulation environment for the efficient design space exploration of heterogeneous embedded systems. Using Sesame, a designer can model embedded applications and MPSoC architectures at the system level, map the former onto the latter, and perform application-architecture cosimulations for rapid performance evaluations. Based on these evaluations, the designer can further refine (parts of) the design, experiment with different hardware/software partitionings, perform simulations at multiple levels of abstraction, or even have mixed-level simulations where architecture model components operate at different levels of abstraction. To achieve this flexibility, the Sesame environment uses separate application and architectures models. According to the Y-chart approach [2], an application model derived from a target application domain describes the functional behavior of an application in an architecture-independent manner. This model correctly expresses the functional behavior, but is free from architectural issues, such as timing characteristics, resource utilization, or bandwidth constraints. Next, a platform architecture model defined with the application domain in mind defines architecture resources and captures their performance constraints. Finally, an explicit mapping step maps an application model onto an architecture model for cosimulation, after which the system performance can

be evaluated quantitatively. The layered infrastructure of Sesame is illustrated in Figure 1.

For application modeling, Sesame uses the Kahn process network (KPN) model of computation [6] in which parallel processes implemented in a high-level language communicate with each other via unbounded FIFO channels. Hence, the KPN model unveils the inherent task-level parallelism available in the application and makes the communication explicit. Furthermore, the code of each Kahn process is instrumented with annotations describing the application's computational actions which allows to capture the computational behavior of an application. The reading from and writing to FIFO channels represent the communication behavior of a process within the application model. When the Kahn model is executed, each process records its computational and communication actions, and generates a trace of *application events*. These application events are an abstract representation of the application behavior and are necessary for driving an architecture model. Application events are generally coarse-grained, such as *read(channel id, pixel block)* or *execute(DCT)*.

An architecture model simulates the performance consequences of the computation and communication events generated by an application model. It solely accounts for architectural (performance) constraints and does not need to model functional behavior. This is possible because the functional behavior is already captured by the application model, which drives the architecture simulation. The timing consequences of application events are simulated by parameterizing each architecture model component with an event table containing operation latencies. The table entries could include, for example, the latency of an *execute(DCT)* event or the latency of a memory access in the case of a memory component. With respect to communication, issues such as synchronization and contention on shared resources are also captured in the architecture model.

To realize trace-driven cosimulation of application and architecture models, Sesame has an intermediate mapping layer with two main functions. First, it controls the mapping of Kahn processes (i.e., their event traces) onto architecture model components by dispatching application events to the correct architecture model component. Second, it makes sure that no communication deadlocks occur when multiple Kahn processes are mapped onto a single architecture model component. In this case, the dispatch mechanism also provides various strategies for application event scheduling.

Extending the Sesame framework to also support power modeling of MPSoCs could be done fairly easily by adding power consumption numbers to the event tables. So, this means that a component in the architecture model not only accounts for the timing consequences of an incoming application event, but also accounts for the power that is consumed by the execution of this application event (which is specified in the event tables now). The power numbers that need to be stored in the event tables can, of course, be retrieved from lower-level power simulators or from (prototype) implementations of components. However, simply adding fixed power numbers to the event tables would be a rigid solution in terms of DSE: these numbers would only be

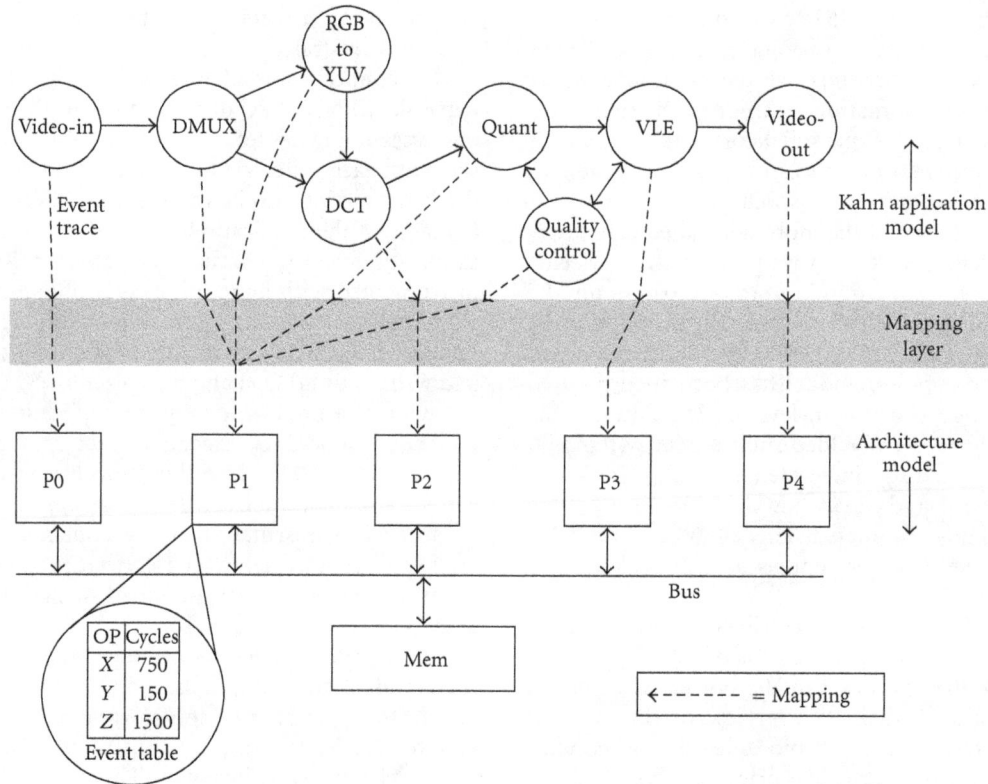

FIGURE 1: The Sesame system-level simulation environment.

valid for the specific implementation used for measuring the power numbers. Therefore, we propose a high-level power estimation method based on so-called event signatures that allows for more flexible power estimation in the scope of system-level DSE. As will be explained in the next sections, signature-based power estimation provides an abstraction of processor activity and communication in comparison to traditional ISS-based power models, while still incorporating an explicit microarchitecture model and thus being able to perform microarchitectural DSE.

3. Event Signatures

An event signature is an abstract execution profile of an application event that describes the computational complexity of an application event (in the case of computational events) or provides information about the data that is communicated (in the case of communication events). Hence, it can be considered as metadata about an application event.

3.1. Computational Events Signatures. A computational signature describes the complexity of computational events in a (micro-)architecture-independent fashion using an abstract instruction set (AIS) [3]. Currently, our AIS is based on a load-store architecture and consists of *instruction classes*, such as Simple Integer Arithmetic, Simple Integer Arithmetic Immediate, Integer Multiply, Branch, Load, and Store. The high level of abstraction of the AIS should allow for capturing the computational behavior of a wide range of

RISC processors with different instruction-set architectures. To construct the signatures, the real machine instructions of the application code represented by an application event (derived from an instruction-set simulator as will be explained here in after) are first mapped onto the various AIS instruction classes, after which a compact execution profile is made. This means that the resulting signature is a vector containing the instruction counts of the different AIS instruction classes. Here, each index in this vector specifies the number of executed instructions of a certain AIS class in the application event. We note that the generation of signatures for each application event is a one-time effort, unless for example, an algorithmic change is made to an application event's implementation.

To generate computational signatures, each Kahn application process is simulated using a particular instruction-set simulator (ISS); depending on the class of target processor, the application will be mapped on. For example, we currently use ISSs from the SimpleScalar simulator suite [7] for the more complex multiple-issue processors, while we deploy the MicroBlaze cycle-accurate instruction-set simulator provided by Xilinx for the more simple soft cores. Taking the signature generation for the MicroBlaze processor as an example in Figure 2, application files are loaded into mb-gdb, which is the GNU C debugger for MicroBlaze. Mb-gdb is used to send instructions of the loaded executable files to the MicroBlaze instruction-set simulator and to perform cycle-accurate simulation of the execution of the software programs, as in [8].

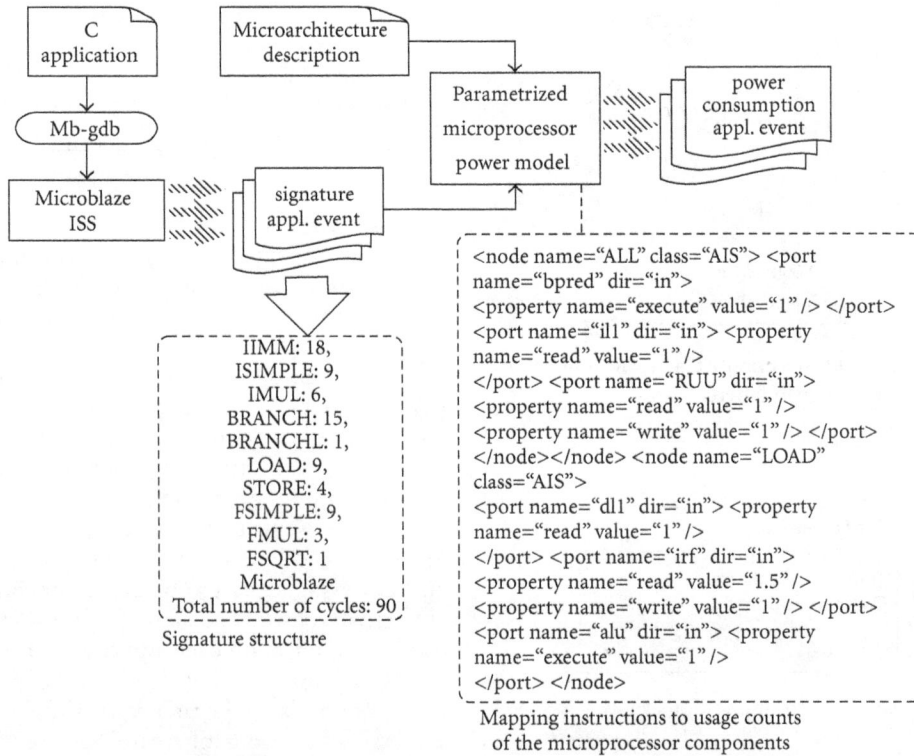

IIMM: 18,
ISIMPLE: 9,
IMUL: 6,
BRANCH: 15,
BRANCHL: 1,
LOAD: 9,
STORE: 4,
FSIMPLE: 9,
FMUL: 3,
FSQRT: 1
Microblaze
Total number of cycles: 90
Signature structure

```
<node name="ALL" class="AIS"> <port
name="bpred" dir="in">
<property name="execute" value="1" /> </port>
<port name="il1" dir="in"> <property
name="read" value="1" />
</port> <port name="RUU" dir="in">
<property name="read" value="1" />
<property name="write" value="1" /> </port>
</node></node> <node name="LOAD"
class="AIS">
<port name="dl1" dir="in"> <property
name="read" value="1" />
</port> <port name="irf" dir="in">
<property name="read" value="1.5" />
<property name="write" value="1" /> </port>
<port name="alu" dir="in"> <property
name="execute" value="1" />
</port> </node>
```

Mapping instructions to usage counts
of the microprocessor components

FIGURE 2: Computational event signature generation for MicroBlaze.

Using these ISSs, the event signatures are constructed—by mapping the executed machine instructions onto the AIS as explained above—for every computational application event that can be generated by the Kahn process in question. The event signatures act as an input to our parameterized microprocessor power model, which will be described in more detail in the next section. For each signature, the ISS may also provide the power model with some additional microarchitectural information, such as cache missrates, and branch misprediction rates. In our case, only instruction and data cache miss-rates are used. As will be explained later on, the microprocessor power model subsequently uses a microarchitecture description file in which the mapping of AIS instructions to usage counts of microprocessor components is described.

The microprocessor power model also uses a microarchitecture description file in which the mapping of AIS instructions to usage counts of microprocessor components is described. An example fragment of this mapping description is shown in Figure 2. It specifies that for every AIS instruction (indicated by the ALL tag), the instruction cache (il1) is read, the register update unit (RUU) is read and written, and branch prediction is performed. Furthermore, it specifies that for the AIS instruction LOAD, the ALU is used (to calculate the address), the level-1 data cache (dl1) is accessed, and the integer register file (irf) is read and written. With respect to the latter, it takes register and immediate addressing modes into account by assuming 1.5 read operations to the irf on average. In addition, the microarchitecture description file also contains the

parameters for our power model, such as, the dimensions and organization of memory structures (caches, register file, etc.) in the microprocessor, clock frequency, and so on. Clearly, this microarchitecture description allows for easily extending the AIS and facilitates the modeling of different microarchitecture implementations.

3.2. Communication Event Signatures. In Sesame, the Kahn processes generate *read* and *write* communication events as a side effect of reading data from or writing data to ports. Hence, communication events are automatically generated. For the sake of power estimation, the communication events are also extended with a signature, as shown in Figure 4. A communication signature describes the complexity of transmitting data through a communication channel (e.g., FIFO, Memory Bus, PLB Bus) based on the dimension of the transmitted data and the statistical distribution of the contents of the data itself.

More specifically, we calculate the average Hamming distance of the data words within the data chunk communicated by a *read* or *write* event (which could be, e.g., a pixel block or even an entire image frame), after which the result is again averaged with Hamming distance of the previous data transaction on the same communication channel. In this way, we can get information about the usage of the channel and the switching factor, which is related to the data distribution. In our transaction-level architecture models, we use the assumption that the communications performed by the KPN application model are not interleaved at the architecture level. For example, if a pixel block is transferred

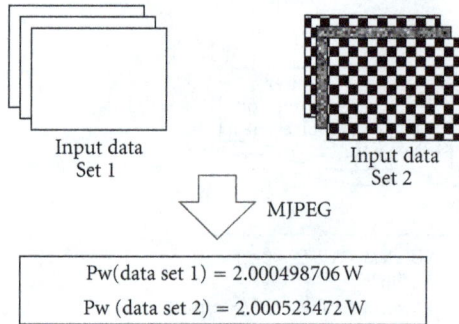

Figure 3: Measured power consumption of MJPEG application mapped on one MicroBlaze using two different input sets.

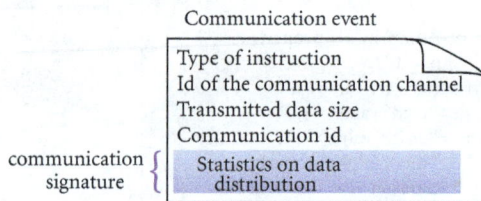

Figure 4: Structure of communication events.

between two KPN processes, then the architecture model simulates the (bus/network) transactions of the consecutive data words in the pixel block, without interleaving these transactions with other ones. In Figure 3, we show the impact of power on a MJPEG application using input sets with different data distribution. In the first input data set picture, the correlation between pixel blocks is very high, and consequently the average Hamming distance of the data will be zero. This results in lower power values with respect to the second input data set picture, which presents a higher Hamming distance distribution.

3.3. Signature-Based, System-Level Power Estimation. In
Figure 5, the entire signature-based power modeling framework is illustrated. First the event traces are generated, together with the communication signatures.

The Kahn application model is used to generate the event traces, which represent the workload that is imposed on the underlying MPSoC architecture model. During this stage, the *average Hamming distance*, as explained in the previous subsection, is computed. This information is then integrated in the trace events, forming the communication signature. The communication signature generation is mapping dependent: communication patterns change with different mappings.

In addition, the computational signatures are generated (Figure 5, left side). In particular, the Kahn application processes for which a power estimation needs to be performed, are simulated using the ISS, constructing the event signatures (as explained in the previous section) for every computational application event that can be generated by the Kahn process in question. After that the computational event signatures are generated, the power consequences of trace

Table 1: Different possibilities of reusing signatures in DSE.

Comp. signatures	Comm. signatures
μ-architectural exploration	μ-architectural exploration
Mapping exploration (limited)	Architectural exploration

events generated by the application model, are computed. As will be explained in the next section, the microprocessor power model uses a microarchitecture description file in which the mapping of AIS instructions to usage counts of microprocessor components is described.

The Sesame architecture model simulates the performance and power consequences of the computation and communication events generated by the application model. To this end, each architecture model component is parameterized with an event table containing the latencies of the application events it can execute (as explained in Section 2). Moreover, each architecture model component now also has an underlying signature-based power model. These models are activity-based. The activity counts are derived from the different application events in the event traces as well as the signature information of the separate events. The total power consumption is then obtained by simply adding the average power contributions of microprocessor(s), memories, and interconnect(s).

The structure of the entire system-level power model is composed by separate and independent modules, which allow for the reuse of the different underlying component models as well as the generated signatures (as shown in Table 1). For example, once computational signatures are generated for application events, it is possible to explore different microarchitectures executing the same application with the same mapping. Moreover, given the computational event signatures, it is also possible to do mapping exploration, limited to the case of homogeneous systems. Communication signatures can be reused for both microarchitectural and architectural exploration.

4. Power Model

We propose a high-level power estimation method based on the previously discussed event signatures that allows for flexible power estimation in the scope of system-level DSE. As will be explained in the subsequent subsections, signature-based power estimation provides an abstraction of processor (and communication) activity in comparison to, for example, traditional ISS-based power models, while still incorporating an explicit microarchitecture model and thus being able to perform microarchitectural DSE. The power models are based on FPGA technology, since we have incorporated these models in our system-level MPSoC synthesis framework Daedalus [5], which targets FPGA-based (prototype) implementations. The MPSoC power model is formed by three main building blocks, modeling the microprocessors, the memory hierarchy, and the interconnections, respectively. The model is based on the activity counts that can be derived from the application events and their signatures as described before and on the power characteristics of the

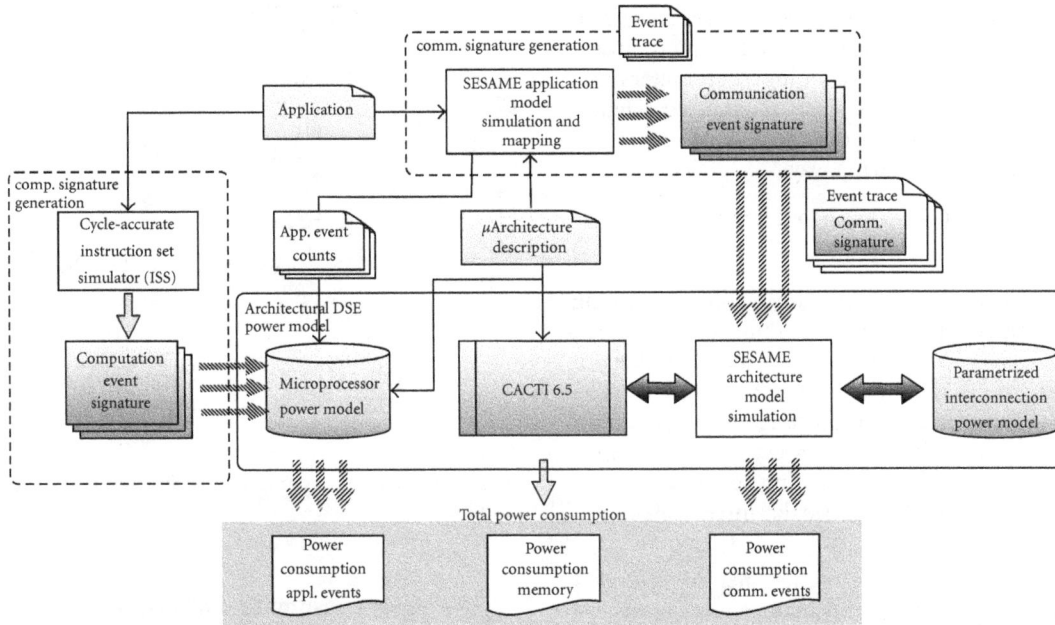

FIGURE 5: System-level power estimation framework.

components themselves, measured in terms of LUTs used. In particular, we estimate through synthesis on FPGA the maximum number of LUTs used for each component. The resulting model is therefore a compositional power model, consisting of the various components (for which the models are described below) used in the MPSoC under study. In the remainder of this paper, we will focus on homogeneous systems, but the used techniques do allow the modeling and simulation of heterogeneous systems as well.

4.1. Interconnection Power Model. In this section, we derive architectural level parameterized, and activity-based power models for major network building blocks within our targeted MPSoCs. These include FIFO buffers, crossbar switches, buses, and arbiters. The currently modeled building blocks—network components as well as processor and memory components—are all part of the IP library of our Daedalus synthesis framework [5], which allows the construction of a large variety of MPSoC systems. Consequently, all our modeled MPSoCs can actually be rapidly synthesized to and prototyped on FPGA, allowing us to easily validate our power models.

Our network power models are composed of models for the aforementioned network building blocks, for which each of them we have derived parameterized power equations. These equations are all based on the common power equation for CMOS circuits:

$$P_{interconnect} = V_{dd}^2 f C \alpha, \quad (1)$$

where f is the clock frequency, V_{dd} the operating voltage, C the capacitance of the component, and α is the average switching activity of the component, respectively. The capacitance values for our component models are obtained

through an estimation of the number of LUTs used for the component in question as well as the capacitance of a LUT itself. Here, we estimate the number of LUTs needed for every component through synthesis, after which the capacitance is obtained using the X-Power tool from Xilinx. The activity rate α is primarily based on the *read* and *write* events from the application event traces that involve the component in question. For example, for an arbiter component of a bus, the total time of read and write transactions to the arbiter (i.e., the number of *read* and *write* events that involve the arbiter) as a fraction of the total execution time is taken as the access rate (i.e., activity rate). Consequently, the power consumption of an arbiter is modeled as follows:

$$P_{arbiter} = \beta \times V_{dd}^2 \times f \times C_{LUT} \times n_{LUTs} \times access\_rate, \quad (2)$$

where C_{LUT}, n_{LUTs}, f, and V_{dd} are, respectively, the estimated capacitance of a LUT, the estimated number of LUTs needed to build the arbiter, the clock frequency, and the operating voltage. β is a scaling factor obtained through precalibration of the model, and

$$access\_rate = \frac{T_{reads} + T_{writes}}{T_{total\_exec}}. \quad (3)$$

Here, T_{reads} and T_{writes} are the total times spend on the execution of read and write transactions, respectively, and T_{total_exec} is the total execution time.

For communication channels like busses, not only the number of *read* and *write* events play a role to determine the activity factor, but also the data that is actually communicated. To this end, we consider the *Hamming distance distribution* between the data transactions, as explained in the previous section on communication signatures. Thus, every communication trace event is carrying the statistical activity-based information of the channel from/to which the data is

read/written. Consequently, for any activity (read/write of
data) in the channel, the dynamic power of the interconnection is calculated according to technology parameters and
the statistical distribution of the data transmitted. Hence,
for every packet transmitted over the channel, the estimated
power is computed in the following way:

$$P_{chan} = \beta \times V_{dd}^2 \times f \times C_{chan} \times n_{LUTs} \times \text{Hamm\_dist}(e), \tag{4}$$

where β, C_{chan}, f, V_{dd}, and n_{LUTs} are again the scaling
factor, estimated capacitance of the communication channel,
clock frequency, the operating voltage, and number of
LUTs needed to build the interconnection channel. The
Hamm_dist(e) parameter is the average Hamming distance
of the data transmitted in the *read/write* events. In our models, leakage power is calculated according to the estimated
look-up tables needed to build a particular interconnection.

4.2. Memory Power Model.
For on-chip memory (level 1
and 2 caches, register file, etc.) and main memory, we use
the analytical energy model developed in CACTI 6.5 [9]
to determine the power consumption of read and write
accesses to these structures. These power estimates include
leakage power. The access rates for the processor-related
memories, such as caches and register file, are derived from
the computational signatures, as will be explained in the next
subsection. Moreover, we use the cache missrate information
provided by the ISS used to generate the computational
signatures to derive the access counts for structures like the
level-2 cache and the processor's load/store queue.

For the main memory and communication buffers, we
calculate the access rate in the same fashion as for a network
arbiter component as explained above: the communication
application events are used to track the number of accesses to
the memory. That is, the total time taken by read and write
accesses (represented by the communication application
events) to a memory as a fraction of the total execution
time is taken as the access rate. Subsequently, the signal
rate represents the switching probability of the signals. For
every read/write event to the memory, the average Hamming
distance contained in the communication event signature is
extracted, and the signal rate is calculated as follows:

$$\text{signal\_rate} = \gamma \times \text{Hamm\_dist}(e), \tag{5}$$

where the γ is again a scaling factor obtained through
precalibration of the model.

4.3. Microprocessor Power Model.
The microprocessor model
that underlies our power model is based on [3]. It assumes
a dynamic pipelined machine, consisting of one arithmetic
logical unit, one floating point unit, a multiplier, and two
levels of caches. However, this model can easily be extended
to other processor models, by simply introducing new units.
For the power model of the clock component, three sub-
components are recognized: the clock distribution wiring,
the clock buffering, and the clocked node capacitance. We
assume a H-tree-based clock network using a distributed
driver scheme (i.e., applying clock buffers) [3].

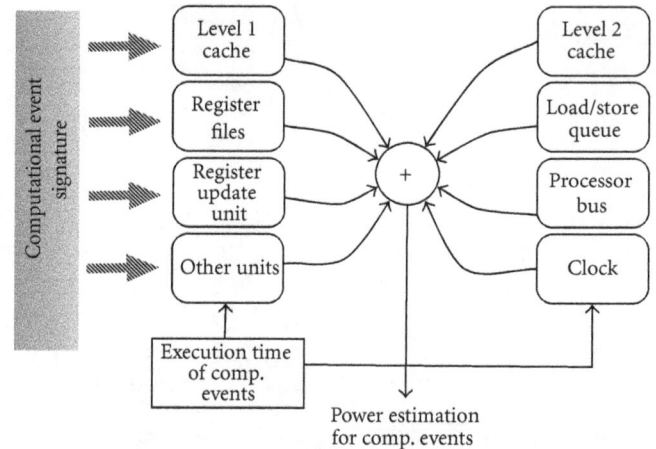

FIGURE 6: Different components in the microprocessor power
model.

The power consumption of a computational application
event is calculated by accumulating the power consumption
of each of the components that constitute the microprocessor
power model, as shown in Figure 6. More specifically, the first
step to calculate an application event's power consumption is
to map its signature to usage counts of the various processor
components. So, here it is determined how often for example,
the ALU (see other units in Figure 6), the register file and the
level-1 instruction and data caches are accessed during the
execution of an application event. The microprocessor power
model uses an XML-based micro-architecture description
file in which the mapping of AIS instructions to usage counts
of microprocessor components is described. This microarchitecture description file also contains the parameters for
our microprocessor power model, such as, the dimensions
and organization of memory structures (caches, register file,
etc.) in the microprocessor, clock frequency, and so on.
Clearly, this micro-architecture description allows for easily
extending the AIS and facilitates the modeling of different
micro-architecture implementations.

The above ingredients (the event signatures, additional
micro-architectural information per signature such as cache
statistics, and the micro-architecture description of the
processor) subsequently allow the power model to produce
power consumption estimates for each computational application event by accumulating the power consumption of the
processor components used by the application event.

5. Validation

As mentioned before, we have integrated our power model
into the Daedalus system-level design flow for the design
of MPSoC-based embedded multimedia systems [4, 5]. This
allows direct validation and calibration of our power model.

5.1. The Daedalus Design Flow.
Daedalus offers a fully
integrated tool flow in which design space exploration (DSE),
system-level synthesis, application mapping, and system
prototyping of MPSoCs are highly automated, which allows

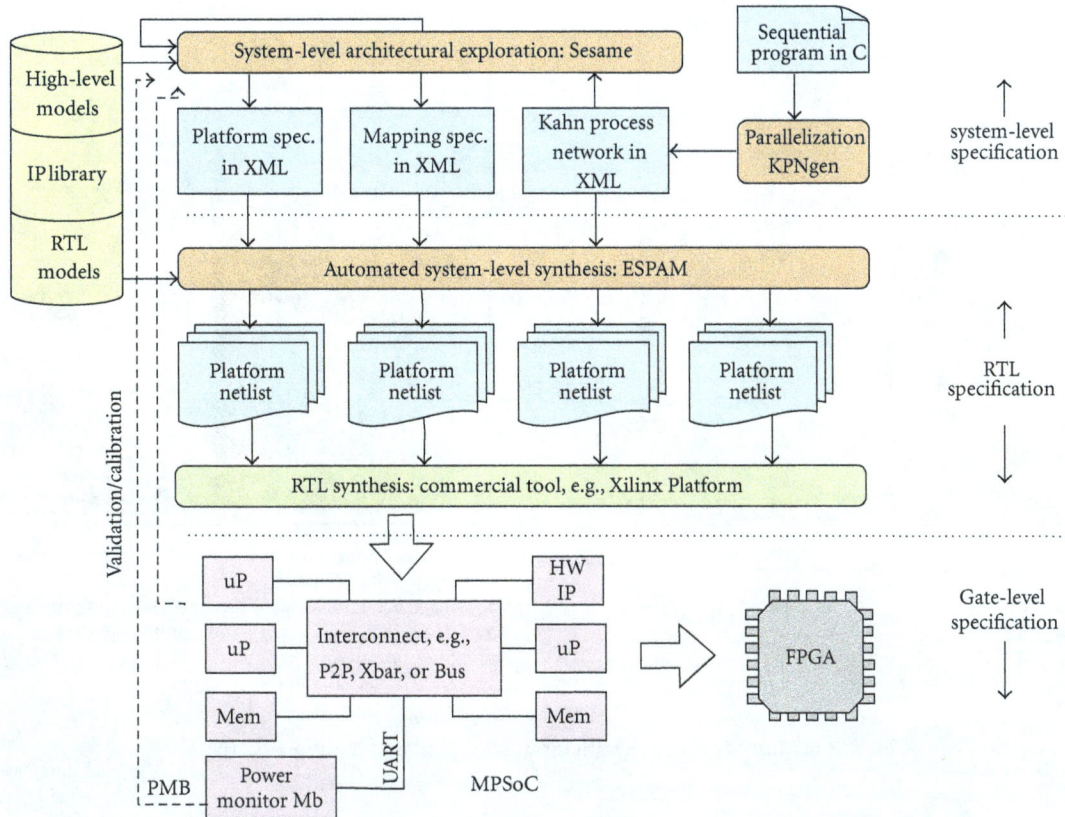

FIGURE 7: The Daedalus design and validation tool flow.

a direct validation and calibration of our power model. In Figure 1, the conceptual design flow of the Daedalus framework is depicted.

A key assumption in Daedalus is that the MPSoCs are constructed from a library of predefined and preverified IP components. These components include a variety of programmable and dedicated processors, memories, and interconnects, thereby allowing the implementation of a wide range of MPSoC platforms. So, this means that Daedalus aims at composable MPSoC design, in which MPSoCs are strictly composed of IP library components. Daedalus consists of three core tools.

Starting from a sequential multimedia application specification in C, the KPNgen tool [6] allows for automatically converting the sequential application into a parallel Kahn process network (KPN) specification. Here, the sequential input specifications are restricted to so-called static affine-nested loop programs, which is an important class of programs in, for example, the scientific and multimedia application domains. The generated or handcrafted KPNs (the latter in the case that, e.g., the input specification did not entirely meet the requirements of the KPNgen tool) are subsequently used by the Sesame modeling and simulation environment [2, 10] to perform system-level architectural design space exploration. To this end, Sesame uses (high-level) architecture model components from the IP component library (see the left part of Figure 7). As discussed before, Sesame allows for quickly evaluating the performance of different application to architecture mappings, HW/SW partitionings, and target platform architectures. Such exploration should result in a number of promising candidate system designs, of which their specifications (system-level platform description, application-architecture mapping description, and application description) acting as an input to the ESPAM tool [4, 5]. This tool uses these system-level input specifications, together with RTL versions of the components from the IP library, to automatically generate synthesizable VHDL that implements the candidate MPSoC platform architecture. In addition, it also generates the C code for those application processes that are mapped onto programmable cores. Using commercial synthesis tools and com pilers, this implementation can be readily mapped onto an FPGA for prototyping. Such prototyping also allows calibrating and validating Sesames system-level models, and as a consequence, improving the trustworthiness of these models.

5.2. Experimental Results. By deploying Daedalus, we have designed several different candidate MPSoC configurations and compared our power estimates for these architectures with the real measurements. The studied MPSoCs contain different numbers of MicroBlaze processors that are interconnected using a crossbar network and also a point-to-point network. The softcores on the FPGA device used in the framework do not use caches at this moment. This is considered to be future work. The validation environment

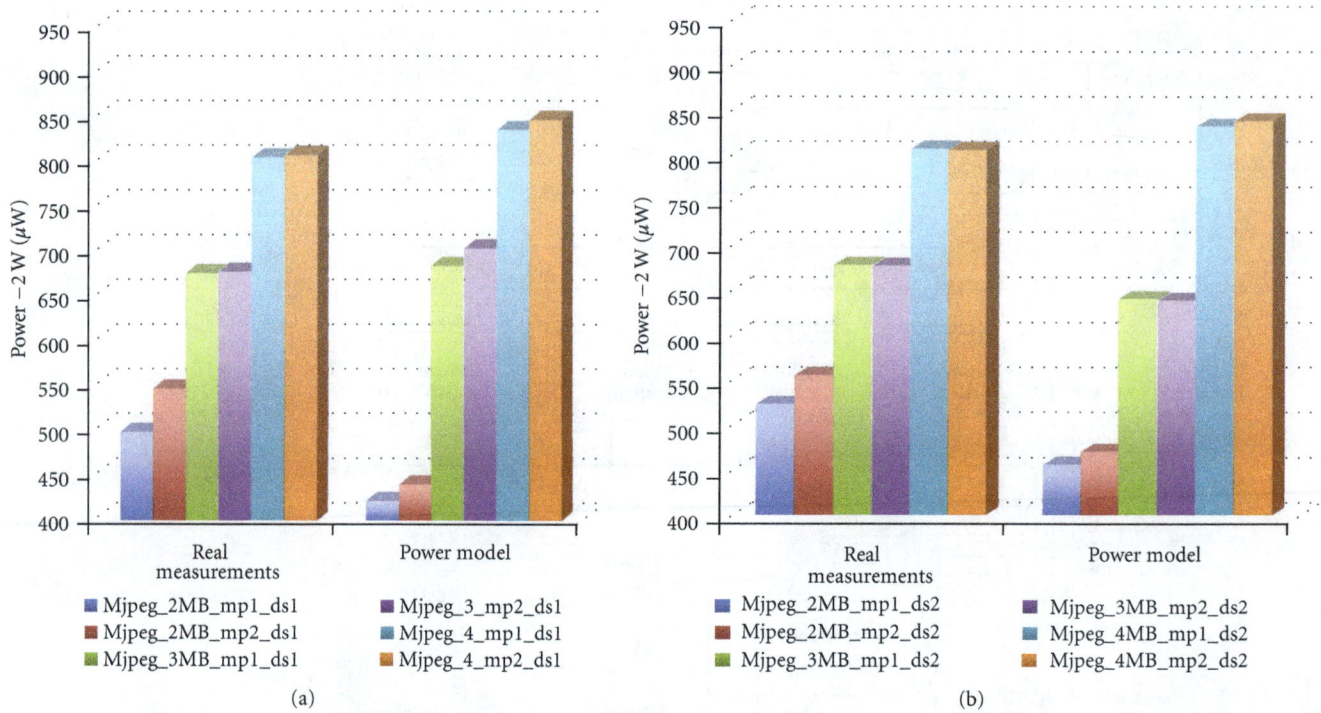

FIGURE 8: Mjpeg application with input set ds1 (a) and input set ds2 (b).

is formed by the architecture itself and an extra MicroBlaze. This extra MicroBlaze polls the power values in the internal measurement registers in our target Virtex-6 FPGA and interfaces an I2C controller in the FPGA design with the I2C interface of the PMBus controller chip [11]. In order to do this, it runs a software driver which implements the PMBus protocol [11]. The extra MicroBlaze "polls" the power values in the Virtex-6 FPGA internal measurement registers and prints out the read values through the UART to the pc, as shown in Figure 7. In this way, we have a fully automated system to register the power values of an architecture running a particular application with a given mapping. As we introduced an extra MicroBlaze in the design, the resulting power consumption of the system is scaled by a fixed factor, which is dependent on the measurement infrastructure. This is, however, not a problem since our primary aim is to provide high fidelity rankings in terms of power behavior (which is key to early design space exploration) rather than obtaining near-perfect absolute power estimations [12]. Evidently, the additional power consumed by the extra MicroBlaze does not affect the fidelity of the rankings (i.e., the extra MicroBlaze exists in every MPSoC configuration), while the power measurements obtained are much more accurate compared to, for example, using a simulator [13].

The results of the validation experiments are shown in Figures 8, 9, 10, and 11. In the experiments, we compare the total power consumption, which is both leakage and dynamic power. In these experiments, we mapped three different parallel multimedia applications onto the target MPSoCs: a Motion-JPEG encoder (Mjpeg), a Periodogram, which is an estimate of the spectral density of a signal, and a Sobel

filter for edge detection in images. In addition, for each of the applications, we also investigated two different task mappings onto the target architectures. Here, we selected one "good" mapping, in terms of task communication, as well as a "poor" one for each application. That is in the "good" mapping we minimize task communications, while in the "poor" one we maximize task communications. The experiments in Figures 8, 9, 10, and 11 apply the following notation: app_{name}-n_{proc}-$mapping_{type}$, where app_{name} is the application considered, n_{proc} indicates the number of processors used in the architecture (e.g., "3mb" indicates an MPSoC with 3 MicroBlaze processors), and $mapping_{type}$ refers to the type of mapping used. With respect to the latter, the tag mp1 indicates the good mapping, while mp2 refers to the poor mapping. For the Motion-JPEG application, we also considered two different data input sets: the first input set (ds1) is characterized by a high data correlation, while the second input set (ds2) has a very low data correlation, in terms of measured average *Hamming distance distribution* of the input data. With respect to our previous work [14], we extend the analysis to a point-to-point architecture based on FIFOs. That is we tested the power model on two different communication architecture configurations: the first one is crossbar-based, while the second one is a point-to-point network based on FIFOs. The power values in Figures 8, 9, 10, and 11 are scaled by a factor of 2 W for the sake of improved visibility. Most charts show a very little difference between the good and bad configurations (mp1 versus mp2) for a number of processors greater than 2; this is explained by the fact that a design with a larger number of processors implies a higher use of the communication channels. Given an application with m tasks and n processors, if $m \gg n$,

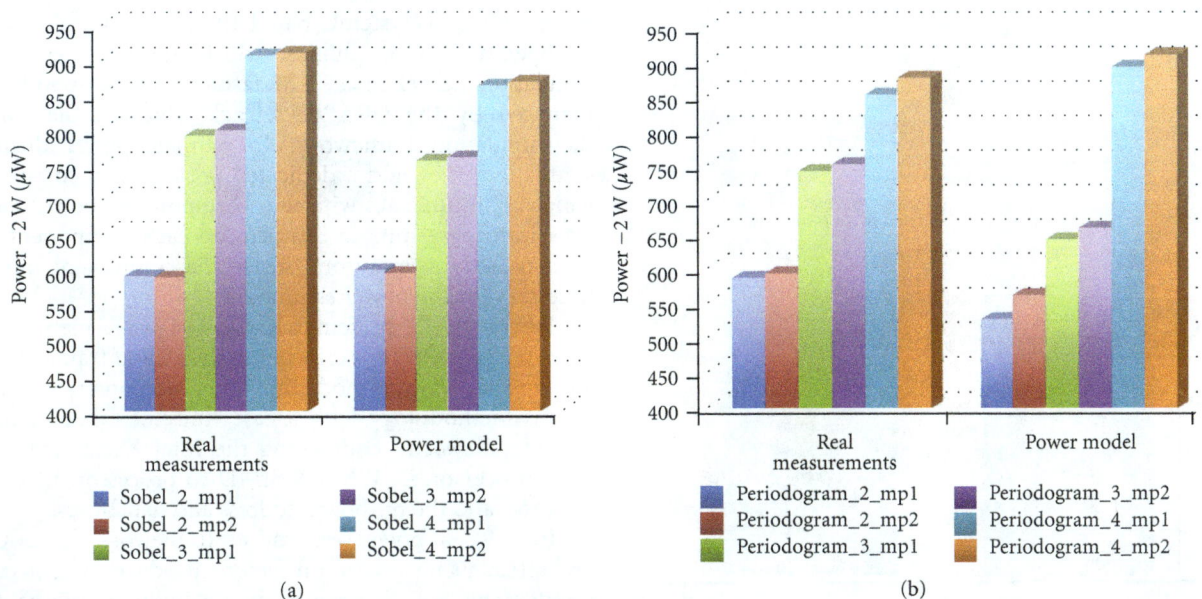

FIGURE 9: Sobel filter (a) and Periodogram application (b).

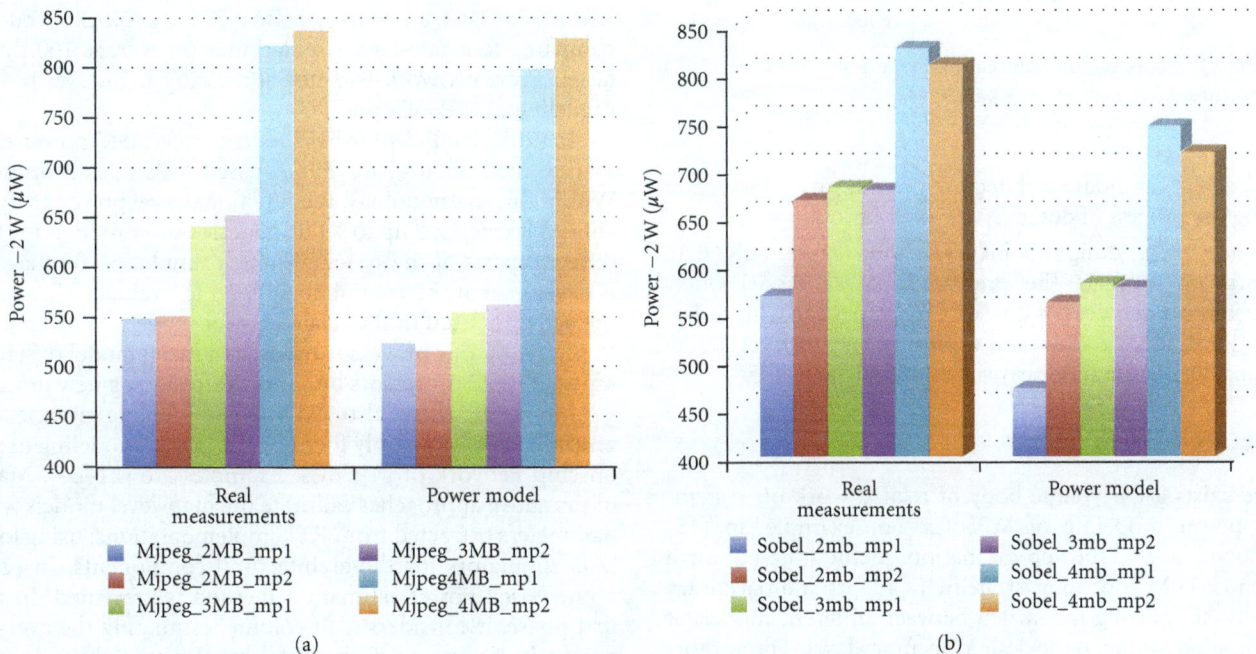

FIGURE 10: Mjpeg (a) and Sobel (b) applications in a point-to-point FIFO architecture.

then this implies that a good mapping can be beneficial for reducing tasks communication. However, in the case of $m = n$, the tasks mapping cannot avoid substantial communications.

The results in Figures 8, 9, 10, and 11 show that our power model performs quite decently in terms of absolute accuracy. We observed an average error of our power estimations of around 7%, with a standard deviation of 5% for the crossbar networks and an average error of our power estimations of around 10%, with a standard deviation of 6% for the point-to-point networks. More important in

the context of early design space exploration, however, is the fact that our power model appears to be very capable of estimating the right power consumption trends for the various MPSoC configurations, applications, and mappings. We explicitly checked the fidelity of our estimations in terms of quality ranking of candidate architectures by ranking all design instances according to their consumed power for a specific application. Our estimates result in a ranking of the power values that is correct for every application we considered, therefore showing a high fidelity. This high-fidelity quality-ranking of candidate architectures thus allows

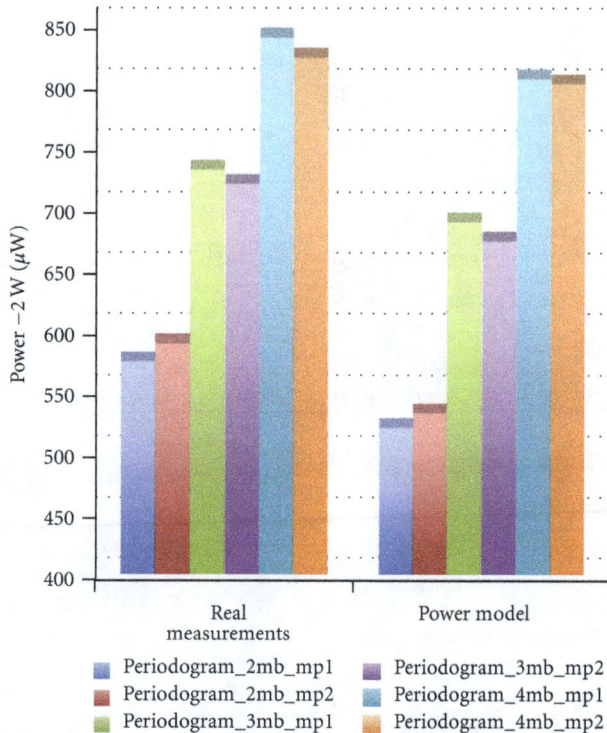

FIGURE 11: Periodogram application in a point-to-point FIFO architecture.

for a correct candidate architecture generation and selection during the process of design space exploration.

Since every design point evaluation takes only 0.16 seconds on average, the presented power model offers remarkable potentials for quickly experimenting with different MPSoC architectures and exploring system-level design options during the very early stages of design.

6. Related Work

There exists a fairly large body of related work on system-level power modeling of MPSoCs. For example, in [15] developed a SoC power estimation method based on a SystemC TLM modeling strategy. It adopts multiaccuracy models, supporting the switch between different models at run time according to the desired accuracy level. The authors validate their model using the STBus NoC and an analytical power model of this NoC. An MPEG4 application was tested, achieving up to 82% speed-up compared to TLM BCA (bus cycle-accurate) simulation.

Atitallah et al. [16] uses a stack of abstract models. The higher-abstraction model, named Timed Programmer View (PVT), omits details related to the computation and communication resources. Such an abstract model enables designers to select a set of solutions to be explored at lower abstraction levels. The second model, CABA (cycle-accurate bit-accurate), is used for power estimation and platform configuration.

In [17], a system-level cycle-based framework to model and design heterogeneous MPSoC (called GRAPES) is

presented. C++/SystemC-based IP system modules can be wrapped to act as plugins, which are managed by the simulation kernel in a TLM fashion. Those modules are managed by the GRAPES kernel, which is the core of the simulation framework. To estimate power during a simulation, they add a dedicated port to each component, which communicates with the corresponding power model. This feature permits to characterize each component with a set of activity monitors (inside the component module) necessary for the power estimation.

Reference [18] presents a simulation-based methodology for extending system performance modeling frameworks to also include power modeling. They demonstrate the use of this methodology with a case study of a real, complex embedded system, comprising the Intel XScale-embedded microprocessor, its WMMX SIMD co processor, L1 caches, SDRAM, and the on-board address and data buses.

In [19], a power estimation framework for SoCs is presented, using power profiles to produce cycle-accurate results. The SoC is divided in its building blocks (e.g., processors, memories, communication, and peripherals), and the power estimation is based on the RTL analysis of each component. The authors validate the framework using an ARM926EJ-S CPU and the AMBA AXI 3.0 as NoC. speed-up compared to a gate-level simulation is on average 100 times faster. Previous work was not addressing high-level power modeling of MPSoCs on FPGA.

In [20], an efficient hybrid system level (HSL) power estimation methodology for FPGA-based MPSoC is proposed. Within this methodology, the functional level power analysis (FLPA) is extended up to set up generic power models for the different parts of the system. Then, a simulation framework is developed at the transactional level to evaluate accurately the activities used in the related power models. With respect to this work, our processor model can easily model different kinds of RISC processors by simply introducing new units.

Moreover, there also exist a considerable number of research efforts that only focus on the power modeling of the on-chip network of MPSoCs. Examples are [21–24]. Many of the above approaches calibrate the high-level models with parameters extracted from RTL implementations, using low-level simulators for the architectural components. In [21], a rate-based power estimation method is presented. In the first phase, it considers data volume, estimating the average power in function of the total transmitted data: in the second phase, it calibrates the model through definition of the consumed power for each transition rate. In particular, the calibration uses RTL model of the NoC, while the latter uses an actor-oriented model. After the calibration, a power dissipation table is generated for each injection rate and router element. Using linear approximation, they determine the power dissipation for each injection rate. In [22], an energy estimation model based on the traffic flow in the NoCs building blocks (routers and interconnection wires) is presented. The authors represents the amount of energy consumed in the transmission of a data bit throughout the NoC (in its routers and interconnection wires). In [23], a NoC power and performance analysis with different traffic models, using analytical models, is presented. The authors target a NoC with a mesh topology. The employed traffic models are: uniform, local, hot-spot, and matrix transpose. Results were compared to Synopsys Power Compiler and ModelSim, showing an error of 2% for power estimation and 3% for throughput. In [24], a methodology for accurate analysis of power consumption of message-passing primitives in a MPSoC is proposed, and, in particular, an energy model which allows to model the traffic-dependent nature of energy consumption through the use of a single, abstract parameter, namely, the size of the message exchanged. The ISS performs cycle-accurate simulation of the cores, while the rest of the system is described in SystemC at signal level. In [25], the authors employ a framework that takes as input message flows and derives a power profile of the network fabric. The authors map the CPU data path as a graph and the application as a set of messages that flow in this graph. Those mapped CPUs are connected into the network fabric, mapping the entire MPSoC as a network. The authors make use of a network power estimation tool, called LUNA, to evaluate the power dissipation of the entire MPSoC.

To the best of our knowledge, none of the existing efforts have incorporated the power models in a (highly automated) system-level MPSoC synthesis framework, allowing for accurate and flexible validation of the models. Instead, most existing works either use simulation-based validation (e.g., [15, 21–23, 25]) or validation by means of measurements on fixed target platforms (e.g., [18, 19]). Consequently, in general, related system-level MPSoC modeling efforts do also not target FPGA technology in their system-level power models.

7. Conclusion

We presented a framework for high-level power estimation of multiprocessor systems-on-chip (MPSoC) architectures on FPGA. The technique is based on abstract execution profiles called "event signatures", and it operates at a higher level of abstraction than, for example, commonly-used instruction-set simulator (ISS)-based power estimation methods and should thus be capable of achieving good evaluation performance. The model is based on the activity counts from the signatures and from the power characteristics of the components themselves, measured in terms of LUTs used. The signature-based power modeling technique has been integrated in our Daedalus system-level MPSoC synthesis framework, which allows a direct validation and calibration of the power model. We compared the results from our signature-based power modeling to those from real measurements on a Virtex-6 FPGA board. These validation results indicate that our high-level power model achieves good power estimates in terms of DSE. As future work, we plan to perform additional experiments (e.g., using different memory hierarchies and different processor components) as well as to deploy the power model in real system-level DSE experiments.

Acknowledgments

This work has been partially supported by the MADNESS STREP-FP7 European Project and the NWO EASY Project. We would like to give special credits to Todor Stefanov and Mohamed Bamakhrama for their support on implementing the MicroBlaze software driver for the PMBus controller.

References

[1] A. B. Kahng, L. Bin, L. S. Peh, and K. Samadi, "ORION 2.0: A fast and accurate NoC power and area model for early-stage design space exploration," in *Proceedings of the Design, Automation and Test in Europe Conference and Exhibition (DATE '09)*, pp. 423–428, April 2009.

[2] A. D. Pimentel, C. Erbas, and S. Polstra, "A systematic approach to exploring embedded system architectures at multiple abstraction levels," *IEEE Transactions on Computers*, vol. 55, no. 2, pp. 99–111, 2006.

[3] P. Stralen and A. D. Pimentel, "A high-level microprocessor power modeling technique based on event signatures," in *Proceedings of the IEEE/ACM/IFIP Workshop on Embedded Systems for Real-Time Multimedia (ESTIMedia '07)*, 2007.

[4] M. Thompson, H. Nikolov, T. Stefanov et al., "A framework for rapid system-level exploration, synthesis, and programming of multimedia MP-SoCs," in *Proceedings of the 5th International Conference on Hardware/Software Codesign and System Synthesis (CODES+ISSS '07)*, pp. 9–14, October 2007.

[5] H. Nikolov, M. Thompson, T. Stefanov et al., "Daedalus: Toward composable multimedia MP-SoC design," in *Proceedings of the 45th Design Automation Conference (DAC '08)*, pp. 574–579, June 2008.

[6] G. Kahn, "The semantics of a simple language for parallel programming," in *Proceedings of the IFIP Congress*, J. L. Rosenfeld, Ed., vol. 74, pp. 471–475, North-Holland Puplishing Company, New York, NY, USA, 1974.

[7] T. Austin, E. Larson, and D. Ernest, "SimpleScalar: An infrastructure for computer system modeling," *Computer*, vol. 35, no. 2, pp. 12–67, 2002.

[8] J. Ou and V. K. Prasanna, "Rapid energy estimation for hardware-software codesign using FPGAs," *EURASIP Journal on Embedded Systems*, vol. 2006, Article ID 98045, 11 pages, 2006.

[9] N. Muralimanohar, R. Balasubramonian, and N. Jouppi, "Optimizing NUCA organizations and wiring alternatives for large caches with CACTI 6.0," in *Proceedings of the 40th IEEE/ACM International Symposium on Microarchitecture (MICRO '07)*, pp. 3–14, December 2007.

[10] C. Erbas, A. D. Pimentel, M. Thompson, and S. Polstra, "A framework for system-level modeling and simulation of embedded systems architectures," *EURASIP Journal on Embedded Systems*, vol. 2007, Article ID 82123, 11 pages, 2007.

[11] http://pmbus.org/specs.html.

[12] P. K. Huang, M. Hashemi, and S. Ghiasi, "System-level performance estimation for application-specific MPSoC interconnect synthesis," in *Proceedings of the Symposium on Application Specific Processors (SASP '08)*, pp. 95–100, June 2008.

[13] J. Becker, M. Huebner, and M. Ullmann, "Power estimation and power measurement of xilinx virtex fpgas: Trade-offs and limitations," in *Proceedings of the 16th Symposium on Integrated Circuits and Systems Design (SBCCI '07)*, p. 283, IEEE Computer Society, Washington, DC, USA, 2003.

[14] R. Piscitelli and A. D. Pimentel, "A high-level power model for mpsoc on fpga," in *Proceedings of the 18th Reconfigurable Architectures Workshop (RAW '11)*, 2011.

[15] D. Sciuto, G. Beltrame, and C. Silvano, "Multi-accuracy power and performance transaction-level modeling," in *Proceedings*

of the Conference on Design, Automation and Test in Europe (DATE '08), 2008.

[16] J. L. Dekeyser, R. B. Atitallah, and S. Niar, "MPSoC power estimation framework at transaction level modeling," in *Proceedings of the 19th International Conference on Microelectronics (ICM '07)*, pp. 245–248, December 2007.

[17] M. Monchiero, G. Palermo, C. Silvano, and O. Villa, "A modular approach to model heterogeneous MPSoC at cycle level," in *Proceedings of the 11th EUROMICRO Conference on Digital System Design Architectures, Methods and Tools (DSD '08)*, pp. 158–164, September 2008.

[18] A. Varma, E. Debes, I. Kozintsev, P. Klein, and B. Jacob, "Accurate and fast system-level power modeling: An XScale-based case study," *Transactions on Embedded Computing Systems*, vol. 7, no. 3, article 25, 2008.

[19] I. Lee, H. Kim, P. Yang et al., "PowerViP: SoC power estimation framework at transaction level," in *Proceedings of the Asia and South Pacific Design Automation Conference (ASP-DAC '06)*, pp. 551–558, IEEE Press, January 2006.

[20] S. Niar, E. Senn, S. K. Rethinagiri, R. B. Atitallah, and J. L. Dekeyser, "Hybrid system level power consumption estimation for fpga-based mpsoc," in *Proceedings of the 29th IEEE International Conference on Computer Design (ICCD '11)*, October 2011.

[21] L. Ost, G. Guindani, L. Indrusiak, S. Maatta, and F. Moraes, "Using abstract power estimation models for design space exploration in NoCbased MPSoC," *IEEE Design and Test of Computers*, vol. 28, no. 2, pp. 16–29, 2011.

[22] J. Hu and R. Marculescu, "Energy-aware mapping for tile-based noc architectures under performance constraints," in *Proceedings of the Asia and South Pacific Design Automation Conference (ASP-DAC '03)*, pp. 233–239, New York, NY, USA, 2003.

[23] S. Koohi, M. Mirza-Aghatabar, S. Hessabi, and M. Pedram, "High-level modeling approach for analyzing the effects of traffic models on power and throughput in mesh-based nocs," in *Proceedings of the 21st International Conference on VLSI Design (VLSID '08)*, 2008.

[24] M. Loghi, L. Benini, and M. Poncino, "Power macromodeling of MPSoC message passing primitives," *ACM Transactions in Embedded Computing Systems*, vol. 6, no. 4, article 31, 2007.

[25] N. Eisley, V. Soteriou, and L. Peh, "High-level power analysis for multi-core chips," in *Proceedings of the International conference on Compilers, Architecture and Synthesis for Embedded Systems (CASES '06)*, pp. 389–400, New York, NY, USA, 2006.

4

Low Complexity Submatrix Divided MMSE Sparse-SQRD Detection for MIMO-OFDM with ESPAR Antenna Receiver

Diego Javier Reinoso Chisaguano and Minoru Okada

Graduate School of Information Science, Nara Institute of Science and Technology, 8916-5 Takayama, Ikoma-shi, Nara 630-0192, Japan

Correspondence should be addressed to Diego Javier Reinoso Chisaguano; diegojavier-r@is.naist.jp

Academic Editor: Mohamed Masmoudi

Multiple input multiple output-orthogonal frequency division multiplexing (MIMO-OFDM) with an electronically steerable passive array radiator (ESPAR) antenna receiver can improve the bit error rate performance and obtains additional diversity gain without increasing the number of Radio Frequency (RF) front-end circuits. However, due to the large size of the channel matrix, the computational cost required for the detection process using Vertical-Bell Laboratories Layered Space-Time (V-BLAST) detection is too high to be implemented. Using the minimum mean square error sparse-sorted QR decomposition (MMSE sparse-SQRD) algorithm for the detection process the average computational cost can be considerably reduced but is still higher compared with a conventional MIMOOFDM system without ESPAR antenna receiver. In this paper, we propose to use a low complexity submatrix divided MMSE sparse-SQRD algorithm for the detection process of MIMOOFDM with ESPAR antenna receiver. The computational cost analysis and simulation results show that on average the proposed scheme can further reduce the computational cost and achieve a complexity comparable to the conventional MIMO-OFDM detection schemes.

1. Introduction

In multipath fading channels, multiple input multiple output (MIMO) antenna systems can achieve a great increase in the channel capacity [1]. MIMO-OFDM combines the advantages of the MIMO systems with orthogonal frequency division multiplexing (OFDM) modulation, achieving a good performance for frequency selective fading channels. Due to these advantages, MIMO-OFDM allows high data rates in wireless communications systems. It is used in the wireless local area network (WLAN) standard IEEE 802.11n [2] and is also considered for the next-generation systems.

One of the limitations of MIMO-OFDM is that it requires one radio frequency (RF) front-end circuit for every receiver and transmitter antenna. Comparing MIMO-OFDM 2Tx-2Rx with MIMO-OFDM 2Tx-4Rx, MIMO-OFDM 2×4 can achieve better diversity gain and bit error rate performance but requires more RF front-end circuits, A/D converters, and FFT blocks for every additional branch.

In [3, 4] a MIMO-OFDM 2×2 scheme with electronically steerable passive array radiator (ESPAR) antenna receiver diversity has been proposed. It utilizes for every receiver a 2-element ESPAR antenna whose directivity is changed at the same frequency of the OFDM symbol rate. Compared to the conventional MIMO-OFDM 2×2 systems, this scheme gives additional diversity gain and improves the bit error rate performance without increasing the number of RF front-end circuits. For the detection the zero forcing (ZF) Vertical-Bell Laboratories Layered Space-Time (V-BLAST) algorithm [5, 6] is used but, due to the large size of the channel matrix, the required computational effort is very high.

In order to reduce the computational cost of the detection process of the scheme proposed in [3, 4], the use of a minimum mean square error sparse-sorted QR decomposition (MMSE sparse-SQRD) algorithm based on the SQRD algorithm introduced in [7, 8] was proposed by the authors in [9]. The computational cost reduction is achieved by exploiting the sparse structure of the channel matrix. This detection algorithm considerably reduces the average computational cost and also improves the bit error rate performance compared to the original scheme [3, 4]. A submatrix divided MMSE sparse-SQRD algorithm for the

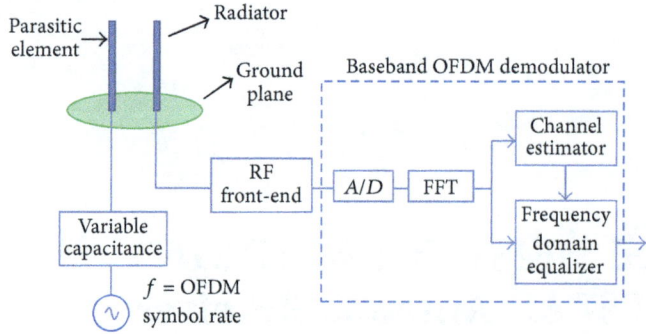

FIGURE 1: Block diagram of OFDM receiver with ESPAR antenna.

detection process of MIMO-OFDM with ESPAR antenna receiver was proposed by the authors in [10] for further reduction in the computational cost. This algorithm divides the channel matrix into k smaller submatrices reducing the computational cost but adding a small degradation in the bit error rate performance.

This paper is an extension of [10] including results of the bit error performance and computational cost for higher order submatrix division schemes. Also, another approach to further reduce the bit error degradation originated by the submatrix division algorithm is introduced.

The rest of this paper is organized as follows. Sections 2 and 3 gives a brief background description about OFDM and MIMO-OFDM with ESPAR antenna receiver. In Section 4, detection algorithms based on QR decomposition are shown. Then in Section 5 a detailed explanation about the MMSE sparse-SQRD algorithm is included. In Section 6 the proposed submatrix divided scheme is described. The computational cost analysis and simulation results are presented in Sections 7 and 8, respectively. And finally, in Section 9 conclusions are included.

2. OFDM with ESPAR Antenna

ESPAR is a small size and low power consumption antenna [11, 12]. It is composed by a radiator element connected to the RF front-end and one or more parasitic (passive) elements terminated by variables capacitances. The beam directivity can be controlled modifying the variables capacitances. This antenna requires only one RF front-end and therefore is known also as single RF port antenna array.

In [13] an OFDM receiver using ESPAR antenna is proposed. In this scheme the directivity of the ESPAR antenna is changed by a periodic wave whose frequency is the OFDM symbol rate. The block diagram of this scheme is shown in Figure 1.

A two-element ESPAR antenna is utilized. The periodic variation of the directivity causes intercarrier interference (ICI) in the received signal. The ICI is caused by the addition of phase shifted components to the received signal. The frequency domain equalizer in Figure 1 uses both the shifted and nonshifted components in the detection. Due to this effect this scheme obtains diversity gain, therefore improving the bit error rate performance.

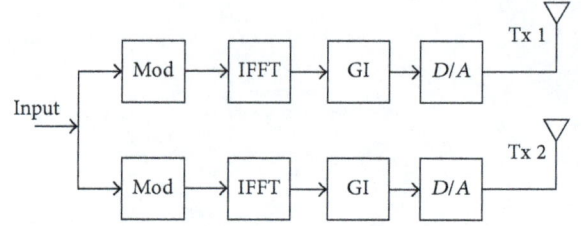

FIGURE 2: Block diagram of the MIMO-OFDM transmitter.

3. MIMO-OFDM with ESPAR Antenna Receiver

Based on [13], a MIMO-OFDM receiver with ESPAR antenna was proposed in [3, 4] and is described in this section. The block diagrams of the receiver and transmitter are shown in Figures 2 and 3, respectively.

The transmitter is based on the WLAN standard IEEE 802.11n [2]. For simplicity forward error correction (FEC) interleaver blocks are not considered in the system. The receiver uses a 2-element ESPAR antenna where the directivity is also periodically changed according to the OFDM symbol rate. An MMSE channel estimator derived in [3, 4] is used and the detection process is carried out by the ZF V-BLAST detector.

3.1. Channel Estimation. For the channel estimation [13], let $\mathbf{P}_1$ be the pilot symbol and its cyclic shifted $\mathbf{P}_2$. The received signal after the FFT processor at the ith Rx is

$$\mathbf{u_i} = \mathbf{P}_1\mathbf{h}_{i,1}^{ns} + \mathbf{GP}_1\mathbf{h}_{i,1}^{s} + \mathbf{P}_2\mathbf{h}_{i,2}^{ns} + \mathbf{GP}_2\mathbf{h}_{i,2}^{s} + \mathbf{z}, \qquad (1)$$

where $\mathbf{h}_{i,l}^{ns}$ and $\mathbf{h}_{i,l}^{s}$ are the channel response between the ith receive antenna and lth transmit antenna for the phase nonshifting (ns) and phase shifting (s) elements respectively. The matrix $\mathbf{G}$ represents the frequency shift due to directivity variation in ESPAR antenna and $\mathbf{z}$ is the additive white Gaussian noise (AWGN) vector.

From (1) the autocorrelation matrix $\mathbf{R_u} = E[\mathbf{u_i}\mathbf{u}_i^H]$ is given by

$$\begin{aligned} \mathbf{R_u} = \mathbf{P}_1\mathbf{R}_h\mathbf{P}_1^H + \mathbf{GP}_1\mathbf{R}_h\mathbf{P}_1^H\mathbf{G}^H \\ + \mathbf{P}_2\mathbf{R}_h\mathbf{P}_2^H + \mathbf{GP}_2\mathbf{R}_h\mathbf{P}_2^H\mathbf{G}^H + \sigma_z^2\mathbf{I}, \end{aligned} \qquad (2)$$

where σ_z^2 is the noise variance and $\mathbf{R}_h$ is the covariance matrix that represents the delay profile of the channel. Considering that the phase nonshifting (ns) and phase shifting (s) elements are spatially separated enough to be uncorrelated, the cross-correlation matrices $\mathbf{B}_i = E[\mathbf{uh}^H]$ are given by

$$\begin{aligned} \mathbf{B}_i^{ns} &= \mathbf{P}_i\mathbf{R}_h, \\ \mathbf{B}_i^{s} &= \mathbf{GP}_i\mathbf{R}_h. \end{aligned} \qquad (3)$$

Using the MMSE criteria the channel response is given by

$$\begin{aligned} \widetilde{\mathbf{h}}_{i,l}^{ns} &= \left(\mathbf{R_u}^{-1}\mathbf{B}_i^{ns}\right)^H\mathbf{u}_i, \\ \widetilde{\mathbf{h}}_{i,l}^{s} &= \left(\mathbf{R_u}^{-1}\mathbf{B}_i^{s}\right)^H\mathbf{u}_i. \end{aligned} \qquad (4)$$

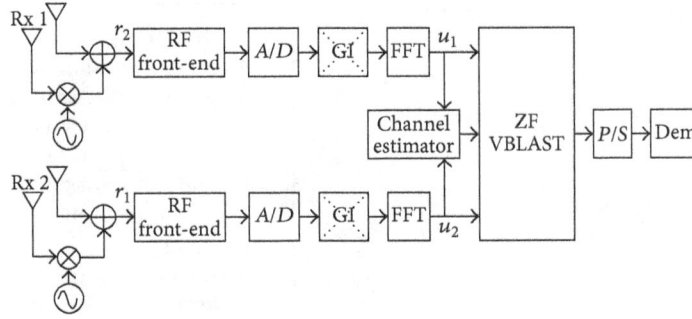

FIGURE 3: Block diagram of the MIMO-OFDM receiver with ESPAR antenna.

The channel matrix $\mathbf{H}$ has a size of $2(N + 2) \times 2N$, where N is the number of data subcarriers.

3.2. Detection. For the detection process the ZF V-BLAST [6] algorithm is used. In this algorithm the received signal vector is multiplied with a filter matrix $\mathbf{G}_{ZF}$, that is, calculated by

$$\mathbf{G}_{ZF} = \mathbf{H}^{\dagger}, \tag{5}$$

where $\mathbf{H}^{\dagger}$ is the Moore-Penrose pseudoinverse of $\mathbf{H}$. The matrix $\mathbf{G}_{ZF}$ is calculated in a recursive way after zeroing one column of the channel matrix $\mathbf{H}$; for this scheme the pseudoinverse is calculated $2N$ times. Due to the large size of the channel matrix $\mathbf{H}$, calculating the pseudoinverse demands a very high computational effort and for this reason the detection process is the main limitation of this scheme.

4. QR Decomposition-Based Detection

Let $\mathbf{x} = [x_1, x_2, \ldots, x_{2N}]^T$ denote the vector of transmitted symbols, let $\mathbf{z} = [z_1, z_2, \ldots, z_{2(N+2)}]^T$ denote the vector of noise components and $\mathbf{u} = [u_1, u_2, \ldots, u_{2(N+2)}]^T$ the vector of received symbols.

4.1. MMSE-QRD. Applying like in [8] the MMSE detector criteria, let us denote the extended channel matrix $\underline{\mathbf{H}}$ and the extended vector of received symbols $\underline{\mathbf{u}}$ by

$$\underline{\mathbf{H}} = \begin{bmatrix} \mathbf{H} \\ \sigma_z \mathbf{I}_{2N} \end{bmatrix},$$

$$\underline{\mathbf{u}} = \begin{bmatrix} \mathbf{u} \\ \mathbf{0}_{2N,1} \end{bmatrix}, \tag{6}$$

where σ_z is the noise standard deviation, $\mathbf{I}_{2N}$ is an identity matrix of size $2N \times 2N$, and $\mathbf{0}_{2N,1}$ is a column vector with $2N$ zero elements.

The QR decomposition of the extended channel matrix $\underline{\mathbf{H}}$ can be expressed by

$$\underline{\mathbf{H}} = \underline{\mathbf{Q}}\underline{\mathbf{R}}, \tag{7}$$

where $\underline{\mathbf{Q}}$ is a unitary matrix and $\underline{\mathbf{R}}$ is an upper triangular matrix. And the extended vector of received symbols $\underline{\mathbf{u}}$ is given by

$$\underline{\mathbf{u}} = \underline{\mathbf{Q}}\underline{\mathbf{R}}\mathbf{x} + \mathbf{z}. \tag{8}$$

Then (8) is multiplied by $\underline{\mathbf{Q}}^H$ to obtain

$$\mathbf{y} = \underline{\mathbf{Q}}^H \underline{\mathbf{u}} = \underline{\mathbf{R}}\mathbf{x} + \boldsymbol{\nu}, \tag{9}$$

where $\underline{\mathbf{Q}}^H$ is the Hermitian transpose of $\underline{\mathbf{Q}}$ and $\nu = \underline{\mathbf{Q}}^H \mathbf{z}$. The statistical properties of ν remain unchanged because $\underline{\mathbf{Q}}$ is a unitary matrix.

4.2. MMSE-SQRD. In [8] an MMSE sorted QRD detection algorithm based on the modified Gram-Schmidt algorithm is introduced. The starting condition is that $\underline{\mathbf{Q}} = \underline{\mathbf{H}}$; then the norms of the column vectors of $\underline{\mathbf{Q}}$ are calculated. For every step the column of $\underline{\mathbf{Q}}$ with the minimum norm is found to maximize $\underline{\mathbf{R}}_{k,k}$ and the columns of $\underline{\mathbf{Q}}$ are exchanged before the orthogonalization process. This algorithm calculates an improved matrix $\underline{\mathbf{R}}$ that reduces the error propagation through the detection layers. During the calculation, a permutation vector $\mathbf{p}$ carries the column exchanging operations for reordering the detected symbols at the end of the algorithm.

After the matrices $\underline{\mathbf{R}}$ and $\underline{\mathbf{Q}}$ are calculated, then $\mathbf{y}$ is obtained according to (9) and the symbols are detected iteratively. After the symbols are detected, they are reordered using the permutation vector $\mathbf{p}$ to find the original sequence of the detected symbols.

5. MMSE Sparse-SQRD Algorithm

The extended channel matrix $\underline{\mathbf{H}}$ whose size is $2(2N + 2) \times 2N$ is shown in (10) and as we can see it is a sparse matrix. The MMSE sparse-SQRD algorithm is based on the MMSE-

SQRD algorithm [8] and exploits the sparse structure of $\underline{\mathbf{H}}$ to reduce the computational cost of the detection process:

$$
\underline{\mathbf{H}} = \begin{pmatrix}
H_{A,-N/2}^{ns} & 0 & \cdots & 0 & H_{B,-N/2}^{ns} & 0 & \cdots & 0 \\
H_{A,-N/2}^{s} & \ddots & \ddots & \vdots & H_{B,-N/2}^{s} & \ddots & \ddots & \vdots \\
0 & \ddots & \ddots & 0 & 0 & \ddots & \ddots & 0 \\
\vdots & \ddots & \ddots & H_{A,N/2}^{ns} & \vdots & \ddots & \ddots & H_{B,N/2}^{ns} \\
0 & \cdots & 0 & H_{A,N/2}^{s} & 0 & \cdots & 0 & H_{B,N/2}^{s} \\
H_{C,-N/2}^{ns} & 0 & \cdots & 0 & H_{D,-N/2}^{ns} & 0 & \cdots & 0 \\
H_{C,-N/2}^{s} & \ddots & \ddots & \vdots & H_{D,-N/2}^{s} & \ddots & \ddots & \vdots \\
0 & \ddots & \ddots & 0 & 0 & \ddots & \ddots & 0 \\
\vdots & \ddots & \ddots & H_{C,N/2}^{ns} & \vdots & \ddots & \ddots & H_{D,N/2}^{ns} \\
0 & \cdots & 0 & H_{C,N/2}^{s} & 0 & \cdots & 0 & H_{D,N/2}^{s} \\
\sigma_z & 0 & \cdots & \cdots & \cdots & \cdots & 0 & 0 \\
0 & \sigma_z & 0 & \cdots & & & 0 & 0 \\
\vdots & \ddots & \ddots & \ddots & & \cdots & \sigma_z & 0 \\
0 & \cdots & \cdots & \cdots & & \cdots & 0 & \sigma_z
\end{pmatrix}
\tag{10}
$$

Analysing $\underline{\mathbf{H}}$ given in (10) we can see that every column has only five nonzero elements so for the norm calculation of the column vectors of $\mathbf{Q}$ only these elements should be used. Also the positions of the nonzero elements are fixed so we have this information contained in a matrix as input. Using this information the norm calculation is shown in lines 5–9 of Algorithm 1.

In the orthogonalization process of the algorithm these two calculations

$$
\underline{r}_{i,l} = \mathbf{q}_i^H \mathbf{q}_l, \tag{11}
$$

$$
\mathbf{q}_l = \mathbf{q}_l - \underline{r}_{i,l}\mathbf{q}_i \tag{12}
$$

are performed in an iterative way. $\underline{r}_{i,l}$ denote the elements of the matrix $\underline{\mathbf{R}}$ and $\mathbf{q}_l$, $\mathbf{q}_i$ are column vectors of the matrix $\mathbf{Q}$.

In (11), the multiplication of the zero elements of the column vectors does not influence the final result so these multiplications can be avoided. A vector containing only the indices of the nonzero elements of the column vectors is obtained in line 14 so the number of operations required to calculate $\underline{r}_{i,l}$ is reduced without influencing the final result. This is shown in lines 19–21 in the algorithm. The same strategy is used also in lines 15–17 and 23–25.

Also, due to the sparse structure of (10), the result of (11) can be zero. In this case calculating (12) is unnecessary because it does not change the value of $\mathbf{q}_l$ so it can be avoided using the condition in line 22.

```
1: Input: H, Hnz
2: cols ← # of columns of H
3: rows ← # of rows of H
4: R = 0, Q = H, p = (1, ..., cols)
5: for i = 1, ..., cols do
6:     for j = 1, ..., 5 do
7:         normᵢ := normᵢ + ‖ q_{Hnz(j,i),i} ‖²
8:     end for
9: end for
10: for i = 1, ..., cols do
11:     kᵢ = arg min_{l=i,...,cols} normₗ
12:     exchange columns i and kᵢ in R, Q, norm, p
13:     r_{i,i} = √normᵢ
14:     nz ← indices of the non-zero elements of q_i
15:     for j = 1, ..., length(nz) do
16:         q_{nz(j),i} := q_{nz(j),i}/r_{i,i}
17:     end for
18:     for l = i + 1, ..., cols do
19:         for j = 1, ..., length(nz) do
20:             r_{i,l} := r_{i,l} + (q*_{nz(j),i}) q_{nz(j),l}
21:         end for
22:         if r_{i,l} ≠ 0 then
23:             for j = 1, ..., length(nz) do
24:                 q_{nz(j),l} := q_{nz(j),l} - r_{i,l}q_{nz(j),i}
25:             end for
26:             normₗ := normₗ - ‖ r_{i,l} ‖²
27:         end if
28:     end for
29: end for
30: Q₁ ← Q(1 : rows − cols, :)
31: y = Q₁ᴴu
32: for k = cols, ..., 1 do
33:     d̂ = Σ_{i=k+1}^{cols} r_{k,i}x̂_i
34:     x̂_k = Q[(y_k − d̂)/r_{k,k}]
35: end for
36: Permutate x̂ according to p
```

ALGORITHM 1: MMSE sparse-SQRD.

Also we can consider that the calculation of (9) can be simplified as

$$
\mathbf{y} = \underline{\mathbf{Q}}^H \underline{\mathbf{u}} = \begin{bmatrix} \mathbf{Q}_1 \\ \mathbf{Q}_2 \end{bmatrix}^H \begin{bmatrix} \mathbf{u} \\ \mathbf{0}_{2N,1} \end{bmatrix} = \mathbf{Q}_1^H \mathbf{u}, \tag{13}
$$

where $\mathbf{Q}_1$ is a matrix with the same size of the channel matrix $\mathbf{H}$. This is shown in lines 30-31 of the algorithm.

Using these analysed criteria the MMSE sparse-SQRD algorithm can achieve the same bit error rate performance of the MMSE-SQRD algorithm but with a considerable computational cost reduction.

6. Submatrix Divided Proposed Algorithm

In order to further reduce the computational cost of the detection process an algorithm based on submatrix division of the channel matrix is proposed. The block diagram of the proposed scheme is shown in Figure 4.

This detection scheme is composed by a submatrix builder block and k MMSE sparse-SQRD detectors. The submatrix builder is fed with the received symbols from the FFT processors and the channel state information obtained in the channel estimator. Its function is to build the submatrices and vectors for the detectors. Every detector is fed with a vector of received symbols $\mathbf{s}_i$ and a channel submatrix $\mathbf{H}_i$.

From now on we consider the number of subcarriers to be $N = 56$ like in the IEEE 802.11n [2] standard. Let $\mathbf{a} = [a_1, a_2, \ldots, a_{58}]^T$ denote the vector of received (transmitted and interfered) symbols from the FFT1 processor and let $\mathbf{b} = [b_1, b_2, \ldots, b_{58}]^T$ denote the vector of received symbols from the FFT2 processor. For simplicity we consider that the extended channel submatrix $\underline{\mathbf{H}}_i$ is created inside the ith detector. Now we will explain in detail the submatrix division case when $k = 4$ considering two variations with 2 or 4-symbol overlapping.

6.1. Quarter-Size Submatrix ($k = 4$) with 2-Symbol Overlapping.

In this case we divide the channel matrix into four submatrices denoted as $\mathbf{H}_{12}$, $\mathbf{H}_{22}$, $\mathbf{H}_{32}$ and $\mathbf{H}_{42}$. These matrices are shown in (14), (15), (16), and (17), respectively,

$$\mathbf{H}_{12} = \begin{pmatrix} H_{A,-28}^{ns} & 0 & \cdots & 0 & H_{B,-28}^{ns} & 0 & \cdots & 0 \\ H_{A,-28}^{s} & \ddots & \ddots & \vdots & H_{B,-28}^{s} & \ddots & \ddots & \vdots \\ 0 & \ddots & H_{A,-15}^{ns} & 0 & 0 & \ddots & H_{B,-15}^{ns} & 0 \\ 0 & \cdots & H_{A,-15}^{s} & H_{A,-14}^{ns} & 0 & \cdots & H_{B,-15}^{s} & H_{B,-14}^{ns} \\ H_{C,-28}^{ns} & 0 & \cdots & 0 & H_{D,-28}^{ns} & 0 & \cdots & 0 \\ H_{C,-28}^{s} & \ddots & \ddots & \vdots & H_{D,-28}^{s} & \ddots & \ddots & \vdots \\ 0 & \ddots & H_{C,-15}^{ns} & 0 & 0 & \ddots & H_{D,-15}^{ns} & 0 \\ 0 & \cdots & H_{C,-15}^{s} & H_{C,-14}^{ns} & 0 & \cdots & H_{D,-15}^{s} & H_{D,-14}^{ns} \end{pmatrix}, \tag{14}$$

$$\mathbf{H}_{22} = \begin{pmatrix} H_{A,-14}^{ns} & 0 & 0 & H_{B,-14}^{ns} & 0 & 0 \\ H_{A,-14}^{s} & \ddots & \vdots & H_{B,-14}^{s} & \ddots & \vdots \\ 0 & \ddots & H_{A,-1}^{ns} & 0 & \ddots & H_{B,-1}^{ns} \\ 0 & \cdots & H_{A,-1}^{s} & 0 & \cdots & H_{B,-1}^{s} \\ H_{C,-14}^{ns} & 0 & 0 & H_{D,-14}^{ns} & 0 & 0 \\ H_{C,-14}^{s} & \ddots & \vdots & H_{D,-14}^{s} & \ddots & \vdots \\ 0 & \ddots & H_{C,-1}^{ns} & 0 & \ddots & H_{D,-1}^{ns} \\ 0 & \cdots & H_{C,-1}^{s} & 0 & \cdots & H_{D,-1}^{s} \end{pmatrix}, \tag{15}$$

$$\mathbf{H}_{32} = \begin{pmatrix} H_{A,+1}^{ns} & 0 & \cdots & 0 & H_{B,+1}^{ns} & 0 & \cdots & 0 \\ H_{A,+1}^{s} & \ddots & \ddots & \vdots & H_{B,+1}^{s} & \ddots & \ddots & \vdots \\ 0 & \ddots & H_{A,+14}^{ns} & 0 & 0 & \ddots & H_{B,+14}^{ns} & 0 \\ 0 & \cdots & H_{A,+14}^{s} & H_{A,+15}^{ns} & 0 & \cdots & H_{B,+14}^{s} & H_{B,15}^{ns} \\ H_{C,+1}^{ns} & 0 & \cdots & 0 & H_{D,+1}^{ns} & 0 & \cdots & 0 \\ H_{C,+1}^{s} & \ddots & \ddots & \vdots & H_{D,+1}^{s} & \ddots & \ddots & \vdots \\ 0 & \ddots & H_{C,+14}^{ns} & 0 & 0 & \ddots & H_{D,+14}^{ns} & 0 \\ 0 & \cdots & H_{C,+14}^{s} & H_{C,+15}^{ns} & 0 & \cdots & H_{D,+14}^{s} & H_{D,+15}^{ns} \end{pmatrix}, \tag{16}$$

$$\mathbf{H}_{42} = \begin{pmatrix} H_{A,+15}^{ns} & 0 & 0 & H_{B,+15}^{ns} & 0 & 0 \\ H_{A,+15}^{s} & \ddots & \vdots & H_{B,+15}^{s} & \ddots & \vdots \\ 0 & \ddots & H_{A,+28}^{ns} & 0 & \ddots & H_{B,+28}^{ns} \\ 0 & \cdots & H_{A,+28}^{s} & 0 & \cdots & H_{B,+28}^{s} \\ H_{C,+15}^{ns} & 0 & 0 & H_{D,+15}^{ns} & 0 & 0 \\ H_{C,+15}^{s} & \ddots & \vdots & H_{D,+15}^{s} & \ddots & \vdots \\ 0 & \ddots & H_{C,+28}^{ns} & 0 & \ddots & H_{D,+28}^{ns} \\ 0 & \cdots & H_{C,+28}^{s} & 0 & \cdots & H_{D,+28}^{s} \end{pmatrix}. \tag{17}$$

The vectors of received symbols applied to the four detectors are denoted as

$$\begin{aligned} \mathbf{s}_{1\_2} &= [a_1, \ldots, a_{15}, b_1, \ldots, b_{15}]^T, \\ \mathbf{s}_{2\_2} &= [a_{15}, \ldots, a_{29}, b_{15}, \ldots, b_{29}]^T, \\ \mathbf{s}_{3\_2} &= [a_{30}, \ldots, a_{44}, b_{30}, \ldots, b_{44}]^T, \\ \mathbf{s}_{4\_2} &= [a_{44}, \ldots, a_{58}, b_{44}, \ldots, b_{58}]^T. \end{aligned} \tag{18}$$

And the vectors of detected symbols obtained from the detectors are denoted as

$$\begin{aligned} \mathbf{x}_{1\_2} &= [x_1, \ldots, x_{14}, x_{15}, x_{57}, \ldots, x_{70}, x_{71}]^T, \\ \mathbf{x}_{2\_2} &= [x_{15}, \ldots, x_{28}, x_{71}, \ldots, x_{84}]^T, \\ \mathbf{x}_{3\_2} &= [x_{29}, \ldots, x_{42}, x_{43}, x_{85}, \ldots, x_{98}, x_{99}]^T, \\ \mathbf{x}_{4\_2} &= [x_{43}, \ldots, x_{56}, x_{99}, \ldots, x_{112}]^T. \end{aligned} \tag{19}$$

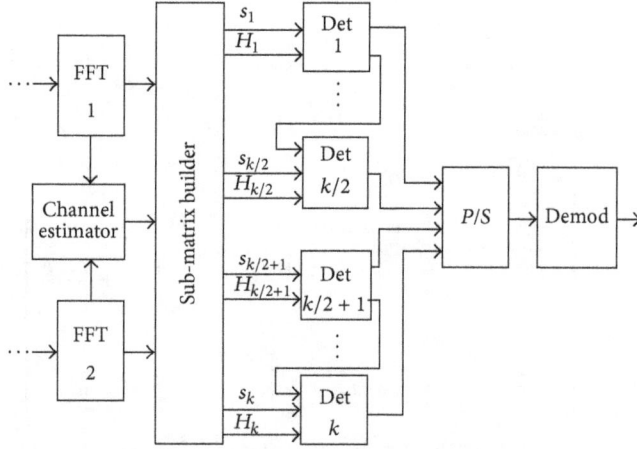

FIGURE 4: Block diagram of the submatrix division proposed scheme.

The submatrix division introduces a degradation in the bit error performance so now we explain the procedure used to minimize this effect. First the channel matrix nonshifted (*ns*) elements associated with the subcarrier −14 ($H_{A,-14}^{ns}$; $H_{B,-14}^{ns}$; $H_{C,-14}^{ns}$; $H_{D,-14}^{ns}$) are included in both $\mathbf{H}_{12}$ and $\mathbf{H}_{22}$. In the same way the nonshifted elements associated with the subcarrier +15 ($H_{A,+15}^{ns}$; $H_{B,+15}^{ns}$; $H_{C,+15}^{ns}$; $H_{D,+15}^{ns}$) are included in $\mathbf{H}_{32}$ and $\mathbf{H}_{42}$.

We overlap the symbols a_{15}, b_{15} in vectors $\mathbf{s}_{1\_2}$ and $\mathbf{s}_{2\_2}$. Using the information from the detection process of the detector 1, the symbols a_{15}, b_{15} in vector $\mathbf{s}_{2\_2}$ are compensated according to (20) where $\tilde{a}_{15}$, $\tilde{b}_{15}$ represent the compensated symbols:

$$
\begin{aligned}
\tilde{a}_{15} &= a_{15} - H_{A,-15}^{s} x_{14} - H_{B,-15}^{s} x_{70}, \\
\tilde{b}_{15} &= b_{15} - H_{C,-15}^{s} x_{14} - H_{D,-15}^{s} x_{70}.
\end{aligned}
\tag{20}
$$

We also overlap symbols a_{44}, b_{44} in $\mathbf{s}_{3\_2}$ and $\mathbf{s}_{4\_2}$. The symbols a_{44}, b_{44} in vector $\mathbf{s}_{4\_2}$ are compensated according to (21) using the information of the detection process of the detector 3. Similarly $\tilde{a}_{44}$ and $\tilde{b}_{44}$ represent the compensated symbols:

$$
\begin{aligned}
\tilde{a}_{44} &= a_{44} - H_{A,+14}^{s} x_{42} - H_{B,+14}^{s} x_{98}, \\
\tilde{b}_{44} &= b_{44} - H_{C,+14}^{s} x_{42} - H_{D,+14}^{s} x_{98}.
\end{aligned}
\tag{21}
$$

During the sorting process of the detector 1, the columns containing the channel matrix nonshifted (*ns*) elements associated with the subcarrier −14 are used first regardless of its norm. It reduces the degradation introduced by these elements in the upper layers during the detection process. The same is performed in the detector 3 with the nonsubcarrier-shifted elements of the subcarrier +15.

In the vectors of detected symbols the overlapped detected elements x_{15}, x_{71} in vector $\mathbf{x}_{1\_2}$ and x_{43}, x_{99} in vector $\mathbf{x}_{3\_2}$ are discarded because they have a higher probability of error.

6.2. Quarter-Size Submatrix ($k = 4$) with 4-Symbol Overlapping. In this subsection another variation with 4-symbol overlapping is introduced. The objective of this idea is to further reduce the degradation in the bit error rate performance created by the submatrix division. Similar to the previous subsection we divide the channel matrix into four submatrices denoted as $\mathbf{H}_{14}$, $\mathbf{H}_{24}$, $\mathbf{H}_{34}$, and $\mathbf{H}_{44}$. These matrices are shown in (23), (24), (25), and (26), respectively. In this case the vectors of received symbols applied to the four detectors are denoted as

$$
\begin{aligned}
\mathbf{s}_{1\_4} &= \left[a_1, \ldots, a_{15}, a_{16}, b_1, \ldots, b_{15}, b_{16} \right]^T, \\
\mathbf{s}_{2\_4} &= \left[a_{15}, a_{16}, \ldots, a_{29}, b_{15}, b_{16} \ldots, b_{29} \right]^T, \\
\mathbf{s}_{3\_4} &= \left[a_{30}, \ldots, a_{44}, a_{45}, b_{30}, \ldots, b_{44}, b_{45} \right]^T, \\
\mathbf{s}_{4\_4} &= \left[a_{44}, a_{45}, \ldots, a_{58}, b_{44}, b_{45}, \ldots, b_{58} \right]^T,
\end{aligned}
\tag{22}
$$

$$
\mathbf{H}_{14} =
\begin{pmatrix}
H_{A,-28}^{ns} & 0 & \cdots & 0 & H_{B,-28}^{ns} & 0 & \cdots & 0 \\
H_{A,-28}^{s} & \ddots & \ddots & \vdots & H_{B,-28}^{s} & \ddots & \ddots & \vdots \\
0 & \ddots & H_{A,-14}^{ns} & 0 & 0 & \ddots & H_{B,-14}^{ns} & 0 \\
0 & \cdots & H_{A,-14}^{s} & H_{A,-13}^{ns} & 0 & \cdots & H_{B,-14}^{s} & H_{B,-13}^{ns} \\
H_{C,-28}^{ns} & 0 & \cdots & 0 & H_{D,-28}^{ns} & 0 & \cdots & 0 \\
H_{C,-28}^{s} & \ddots & \ddots & \vdots & H_{D,-28}^{s} & \ddots & \ddots & \vdots \\
0 & \ddots & H_{C,-14}^{ns} & 0 & 0 & \ddots & H_{D,-14}^{ns} & 0 \\
0 & \cdots & H_{C,-14}^{s} & H_{C,-13}^{ns} & 0 & \cdots & H_{D,-14}^{s} & H_{D,-13}^{ns}
\end{pmatrix},
\tag{23}
$$

$$
\mathbf{H}_{24} =
\begin{pmatrix}
H_{A,-14}^{ns} & 0 & 0 & H_{B,-14}^{ns} & 0 & 0 \\
H_{A,-14}^{s} & \ddots & \vdots & H_{B,-14}^{s} & \ddots & \vdots \\
0 & \ddots & H_{A,-1}^{ns} & 0 & \ddots & H_{B,-1}^{ns} \\
0 & \cdots & H_{A,-1}^{s} & 0 & \cdots & H_{B,-1}^{s} \\
H_{C,-14}^{ns} & 0 & 0 & H_{D,-14}^{ns} & 0 & 0 \\
H_{C,-14}^{s} & \ddots & \vdots & H_{D,-14}^{s} & \ddots & \vdots \\
0 & \ddots & H_{C,-1}^{ns} & 0 & \ddots & H_{D,-1}^{ns} \\
0 & \cdots & H_{C,-1}^{s} & 0 & \cdots & H_{D,-1}^{s}
\end{pmatrix},
\tag{24}
$$

$$\mathbf{H}_{34} = \begin{pmatrix} H^{ns}_{A,+1} & 0 & \cdots & 0 & H^{ns}_{B,+1} & 0 & \cdots & 0 \\ H^{s}_{A,+1} & \ddots & \ddots & \vdots & H^{s}_{B,+1} & \ddots & \ddots & \vdots \\ 0 & \ddots & H^{ns}_{A,+15} & 0 & 0 & \ddots & H^{ns}_{B,+15} & 0 \\ 0 & \cdots & H^{s}_{A,+15} & H^{ns}_{A,+16} & 0 & \cdots & H^{s}_{B,+15} & H^{ns}_{B,16} \\ H^{ns}_{C,+1} & 0 & \cdots & 0 & H^{ns}_{D,+1} & 0 & \cdots & 0 \\ H^{s}_{C,+1} & \ddots & \ddots & \vdots & H^{s}_{D,+1} & \ddots & \ddots & \vdots \\ 0 & \ddots & H^{ns}_{C,+15} & 0 & 0 & \ddots & H^{ns}_{D,+15} & 0 \\ 0 & \cdots & H^{s}_{C,+15} & H^{ns}_{C,+16} & 0 & \cdots & H^{s}_{D,+15} & H^{ns}_{D,+16} \end{pmatrix}, \tag{25}$$

$$\mathbf{H}_{44} = \begin{pmatrix} H^{ns}_{A,+15} & 0 & 0 & H^{ns}_{B,+15} & 0 & 0 \\ H^{s}_{A,+15} & \ddots & \vdots & H^{s}_{B,+15} & \ddots & \vdots \\ 0 & \ddots & H^{ns}_{A,+28} & 0 & \ddots & H^{ns}_{B,+28} \\ 0 & \cdots & H^{s}_{A,+28} & 0 & \cdots & H^{s}_{B,+28} \\ H^{ns}_{C,+15} & 0 & 0 & H^{ns}_{D,+15} & 0 & 0 \\ H^{s}_{C,+15} & \ddots & \vdots & H^{s}_{D,+15} & \ddots & \vdots \\ 0 & \ddots & H^{ns}_{C,+28} & 0 & \ddots & H^{ns}_{D,+28} \\ 0 & \cdots & H^{s}_{C,+28} & 0 & \cdots & H^{s}_{D,+28} \end{pmatrix}. \tag{26}$$

And the vectors of detected symbols obtained from the detectors are denoted as

$$\mathbf{x}_{1\_4} = \left[x_1, \ldots, x_{14}, x_{15}, x_{16}, x_{57}, \ldots, x_{70}, x_{71}, x_{72} \right]^T,$$

$$\mathbf{x}_{2\_4} = \left[x_{15}, \ldots, x_{28}, x_{71}, \ldots, x_{84} \right]^T,$$

$$\mathbf{x}_{3\_4} = \left[x_{29}, \ldots, x_{42}, x_{43}, x_{44}, x_{85}, \ldots, x_{98}, x_{99}, x_{100} \right]^T, \tag{27}$$

$$\mathbf{x}_{4\_4} = \left[x_{43}, \ldots, x_{56}, x_{99}, \ldots, x_{112} \right]^T.$$

In this variation 4 symbols a_{15}, a_{16}, b_{15}, b_{16} in vectors $\mathbf{s}_{1\_4}$ and $\mathbf{s}_{2\_4}$ are overlapped. Similar to the previous subsection the symbols a_{15}, b_{15} in vector $\mathbf{s}_{2\_4}$ are compensated according to (20) using the elements of $\mathbf{H}_{14}$ and $\mathbf{x}_{1\_4}$. We also overlap symbols a_{44}, a_{45}, b_{44}, b_{45} in $\mathbf{s}_{3\_4}$ and $\mathbf{s}_{4\_4}$. In the same way the symbols a_{44}, b_{44} in vector $\mathbf{s}_{4\_4}$ are compensated according to (21) using the elements of $\mathbf{H}_{34}$ and $\mathbf{x}_{3\_4}$.

Also the channel matrix elements associated with the subcarriers −14 and −13 are included in both $\mathbf{H}_{14}$ and $\mathbf{H}_{24}$. In the same way the elements associated with the subcarriers +15 and +16 are included in $\mathbf{H}_{34}$ and $\mathbf{H}_{44}$. During the sorting process of the detector 1, the columns containing the channel

matrix elements associated with the subcarriers −14 and −13 are used first regardless of its norm. The same is performed in the detector 3 with the elements of the subcarriers +15 and +16.

In the vectors of detected symbols the overlapped elements x_{15}, x_{16}, x_{71}, x_{72} in vector $\mathbf{x}_{1\_4}$ and x_{43}, x_{44}, x_{99}, x_{100} in vector $\mathbf{x}_{3\_4}$ are discarded because they have a higher probability of error.

7. Computational Cost

The computational cost is analysed in terms of the number of complex floating point operations (flops) $\mathscr{F}$ required. As in [8], for simplicity we consider each complex addition as one flop and each complex multiplication as three flops. We cannot obtain a formula for the number of flops for the submatrix divided proposed algorithm because this number depends on the random sorting, so we obtained an average of the number of flops from the simulation results. Also, for comparison, the number of flops required by the ML detector [14] is

$$\mathscr{F}_{\mathrm{ML}} = M^C \left(4C^2 + 1.5C + 0.5 \right), \tag{28}$$

where M is the constellation size. The ZF-VBLAST algorithm like in [15] requires

$$\mathscr{F} = 9C^4 + \frac{16}{3}C^3 D + \frac{56}{3}C^3 + 10C^2 D + 12C^2 + \frac{26}{3}CD, \tag{29}$$

where C is the number of columns and D is the number of rows of the channel matrix $\mathbf{H}$.

In Tables 1 and 2 a computational cost comparison in terms of the average number of flops per subcarrier is presented for the case of 2- and 4-symbol overlapping, respectively. The tables show the number of flops per subcarrier for different submatrix sizes using different modulation schemes. The tables also include the number of flops for a full size channel matrix when the submatrix division scheme is not utilized. We can see that when the submatrix division order k increases the average number of flops per subcarrier is reduced. For the eighteen ($k = 18$) submatrix size, that is, the maximum achievable division of the scheme, we obtain the minimum average computational cost. Also we can see that the average number of flops is similar for the different modulation schemes. And, the number of flops for the 4-symbols overlapping option is bigger compared with the other 2-symbols overlapping option.

Table 3 shows as reference the number of flops per subcarrier of the conventional MIMO 2 × 2 VBLAST and MIMO 2 × 2 MLD both without ESPAR antenna receiver. Also the computational cost using eighteenth-size ($k = 18$) submatrix division MMSE sparse-SQRD algorithm with 2 and 4-symbols overlapping is included. We can see that the average number of flops per subcarrier of the proposed submatrix division based algorithm is similar to the flops of MIMO 2 × 2 VBLAST and better than MIMO 2 × 2 MLD scheme for 16-QAM and 64-QAM modulation.

TABLE 1: Average number of flops per subcarrier of the proposed algorithm with 2-symbol overlapping.

Submatrix size	QPSK	16-QAM	64-QAM
Full w/o division	7005	6832	6751
Quarter ($k = 4$)	2645	2580	2552
Eighth ($k = 8$)	1078	1060	1050
Eighteenth ($k = 18$)	519	518	516

TABLE 2: Average number of flops per subcarrier of the proposed algorithm with 4-symbol overlapping.

Submatrix size	QPSK	16-QAM	64-QAM
Full w/o division	7005	6832	6751
Quarter ($k = 4$)	2792	2725	2696
Eighth ($k = 8$)	1277	1272	1269
Eighteenth ($k = 18$)	770	768	765

TABLE 3: Flops per subcarrier comparison.

Algorithm	QPSK	16-QAM	64-QAM
MIMO 2×2 VBLAST w/o ESPAR	542	542	542
MIMO 2×2 MLD w/o ESPAR	312	4992	79872
($k = 18$) with 2-sym overlapping	519	518	516
($k = 18$) with 4-sym overlapping	770	768	765

For calculating the total computational cost required by the receiver, based on [16] the number of flops required by the two FFT blocks considering the data symbol and pilot symbol is

$$\mathscr{F}_{\text{FFT}} = 10 N_{\text{FFT}} \log_2 N_{\text{FFT}}, \tag{30}$$

where N_{FFT} is the FFT size. Also the flops required by the channel estimator used for the ESPAR antenna receiver, that was presented in Section 3.1, are given by

$$\mathscr{F}_{\text{CE}} = 32N(N+2). \tag{31}$$

Table 4 presents the total flops per subcarrier required by the receiver using QPSK modulation. Also the complexity of the FFT, channel estimator, and detection blocks is included for the different systems. The MIMO-OFDM systems that are analysed in this table are the original system with ESPAR antenna receiver using ZF-VBLAST detector [3, 4], the system using full-size channel matrix detection, the system using the proposed submatrix divided ($k = 18$) with 4-symbol overlapping detection and the 2×2 VBLAST system without ESPAR antenna receiver. We can observe that using the proposed submatrix divided scheme ($k = 18$) with 4-symbols overlapping the computational cost required for the detection and also the total number of flops per subcarrier required by the receiver are reduced.

8. Simulation Results

To determine the bit error rate performance of the proposed algorithm, a software simulation model of MIMO-OFDM

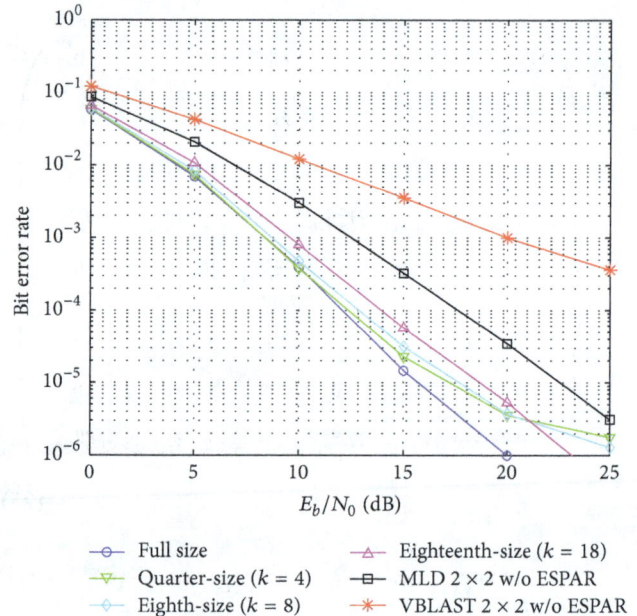

FIGURE 5: Proposed scheme with QPSK and 2-symbol overlapping.

with ESPAR antenna receiver was developed in c++ using the it++ [17] communications library. It is important to note that the system does not include FEC and interleaver. In the simulation the proposed low complexity submatrix divided MMSE sparse-SQRD detection is implemented with quarter-size, eighth-size and eighteenth-size, submatrices. Both options, with 2- and 4-symbol overlapping, are implemented for the previous mentioned submatrix sizes. The configuration settings of the simulation are shown in Table 5.

In Figures 5 and 6 the bit error rate performance using QPSK modulation, for the cases of 2- and 4-symbol overlapping, respectively, is shown. In these figures the performance of the proposed algorithm for quarter-size ($k = 4$), eighth-size ($k = 8$), and eighteenth-size ($k = 18$) submatrices is included. To compare the degradation in the bit error performance created by the algorithm, the performance in the case of a full-size channel matrix without division is included. And also the performance of conventional MIMO-OFDM 2×2 VBLAST and MIMO-OFDM 2×2 MLD systems without ESPAR antenna receiver is shown. As we can see in Figure 6, with QPSK modulation and 4-symbol overlapping, the bit error rate performance degradation is minimum even for the case of eighteenth-size ($k = 18$) submatrix size. Also for a BER of 10^{-3}, the proposed scheme with eighteenth-size ($k = 18$) submatrix size that achieves the minimum computational cost obtains an additional gain of about 11 dB compared to a conventional MIMO-OFDM 2×2 VBLAST system without ESPAR antenna receiver.

In the same way the bit error rate using 16-QAM modulation is shown in Figures 7 and 8. With 16-QAM modulation the degradation in the bit error rate performance is bigger compared with the QPSK results. In this case also the degradation is smaller in the case of 4-symbols overlapping. With 16-QAM for a BER of 10^{-3}, the proposed scheme with

TABLE 4: Total flops per subcarrier of the receiver using QPSK mod.

MIMO-OFDM system	FFT	Channel estimator	Detection	Total
ZF-VBLAST with ESPAR [3, 4]	70	1856	4.1×10^7	4.1×10^7
Full-size w/o division	70	1856	7005	8931
Proposed ($k = 18$) 4-sym. over.	70	1856	770	2696
2×2 VBLAST w/o ESPAR	70	173	542	785

TABLE 5: Simulation settings.

Tx	Modulation	QPSK, 16-QAM, 64-QAM
	Pilot sequence	HTLTF
	Number of subcarriers	56
	FFT size	64
	GI	1/4
Channel	Rayleigh fading	2 rays
	Noise type	AWGN
	Bandwidth	20 MHz
Rx	Synchronization of symbols	Perfect
	Channel estimation	MMSE

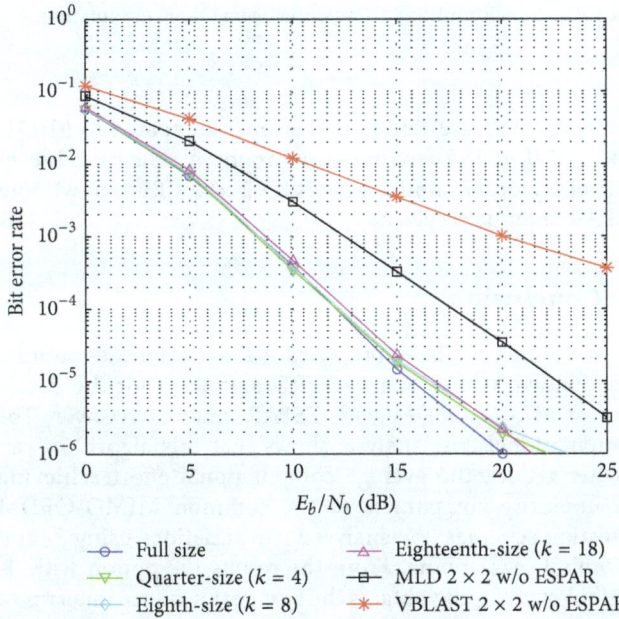

FIGURE 6: Proposed scheme with QPSK and 4-symbol overlapping.

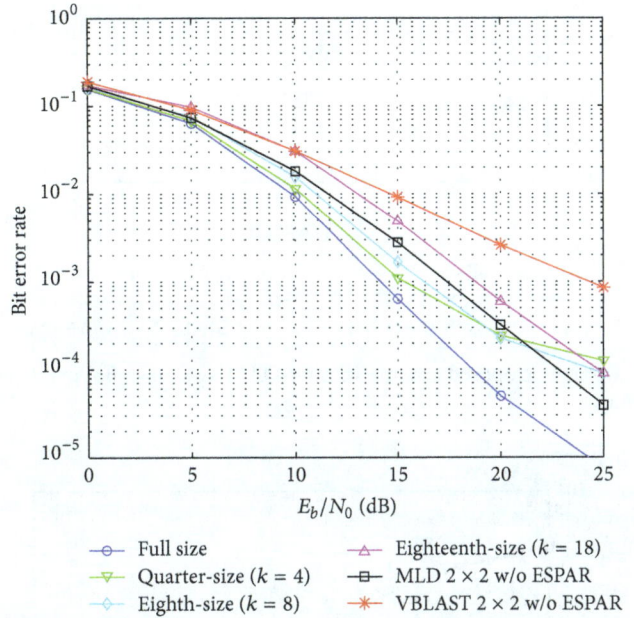

FIGURE 7: Proposed scheme with 16-QAM and 2-symbol overlapping.

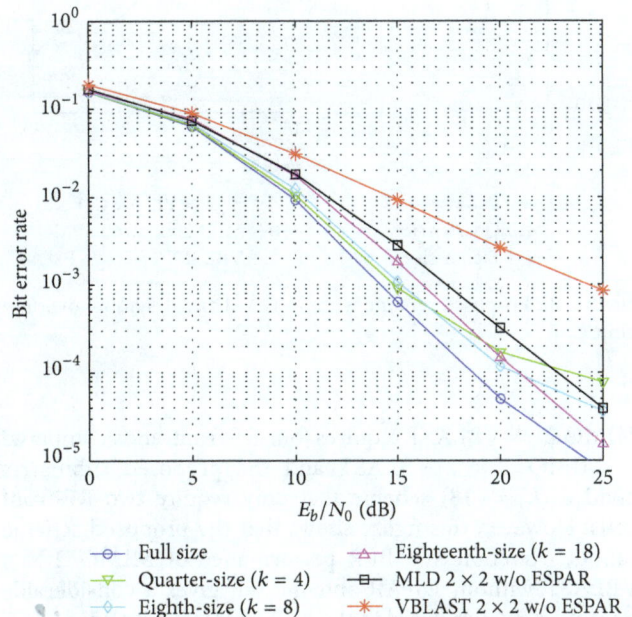

FIGURE 8: Proposed scheme with 16-QAM and 4-symbol overlapping.

eighteenth-size ($k = 18$) submatrix size, obtains an additional gain of about 8.5 dB compared to a conventional MIMO-OFDM 2×2 VBLAST system.

The results for 64-QAM are shown in Figures 9 and 10. In this case the degradation is much bigger and the best result is obtained with the 4-symbol overlapping option.

In Figure 11 the BER performance of the proposed submatrix divided ($k = 18$) scheme with 4-symbol overlapping is compared with the conventional MIMO 2×2 VBLAST and MIMO 2×4 VBLAST without ESPAR antenna using QPSK modulation. This is not a fair comparison in terms of the number of RF front-ends in the receiver side because

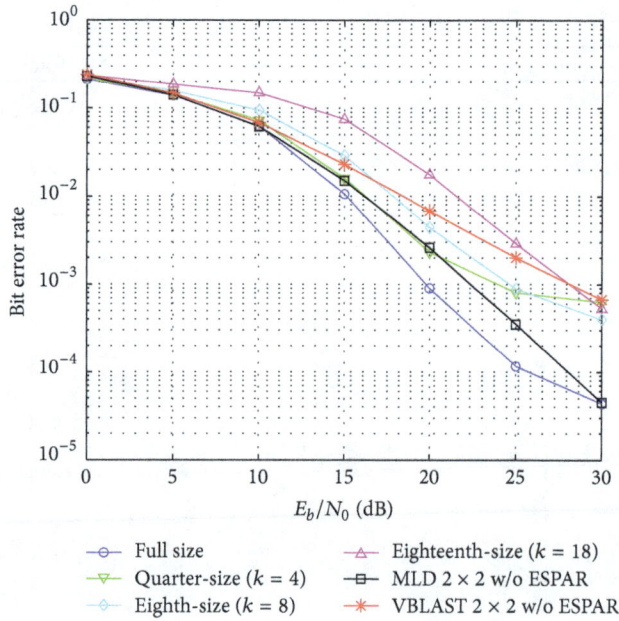

FIGURE 9: Proposed scheme with 64-QAM and 2-symbol overlapping.

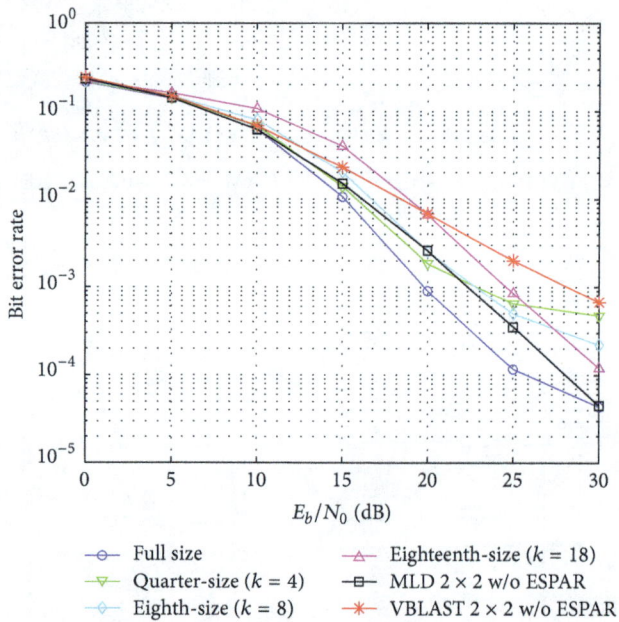

FIGURE 10: Proposed scheme with 64-QAM and 4-symbol overlapping.

MIMO 2×4 VBLAST requires four RF front-ends compared to MIMO 2×2 VBLAST and the proposed submatrix divided ($k = 18$) scheme that only require two RF-front ends. However, this figure shows that the proposed scheme cannot overcome the BER performance of MIMO 2×4 VBLAST without ESPAR antenna but gives a considerable improvement compared to the BER of MIMO 2×2 VBLAST without ESPAR antenna receiver. Also in this figure we can observe that the slope of the proposed scheme and MIMO

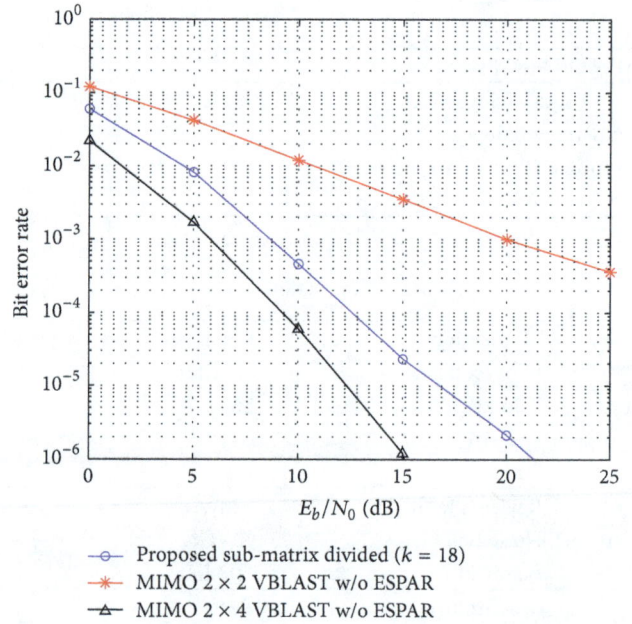

FIGURE 11: Proposed submatrix divided ($k = 18$) scheme with ESPAR antenna receiver versus MIMO 2×2 VBLAST and MIMO 2×4 VBLAST without ESPAR antenna using QPSK modulation.

2×4 VBLAST are similar and steeper compared to MIMO 2×2 VBLAST. Therefore, our proposed scheme achieves a diversity order similar to MIMO 2×4 VBLAST without ESPAR antenna receiver.

9. Conclusion

In this paper, we have proposed a low complexity submatrix divided MMSE Sparse-SQRD algorithm for the detection of MIMO-OFDM with ESPAR antenna receiver. The computational cost analysis shows that this algorithm can further reduce the average computational effort achieving a complexity comparable to the common MIMO-OFDM detection schemes. We analysed two variations using 2- and 4-symbol overlapping. From the results the option with 4-symbol overlapping obtains the best performance in terms of bit error rate, yet increasing the computational cost compared with the other option. The proposed detection scheme is flexible, so the best trade-off between computational cost and bit error rate can be selected depending on the design constraints.

The main application of MIMO-OFDM with ESPAR antenna receiver is to improve the bit error rate performance and diversity gain without increasing the number of RF front-end circuits. And utilizing the proposed low complexity detection scheme we can obtain this improvement in the performance with a low computational cost. The proposed detection scheme is specifically designed to reduce the computational cost of the detection of MIMO-OFDM with ESPAR antenna receiver but it can be also applied in the detection of similar systems that have a large size channel matrix.

In future research we will work in the channel estimator because it is necessary to reduce its computational cost. Also, we will add FEC and interleaver to the system for further improvement in the bit error rate performance.

References

[1] E. Telatar, "Capacity of multi-antenna Gaussian channels," *European Transactions on Telecommunications*, vol. 10, no. 6, pp. 585–595, 1999.

[2] IEEE Computer Society, *IEEE Standard for Information Technology Telecommunication and Information Exchange between Systems Local and Metropolitan Area Networks Specific Requirements*, IEEE Computer Society, New York, NY, USA, 2009.

[3] I. G. P. Astawa and M. Okada, "ESPAR antenna-based diversity scheme for MIMO-OFDM systems," in *Proceedings of the 2009 Thainland—Japan MicroWave*, pp. 1–4, February 2010.

[4] I. G. P. Astawa and M. Okada, "An RF signal processing based diversity scheme for MIMO-OFDM systems," *IEICE Transactions on Communications*, vol. 95, no. 2, pp. 515–524, 2012.

[5] G. J. Foschini, G. D. Golden, R. A. Valenzuela, and P. W. Wolniansky, "Simplified processing for high spectral efficiency wireless communication employing multi-element arrays," *IEEE Journal on Selected Areas in Communications*, vol. 17, no. 11, pp. 1841–1852, 1999.

[6] P. W. Wolniansky, G. J. Foschini, G. D. Golden, and R. A. Valenzuela, "V-BLAST: an architecture for realizing very high data rates over the rich-scattering wireless channel," in *Proceedings of the URSI International Symposium on Signals, Systems, and Electronics (ISSSE '98)*, pp. 295–300, October 1998.

[7] D. Wubben, J. Rinas, R. Bohnke, V. Kuhn, and K. D. Kammeyer, "Efficient algorithm for detecting layered space-time codes," in *Proceedings of the ITG Conference on Source and Channel Coding*, pp. 399–405, Berlin, Germany, January 2002.

[8] D. Wubben, R. Bohnke, V. Kuhn, and K. D. Kammeyer, "MMSE extension of V-BLAST based on sorted QR decomposition," in *Proceedings of the IEEE 58th Vehicular Technology Conference (VTC '03-Fall)*, vol. 1, pp. 508–512.

[9] D. J. Reinoso Ch and M. Okada, "Computational cost reduction of MIMOOFDM with ESPAR antenna receiver using MMSE Sparse-SQRD detection," in *Proceedings of the 27th International Technical Conference on Circuit/Systems, Computers and Communications*, Sapporo, Japan, July 2012.

[10] D. J. R. Chisaguano and M. Okada, "ESPAR antenna assisted MIMO-OFDM receiver using sub-matrix divided MMSE sparse-SQRD detection," in *Proceedings of the International Symposium on Communications and Information Technologies (ISCIT '12)*, pp. 198–203, Gold Coast, Australia, October 2012.

[11] T. Ohira and K. Iigusa, "Electronically steerable parasitic array radiator antenna," *Electronics and Communications in Japan II*, vol. 87, no. 10, pp. 25–45, 2004.

[12] T. Ohira and K. Gyoda, "Electronically steerable passive array radiator antennas for low-cost analog adaptive beamforming," in *Proceedings of the IEEE International Conference on Phased Array Systems and Technology*, pp. 101–104, Dana Point, Calif, USA, May 2000.

[13] S. Tsukamoto and M. Okada, "Single-RF diversity for OFDM system using ESPAR antenna with periodically changing directivity," in *Proceedings of the 2nd International Symposium on Radio Systems and Space Plasma*, pp. 1–4, Sofia, Bulgaria, August 2010.

[14] M. Chouayakh, A. Knopp, and B. Lankl, "Low complexity two stage detection scheme for MIMO systems," in *Proceedings of the IEEE Information Theory Workshop on Information Theory for Wireless Networks (ITW '07)*, pp. 1–5, Solstrand, Norway, July 2007.

[15] J. Benesty, Y. Huang, and J. Chen, "A fast recursive algorithm for optimum sequential signal detection in a BLAST system," *IEEE Transactions on Signal Processing*, vol. 51, no. 7, pp. 1722–1730, 2003.

[16] S. G. Johnson and M. Frigo, "A modified split-radix FFT with fewer arithmetic operations," *IEEE Transactions on Signal Processing*, vol. 55, no. 1, pp. 111–119, 2007.

[17] "Welcome to IT++!," 2010, http://itpp.sourceforge.net/devel/index.html.

5

Optimized Architecture Using a Novel Subexpression Elimination on Loeffler Algorithm for DCT-Based Image Compression

Maher Jridi,[1] Ayman Alfalou,[2] and Pramod Kumar Meher[3]

[1] *Vision Department, L@bIsen, ISEN–Brest, CS 42807, 29228 Brest Cedex 2, France*
[2] *Vision Department, L@bIsen, ISEN–Brest, CS 42807, 29228 Brest Cedex2, France*
[3] *Department of Embedded Systems, Institute for Infocomm Research, Singapore 138632*

Correspondence should be addressed to Maher Jridi, maher.jridi@isen.fr

Academic Editor: Muhammad Shafique

The canonical signed digit (CSD) representation of constant coefficients is a unique signed data representation containing the fewest number of nonzero bits. Consequently, for constant multipliers, the number of additions and subtractions is minimized by CSD representation of constant coefficients. This technique is mainly used for finite impulse response (FIR) filter by reducing the number of partial products. In this paper, we use CSD with a novel common subexpression elimination (CSE) scheme on the optimal Loeffler algorithm for the computation of discrete cosine transform (DCT). To meet the challenges of low-power and high-speed processing, we present an optimized image compression scheme based on two-dimensional DCT. Finally, a novel and a simple reconfigurable quantization method combined with DCT computation is presented to effectively save the computational complexity. We present here a new DCT architecture based on the proposed technique. From the experimental results obtained from the FPGA prototype we find that the proposed design has several advantages in terms of power reduction, speed performance, and saving of silicon area along with PSNR improvement over the existing designs as well as the Xilinx core.

1. Introduction

Many applications such as video surveillance and patient monitoring systems require many cameras for effective tracking of living and nonliving objects. To manage the huge amount of data generated by several cameras, we proposed an optical implementation of an image compression based on DCT algorithm in [1]. But this solution suffers from bad image quality and higher material complexity. After this optical implementation, in this paper we propose a digital realization of an optimized VLSI for image compression system. This paper is an extension of our prior work [2–4] with a new compression scheme along with supplementary simulations and FPGA implementation followed by performance analysis.

More recent video encoders such as H.263 [5] and MPEG-4 Part 2 [6] use the DCT-based image compression along with additional algorithms for motion estimation (ME). A simplified block diagram of the encoder is presented in Figure 1. The 2D DCT of 8×8 blocks of the image is performed to decorrelate each block of input pixels. The DCT

coefficients are then quantized to represent them in a reduced range of values using a quantization matrix. Finally, the quantized components are scanned in a zigzag order, and the encoder employs run-length encoding (RLE) and Huffman coding/binary arithmetic coding (BAC-) based algorithms for entropy coding.

Since the DCT computation and quantization processes are computation intensive, several algorithms are proposed in literature for computing them efficiently in dedicated hardware. Research in this domain can be classified into three parts. The first part is the earliest and concerns the reduction of the number of arithmetic operators required for DCT computation [7–13]. The second research thematic relates to the computation of DCT using multiple constant multiplication schemes [14–25] for hardware implementation. Some other works on design of architectures for DCT make use of convolution formulation. They are efficient but can be used only for prime-length DCT and not suitable for video processing applications [26, 27]. Finally, the third part is about the optimization of the DCT computation in the

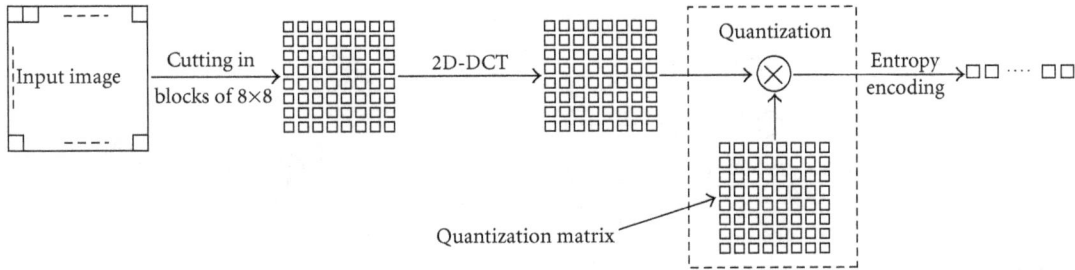

FIGURE 1: Simplified block diagram of the encoder [2].

context of image and video encoding [28–32]. In this paper, we are interested in the last research thematic.

In this paper, we propose a novel architecture of the DCT based on the canonical signed digit (CSD) encoding [33, 34]. Hartley in [35] has used CSD-based encoding and common subexpression elimination (CSE) for efficient implementation of FIR filter. The use of similar CSD and CSE technique for DCT implementation is not suitable. To improve the efficiency of implementation, we identify multiple subexpression occurrences in intermediate signals (but not in constant coefficients as in [35]) in order to compute DCT outputs. Since the calculation of multiple identical subexpression needs to be implemented only once, the resources necessary for these operations can be shared and the total number of required adders and subtractors can be reduced.

The second contribution of the paper is an introduction of a new schema of image compression where the second stage of 1D DCT (DCT on the columns) is configured for joint optimization of the quantization and the 2D DCT computation. Moreover, tradeoffs between image visual quality, power, silicon area, and computing time are analysed.

The remainder of the paper is organized as follows: an overview of fundamental design issues is given in Section 2. Proposed DCT optimization based on CSD and subexpression sharing is described in Section 3. An algorithm based on joint optimization of quantization and 2D DCT computation is proposed in Section 4. Finally, the experimental results are detailed in the Section 5 before the conclusion.

2. Background

Given an input sequence $\{x(n)\}$, $n \in [0, N-1]$, the N-point DCT is defined as:

$$X(n) = \sqrt{\frac{2}{N}} C(n) \sum_{k=0}^{N-1} x(k) \cos \frac{(2k+1)n\pi}{2N}, \quad (1)$$

where $C(0) = 1/\sqrt{2}$ and $C(n) = 1$ if $n \neq 0$.

As stated in the introduction, we find two main types of algorithms for DCT computation. One class of algorithms is focused on reducing the number of required arithmetic operators, while the other class of algorithms are designed for hardware implementation of DCT. In this Section, we provide a brief review of the major developments of different types of algorithms.

2.1. Fast DCT Algorithm. In literature, many fast DCT algorithms are reported. All of them use the symmetry of the cosine function to reduce the number of multipliers. In [36] a summary of these algorithms is presented. In Table 1, we have listed the number of multipliers and adder involved in different DCT algorithms. In [13], the authors show that the theoretical lower limit of 8-point DCT algorithm is 11 multiplications. Since the number of multiplications of Loeffler's algorithm [12] reaches the theoretical limit, our work is based on this algorithm.

Loeffler et al. in [12] proposed to compute DCT outputs on four stages as shown in Figure 2. The first stage is performed by 4 adders and 4 subtractors while the second one is composed of 2 adders, 2 subtractors, and 2 MultAddSub (multiplier, adder and subtractor) blocks. Each MultAddSub block uses 4 multiplications and can be reduced to 3 multiplications by constant arrangements. The fourth stage uses 2 MultSqrt(2) blocks to perform multiplication by $\sqrt{2}$.

2.2. Multiplierless DCT Architecture. The DCT given by (1) can be expressed in inner product form as:

$$Y = \sum_{k=0}^{N-1} x(k) \cdot c(k), \quad (2)$$

where $c(k)$ for $0 \leq k \leq N-1$ are fixed coefficients and equal to $\cos((2k+1)\pi/2N)$, and $x(k)$ for $0 \leq k \leq N-1$ are the input image pixels.

One possible implementation of the inner product in programmable devices uses embedded multipliers. However, these IPs are not designed for constant multipliers. Consequently, they are not power efficient and consume a larger silicon area. Moreover, such a design is not portable for efficient implementation in FPGAs and ASICs. Many multiplierless architectures have, therefore, been introduced for efficient implementation of constant multiplications for the inner product computation. All those methods can be classified as: the ROM-based design [14], the distributed arithmetic (DA-) based design [15], the New distributed

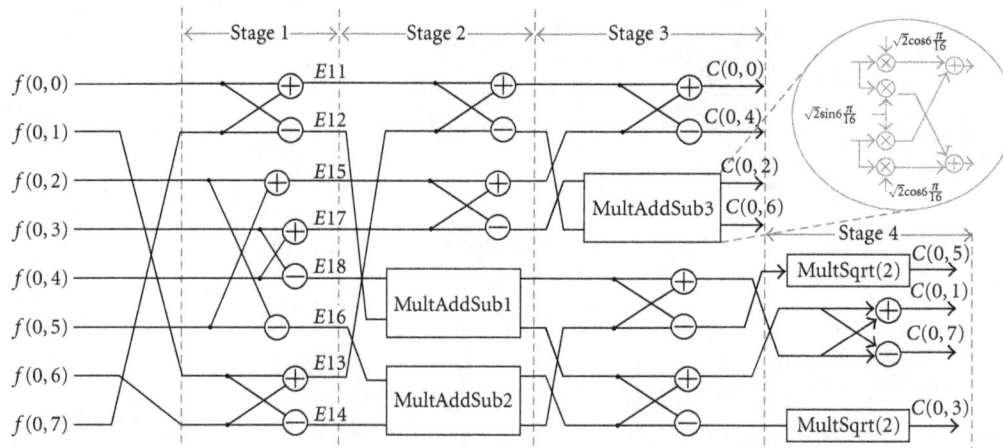

FIGURE 2: Loeffler architecture of 8-point DCT algorithm.

arithmetic (NEDA-) based design [22], and the CORDIC-based design [23].

2.2.1. ROM Multiplier-Based Implementation. This solution is presented in [14] to design a special-purpose VLSI processor of 8×8 2D DCT/IDCT chip that can be used for high-speed image and video coding. Since the DCT coefficient matrix is fixed, the authors of [14] precompute all possible product values and store them in a ROM rather than computing them by any combinational logic. Since the dynamic range of input pixels is 2^8 for gray scale images, the number of stored values in the ROM is equal to $N = 2^8$. Each value is encoded using 16 bits. For example, for an 8-point inner product, the ROM size is about $8 * 2^8 * 16$ bits which is equivalent to 32.768 kbits. To obtain 8-point DCT, 8-point inner products are required and consequently, the ROM size becomes exorbitant for realization of image compression.

2.2.2. Distributed Arithmetic (DA). Distributed arithmetic (DA) [15] is a well-known technique for computing inner products which outperforms the ROM-based design. Authors of [16–18] use the recursive DCT algorithm to derive a design that requires less area than conventional algorithms. By precomputing all the partial inner products corresponding to all possible bit vectors and storing these values in a ROM, the DA method speeds up the inner product computation over the multiplier-based method. Unfortunately, in this case also the size of ROM grows exponentially with the number of inputs and internal precision. This is inherent to the DA technique where a great amount of redundancy is introduced into the ROM to accommodate all possible combinations of bit patterns in the input signal.

2.2.3. New Distributed Arithmetic (NEDA). The New Distributed Arithmetic (NEDA) is adder-based optimization of DA implementation. The NEDA architecture does not require ROMs and multipliers. It provides reduced complexity solution by sharing of common subexpression of input vector to generate optimal shift-add network for

DCT implementation. This results in a low-power, high-throughput architecture for the DCT. Nevertheless, the implementation of NEDA has two main disadvantages, [22]:

(i) the parallel data input leads to higher scanning rate which severely limits the operating frequency of the architecture;

(ii) the assumption of serial data input leads to lower hardware utilization.

2.2.4. CORDIC. COordinate Rotation DIgital Computer (CORDIC) provides a low-cost technique for DCT computation. The CORDIC-based DCT algorithm in [23] utilizes dynamic transformation rather than static ROM addressing. The CORDIC method can be employed in two different modes: the rotation mode and the vectoring mode. Sun et al. in [24] have presented an efficient Loeffler DCT architecture based on the CORDIC algorithm. However, the use of dynamic computation of cosine function in iterative way involves long latency, high-power consumption and involves a costly scale compensation circuit.

2.2.5. CSD. Vinod and lai in [25] have proposed an algorithm to reduce the number of operations by using CSD and CSE techniques, where they have minimized the number of switching events in order to reduce the power consumption. In CSD encoding, the constant multiplications are replaced by additions. Hence, there are two types of additions: interstructural adders to compute the summation terms of the inner product of (2) and intrastructural adder required to replace the constant multipliers. The authors of [25] applied the CSD encoding on the constant multiplier of the conventional DCT computation. Consequently, the number of intrastructural adders is reduced, but the number of interstructural adders is increased to 56 for 8-point DCT. However, as it is reported in Table 1, there are some fast DCT algorithms which use the cosine symmetry to reduce the number of adders (from 26 to 29). Moreover, the authors of [25] have used CSE technique to reduce the number of intrastructural adders. Indeed, since each data (image

TABLE 1: Complexity of different DCT algorithms.

Reference	[7]	[8]	[9]	[10]	[11]	[12]
Multipliers	16	12	12	12	12	11
Adders	26	29	29	29	29	29

TABLE 2: 8-point DCT fixed coefficient representation.

Real value	Decimal	Natural binary	Partial products	CSD	Partial products
$\cos(3\pi/16)$	106	01101010	4	+0−0+0+0	4
$\sin(3\pi/16)$	71	01000111	4	0+00+00−	3
$\cos(\pi/16)$	126	01111110	6	+00000−0	2
$\sin(\pi/16)$	25	00011001	3	00+0−00+	3
$\cos(6\pi/16)$	49	00110001	3	0+0−000+	3
$\sin(6\pi/16)$	118	01110110	5	+000−0−0	3
$\sqrt{(2)}$	181	10110101	5	+0−0−0+0+	5
Total partial products			30		23

pixel) is multiplied with distinct constant element (cosine coefficient), Vinod and lai have proposed to reformulate the DCT matrix for efficient substitution of CSE. With this optimization, the total number of intrastructural adders substantially reduced.

2.3. Joint Optimization. Recent research on DCT implementation uses the optimization to adapt the implementation of the DCT to the specific compression standards in order to reduce the chip size and power consumption. All these recent works in this area exploit the context in which the DCT is used to reduce the computational complexity. Some of them use the characteristics of input signals and the others simplify the DCT architecture. Xanthopoulos and Chandraksan in [28] have exploited the signal correlation property to design a DCT core with low-power dissipation. Yang and Wang have investigated the joint optimization of the Huffman tables, quantization, and DCT [31]. They have tried to find the performance limit of the JPEG encoder by proposing an iterative algorithm to find the optimal DCT bit width for a given Huffman tables and quantization step sizes.

A prediction algorithm is developed in [32] by Hsu and Cheng to reduce the computation complexity of the DCT and the quantization process of H264 standard. They have built a mathematical model based on the offset of the DCT coefficients to develop a prediction algorithm.

In this paper, we propose a model for combined optimization of DCT and quantization to implement them in the same architecture to save the computational complexity for image and video compression.

3. Proposed Algorithm for DCT Computation

In this Section, we present a new multiplierless DCT based on CSD encoding.

3.1. Principle of CSD. The CSD representation was first introduced by Avizienis in [33] as a signed-digit representation of numbers. This representation was created originally to eliminate the carry propagation chains in arithmetic operations. It is a unique signed-digit representation containing the fewest number of nonzero bits. It is therefore used for the implementation of constant multiplications with the minimum number of additions and subtractions. The CSD representation of any given number c is given by:

$$c = \sum_{i=0}^{N-1} c_i \cdot 2^i, \quad c_i = \{-1, 0, 1\}, \tag{3}$$

CSD numbers have two basic properties:

(i) no two consecutive digits in a CSD number are nonzero;

(ii) The CSD representation of a number contains the minimum possible number of nonzero bits and thus the name canonic.

The CSD values of the constants used in 8-point DCT by Loeffler algorithm are listed in Table 2. The digits $1, -1$ are, respectively, represented by $+, -$. For 8 bit width, the saving in term of partial products is about 24%. A generalized statistical study about the average number of nonzero elements in N-bit CSD numbers is presented in [33], and it is proved that this number tends asymptotically to $N/3 + 1/9$. Hence, on average, CSD numbers contain about 33% fewer nonzero bits than 2's complement numbers. Consequently, for multiplications by a constant (where the bit pattern is fixed and known a priori), the numbers of partial products are reduced by nearly 33% in average.

3.2. New CSE Technique for DCT Implementation. To minimize the number of adders, subtractors, and shift operators for DCT computation, we can use the common subexpression elimination (CSE) technique over the CSD representation of constants. CSE was introduced in [35] and applied to digital filters in transpose form. Contrary to transpose form FIR filters, constant coefficients of DCT (shown in Table 2) multiply 8 *different input data* since the DCT consists in transforming 8-point input sequence to 8-point output coefficient. For this reason we cannot exploit the redundancy among the constants for subexpression elimination as in case of FIR filter. Moreover, for bit patterns in the same constant, Table 2 shows that only the constant $\sqrt{2}$ presents one common subexpression which is +0− repeated once with an opposite sign. Consequently, we cannot use the conventional CSE technique in the same manner as in the case of multiple constant multiplication in FIR filters.

We have proposed here a new CSE approach for DCT optimization where we do not consider occurrences in CSD coefficients, but we consider the interaction of these codes. On the other hand, according to our compression method (detailed in the next Section) we use only some of the DCT coefficients (1 to 5 among 8). Hence, it is necessary to compute specific outputs separately. To emphasize the advantage of CSE, we take the example of

$X(2)$ $(X(2) = E35 + E37)$. According to Figure 2, we can express $E35$ as follows:

$$E35 = (E25 + E28)$$

$$= \left(E18 * \cos\left(\frac{3\pi}{16}\right) + E12 * \sin\left(\frac{3\pi}{16}\right)\right) \quad (4)$$

$$+ \left(E14 * \cos\left(\frac{\pi}{16}\right) - E16 * \sin\left(\frac{\pi}{16}\right)\right).$$

Using CSD encoding of Table 2, (7) is equivalent to:

$$E35 = E18(2^7 - 2^5 + 2^3 + 2^1) + E12(2^6 + 2^3 - 2^0)$$
$$- E16(2^5 - 2^3 + 2^0) + E14(2^7 - 2^1). \quad (5)$$

After rearrangement (8) is equivalent to:

$$E35 = 2^7(E18 + E14) + 2^6 E12 - 2^5(E16 + E18)$$
$$+ 2^3(E12 + E16 + E18) + 2^1(E18 - E14) \quad (6)$$
$$- 2^0(E12 + E16).$$

In the same way, we can determine $E37$:

$$E37 = 2^7(E12 + E16) - 2^6 E18 + 2^5(E14 - E12)$$
$$+ 2^3(E12 - E14 - E18) + 2^1(E12 - E16) \quad (7)$$
$$+ 2^0(E14 + E18).$$

Equations (10) and (11) give

$$X(2) = 2^7\left(\overbrace{(E16 + E18)}^{CS1} + E12 + E14\right) + 2^6(E12 - E18)$$

$$- 2^5\left(\overbrace{(E12 - E14)}^{CS2} + \overbrace{(E16 + E18)}^{CS1}\right)$$

$$+ 2^3\left(\overbrace{(E12 - E14)}^{CS2} + E12 + E16\right)$$

$$+ 2^1\left(\overbrace{(E18 - E16)}^{CS3} + \overbrace{(E12 - E14)}^{CS2}\right)$$

$$+ 2^0\left(\overbrace{(E14 - E12)}^{CS2} + \overbrace{(E18 - E16)}^{CS3}\right), \quad (8)$$

where CS1, CS2, and CS3 denote 3 common subexpressions. In fact, the identification of common subexpressions results in significant reduction of hardware and power consumption reductions. For example, CS2 appears 4 times in $X(2)$. This subexpression is implemented only once and resources needed to compute CS2 are shared. An illustration of resources sharing is given in Figure 3.

TABLE 3: Statistics of X(2) calculation.

Components/methods	Multiplier based	CSD	CSD-CSE
Adders/subtractors	11	23	16
Registers	125	188	119
MULT18x18SIOs	4	0	0
Equivalent no of LUT	305	221	200
Latency	$3T_A + T_M$	$6T_A$	$4T_A$
Maximum frequency[1]	143.451	121.734	165.888

[1] Maximum frequency is measured in MHz.

Symbols $\ll n$ denote left shift operation by n-bit positions. It is important to notice that nonoverbraced terms in (12) are potential common subexpressions which could be shared with other DCT coefficients such as $X(4)$, $X(6)$, and $X(8)$. According to this analysis, $X(2)$ is computed by using 11 adders and 4 embedded multipliers. If CSD encoding, is applied 23 adders/subtractors, are required. The proposed method enables to compute $X(2)$ by using only 16 add/subtract operations. This improvement allows to save silicon area and reduces the power consumption without any decrease in the maximum operating frequency.

To emphasize the common subexpression sharing, a VHDL model of calculation of $X(2)$ is developed using three techniques: embedded multipliers, CSD encoding, and CSE of CSD encoding. It is shown in Table 3 that the CSD encoding uses more adders, subtractors, and registers than the proposed combined CSD-CSE technique to replace the 4 embedded multipliers MULT18x18SIOs. Also, we have included the equivalent number of LUT if $X(2)$ is synthesized on Xilinx FPGA without any arithmetic DSP core (without MULT18x18). Total number of LUTs is found to be 305, 221, and 200, respectively, for the multiplier-based design, the CSD-based design, and the combined CSD-CSE-based design. Moreover, it can be observed that the time required to get $X(2)$ coefficient is equal to $3T_A + T_M$, $6T_A$, and $4T_A$, respectively, for the multiplier-based design, the CSD-based design, and the combined CSD-CSE-based design, where T_A is the addition time and T_M is the multiplication time. Consequently, the area-delay product is decreased by sharing subexpression.

4. Joint Optimization

4.1. Principle. As discussed earlier, the 2D DCT is computed in two stages by row/column decomposition using row-wise 1D DCT of input in stage 1, followed by column-wise 1D DCT of intermediate result in stage 2. If we consider an input block $f(i, j)$ of 8×8 samples, the row-wise transform calculates $X(i, v)$ as 1D DCT of $f(i, j)$, and the column-wise transform gives $Y(u, v)$ which are the 1D DCT coefficients applied to $X(i, v)$ for $i, j, u, v \in [1 : 8]$. Hence, for a given 8×8 block of pixels, we obtain 64 DCT coefficients of different frequencies. Unlike the high-frequency coefficients, the low-frequency coefficients have a greater effect on image reconstruction. Moreover, after quantization process, most of the high-frequency coefficients are likely to be zero as

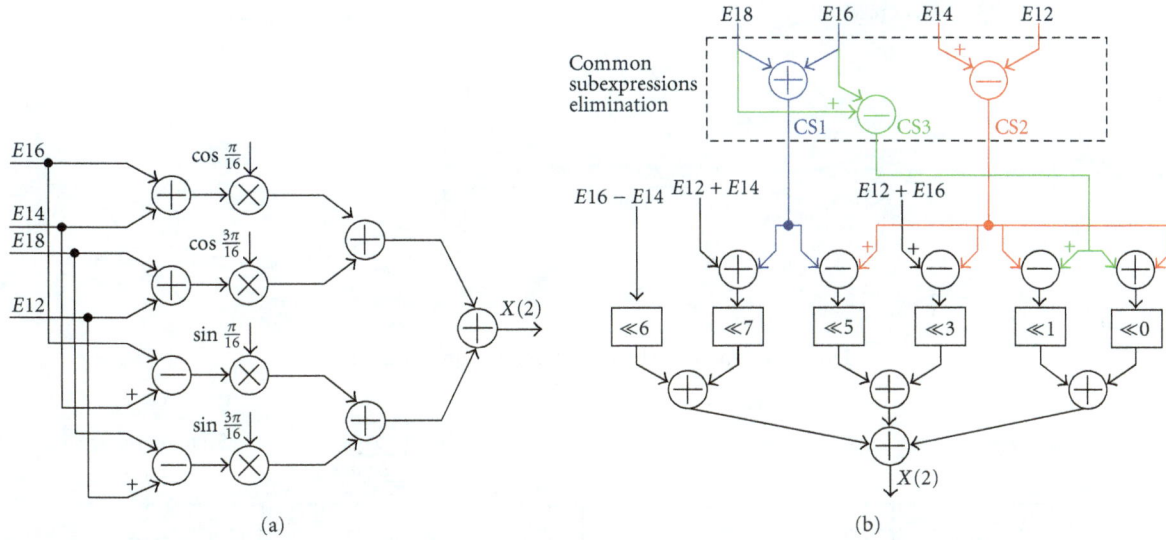

FIGURE 3: $X(2)$ calculation (a) conventional method, (b) shared subexpression using CSD encoding.

shown in Figure 4. Since these coefficients are likely to be zero after quantization, to save computation time and resources we avoid computing these DCT coefficients. In fact, for an 8×8 block of pixels, we compute 64 1D DCT coefficients of the first stage and then we compute only low frequency components for 1D DCT of second stage. With this method, the quantization is done on the fly with the DCT algorithm and consequently we save computational resources. Another advantage of the proposed method is the latency improvement. Since the high-frequency coefficients are to be eventually discarded, the time used for their computation is saved. In fact, for the second 1D DCT algorithm, at least 3 rows do not need to be computed as illustrated in the Figure 5. This gives a saving of at least 12 clock cycles, since the latency of calculation of each row is 4 clock cycles.

4.2. Quantization Levels. For an 8×8 block of pixels, the 1D DCT is calculated for each of the 8 input rows as mentioned in Figure 5. For each row, the first 1D DCT coefficient $X(i, 1)$ is encoded using 11 bits which is the estimated word length without truncation or rounding for $i \in [1 : 8]$. Coefficients $X(i, 2)$ to $X(i, 8)$ are truncated using 8 bits to trade accuracy for compression ratio and computational complexity.

In the second stage, 1D DCT is calculated selectively since the higher frequency components need not be computed. It is known that the lower frequencies tend to spread across either the first row or the first column of the 2D DCT coefficient matrix. However, the computation of an entire row and entire column leads to the computation of all DCT coefficients. For this purpose, we propose an efficient and simple computing scheme by creating 4 DCT zones (shown in Figure 5) where each zone corresponds to a specific compression ratio. For an 8×8 block of pixels, the row-wise transform is applied to compute 64 DCT coefficients while the column-wise transform is applied partially to reduce the computational complexity. Indeed, the proposed

quantization zones are chosen to be square in order to avoid redundancy of computation. In Zone 1, only 4 coefficients $Y(1, 1)$, $Y(1, 2)$, $Y(2, 1)$, and $Y(2, 2)$ are calculated by 2 1D DCT operations. The first one is applied to the first column of intermediate result (output of the row-wise transform) which gives $Y(1, 1)$ and $Y(1, 2)$ while the second one is applied to the second column of intermediate result to compute $Y(2, 1)$ and $Y(2, 2)$. For this quantization mode, all the others DCT coefficients are set to zero. Similarly, in Zone 4, 25 DCT coefficients are calculated by 5 1D DCT operations to compute coefficients $Y(u, v)$, $u, v \in [1 : 5]$.

The compression ratio depends on the zone selection. In Zone 1, $Y(1, 1)$ is encoded using 14 bits. Indeed, the first 1D-DCT coefficient is encoded using 11 bits since in Loeffler DCT algorithm (shown in Figure 2), the DC output is obtained by three cascaded adder stages applied to 8-bit image pixels. Since the 11-bit DC output is fed to the second stage of 1D DCT the bit width of the first output of 2D-DCT is equal to 14 bits. This bit width is taken as reference for encoding the AC coefficients which have less influence than DC coefficient on the quality of reconstructed images. To estimate the bit width of AC coefficients, we were referred to image and video quantization tables (Q) in JPEG standard for Luminance image component and in MPEG-4 standard for intraframe video coding. It is found that for image quality of 50%, $Y(2, 2)$ is divided by $Q(2, 2) = 24$. Then, in order to have a unique bit width by quantization zone, AC coefficients of zone 1 are encoded with 5 bits under the DC bit width (i.e., 9 bits). Likewise, AC coefficients of zone 2 need coefficients of zone 1 along with 5 other coefficients $Y(1, 3)$, $Y(2, 3)$, $Y(3, 1)$, $Y(3, 2)$, and $Y(3, 3)$. All these data are encoded using 8 bits. The additional coefficients of zone 3 are encoded using 7 bits. Finally, the remaining coefficients of zone 4 are encoded using 6 bits since $Y(5, 5)$ will be divided by $Q(5, 5) = 136$.

Hence, by selecting zone 1, the total number of bits is equal to $14 + 9 * 3 = 41$ bits and the compression ratio (CR)

8 × 8 block pixel

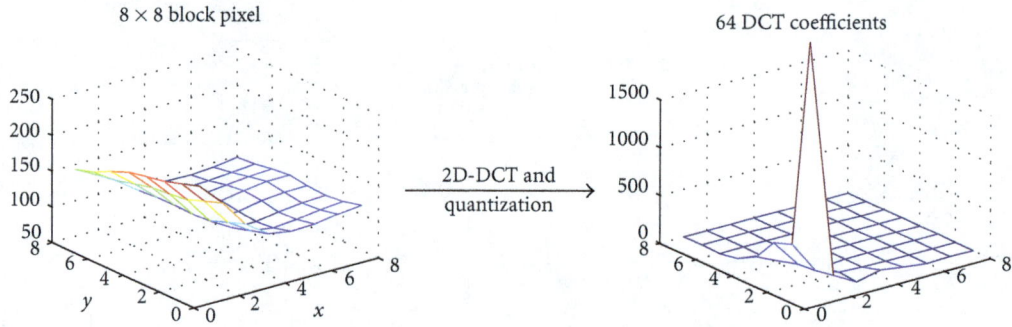

64 DCT coefficients

FIGURE 4: Principle of the DCT and the quantization.

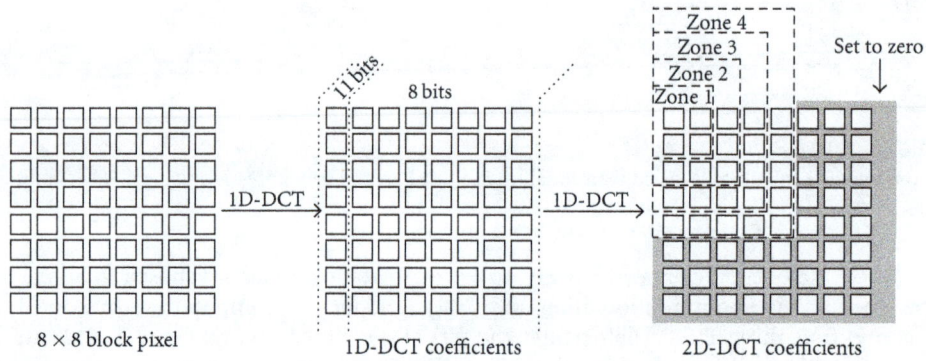

FIGURE 5: Quantization zones.

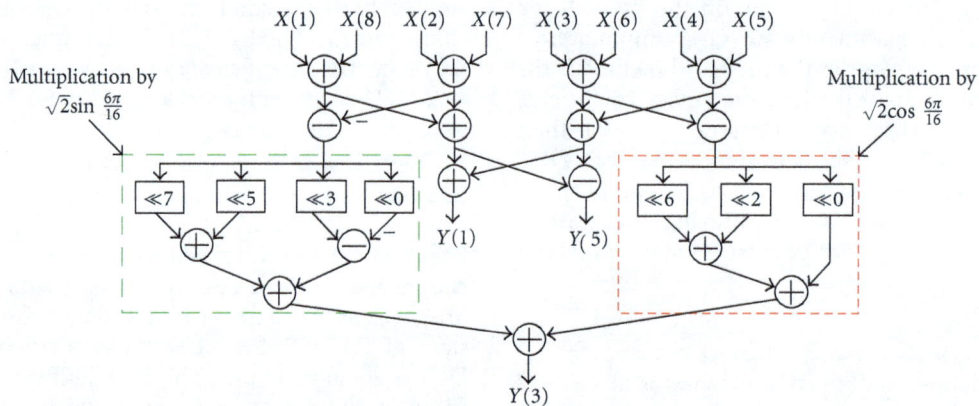

FIGURE 6: $Y(1, v)$, $Y(3, v)$, and $Y(5, v)$ calculation with CSE and CSD.

which is the ratio of the original image size to the compressed image size is equal to 8 bits $*64/14$ bits = 12.48. For the zone 2, the compression ratio is equal to 8 bits $* 64/(14 + 9 * 3 + 8 * 5$ bits) = 6.32. Similarly, for zone 3 and 4, the CRs are, respectively, equal to 3.95 and 2.78.

4.3. DCT Calculation. An example of the calculation of the 1D DCT coefficient ($X(i, 2)$ for $i \in [1 : 8]$) is given in Section 3.2. To compute the DCT coefficients of the 1D DCT of second stage, we use the same method by replacing the inputs by $X(i, v)$ and the outputs by $Y(u, v)$ for $u, v \in [1 : 8]$. According to the algorithm illustrated in Figure 2, for a given column, $Y(1, v)$ and $Y(5, v)$ are calculated using adders and

subtractors while $Y(3, v)$ uses a multiplicative constant and is given by:

$$Y(3) = \sqrt{2} \cos\left(\frac{6\pi}{16}\right) E24 + \sqrt{2} \sin\left(\frac{6\pi}{16}\right) E22. \quad (9)$$

Intermediate results $E22$ and $E24$ in (9) are shown in Figure 2 for a given column. Now, constants $\sqrt{2} \cos(6\pi/16)$ and $\sqrt{2} \sin(6\pi/16)$ are converted to CSD format and given, respectively, by 0 + 000 + 0+ and 0 + 0 + 0 + 00−. $Y(1, v)$, $Y(3, v)$ and $Y(5, v)$ calculations are given in Figure 6.

For $Y(2, v)$ and $Y(4, v)$ also the CSD-CSE techniques are used. $Y(2, v)$ is calculated as in (8), and the common subexpression of $Y(4, v)$ calculation is determined by increasing the number of common subexpressions shared between $Y(2, v)$ and $Y(4, v)$.

According to Figure 2, $Y(4, v) = \sqrt{2}(E26 - E27)$ which is equivalent to:

$$
Y(4, v) = \sqrt{2}\left(E12 * \cos\left(\frac{3\pi}{16}\right) - E18 * \sin\left(\frac{3\pi}{16}\right)\right) \\
- \sqrt{2}\left(E16 * \cos\left(\frac{\pi}{16}\right) - E14 * \sin\left(\frac{\pi}{16}\right)\right). \quad (10)
$$

Using CSD encoding of the constant coefficient, (10) is equivalent to:

$$
Y(4, v) = E12(2^7 + 2^5 - 2^3 - 2^0) \\
- E18(2^7 - 2^5 + 2^2 + 2^0) \\
- E16(2^8 - 2^6 - 2^4 + 2^1) \\
+ E14(2^5 + 2^2 - 2^0). \quad (11)
$$

After rearrangement (11) is equivalent to:

$$
Y(4, v) = 2^7\left(-\overbrace{(E16 + E18)}^{CS1} + \overbrace{(E12 - E16)}^{CS4}\right) \\
+ 2^5\left(-\overbrace{(E16 + E18)}^{CS1} + \overbrace{(E12 + E14)}^{CS5} + E14\right) \\
- 2^3\left(\overbrace{(E12 - E16)}^{CS4}\right) \\
+ 2^2\left(E16 + E14 - \overbrace{(E18 - E16)}^{CS3}\right) \\
- 2^0\left(\overbrace{(E16 + E18)}^{CS1} + \overbrace{(E12 + E14)}^{CS5} + E16\right), \quad (12)
$$

For $Y(4, v)$ calculation, the common subexpression CS1 and CS3 defined for $Y(2, v)$ calculation is used. Two new subexpressions CS4 and CS5 are introduced to further reduce the arithmetic operators. It is important to mention that the equations listed before are expressed to create several occurrences of common subexpression such as CS1, CS3, and CS4 those are used for $Y(2, v)$ calculation. The Signal flow graphs of $Y(2, v)$ and $Y(4, v)$ are shown in Figure 7.

5. Simulation Results

We have coded the proposed method and the existing competing algorithms in VHDL and synthesized them using Xilinx ISE tool.

TABLE 4: Macrostatistics of 1D DCT calculation.

Method	DA [16]	DA [18]	NEDA [19]	CSD [25]	Proposed
Adders	136	144	85	123	72

TABLE 5: Microstatistics of 1D DCT calculation.

Method	NEDA [19]	CORDIC [37]	Xilinx's core	Proposed
Slices	1031	780	531	369[2]	454

[2] Apart from the number 369 slices Xilinx core uses 4 embedded multipliers.

5.1. Synthesis Results. From high-level synthesis results we obtain the number of adders used for different DA-based 1D DCT design and listed in Table 4. It is found that our design uses fewer adders than the other. The direct realization of DA-based DCT design requires 308 adders. Optimizations presented in [19] reduce the number of adders to 85. Regarding the CSD-based design [25], for 8-bit constant width, we found that design of [25] consumes 123 adders (67 intrastrucutral adders + 56 interstructural adders) while the proposed design involves the DCT with 72 adders. We have listed the number of slices occupied by the 1D DCT of [37] and proposed design in the Table 5. The proposed method is compared favorably with the conventional multiplierless architectures using Xilinx XC2VP50 FPGA, the same device employed in [37].

Note that the Xilinx's core uses the Chen's algorithm [7] and requires 369 Slices along with 4 embedded multipliers 18x18SIOs. Besides, the number of slices required by the Xilinx core is relatively low compared with other designs because, the adder/subtractor module of the Xilinx's design alternatively chooses addition and subtraction by using a toggle flop. However, the slice-delay product of proposed design is significantly less than that of the Xilinx DCT IP core since the later has a maximum usable frequency (MUF) of 101 MHz on Spartan3E device while the proposed design provides MUF of 119 MHz.

Regarding the timing analysis derived from synthesis results obtained by cadence 0.18μ library, we find that the proposed design involves a delay about 14.4 ns which represents nearly 15% less than the Xilinx core and 60% less than optimized NEDA-based design [20]. We should underline that in [20] an optimized architecture of NEDA-based design [19] where compressor trees are used to decrease the delay.

Moreover, we have used the XPower tool of Xilinx ISE suite to estimate the dynamic power consumption. The power dissipation of the proposed 1D DCT design and Xilinx's core is about 39 mW and 62 mW respectively.

In order to highlight the effect of subexpression sharing, 1D DCT structure of Loeffler algorithm is implemented with different multiplier designs. The power-delay product in nJ is computed as the product of the DCT computation time (ns) and the power dissipation (W) for the proposed design and the Xilinx core. The power-delay product for different number of DCT coefficients is calculated and plotted in Figure 8. The CSD-based design and the proposed design using CSE involve nearly 43% and 33% of power-delay

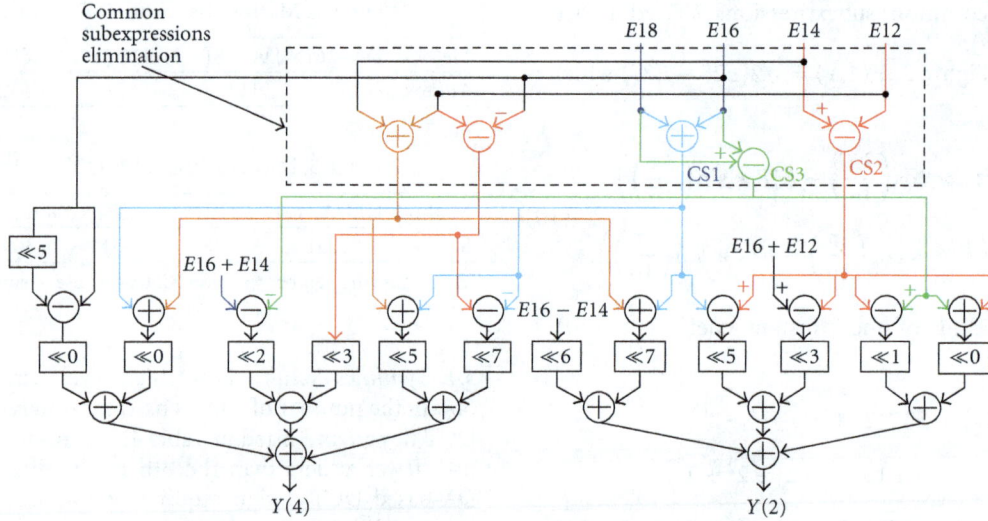

FIGURE 7: $Y(2, v)$ and $Y(4, v)$ calculation with CSE and CSD.

product of the Xilinx's multiplier-based design, respectively. It can be seen in Figure 8 that the computation of only the first 1D DCT coefficient involves the same power-delay product since this coefficient does not require any multiplier. Note that the computation of 4th DCT coefficient requires nearly the same power-delay product as that for 5th DCT coefficient. Indeed, the computation of the fifth DCT coefficient requires only one more subtrator.

For the 2D DCT architecture using the Loeffler algorithm a performance analysis is presented in Table 6 in order to highlight the effects of CSD coding, subexpression sharing, and quantization. The multiplier-based structures considered in the comparison are the Xilinx's embedded multiplier synthesized as multiplier block IP and in LUTs. It should be indicated that the input bit width is of 8 bits, the DC coefficient bit width of the first and second 1D DCT stages are 11 bits and 14 bits, respectively and the constant cosine coefficient bit width is 8 bits. The implementation of 2D DCT is realized by decomposing the 2D DCT into two 1D DCT computations together with a transpose memory. It can be observed in Table 6 that the area-delay complexity of Xinlinx's multiplier-based 2D DCT design (synthesized in block) is nearly the same as that of the combined CSD-CSE design but has nearly twice the power consumption. On the other hand, when the Xilinx's multipliers are synthesized as LUT, the 2D DCT structure has less power-delay product but involves twice the area compared with the combined CSD-CSE structure.

The average computation time (ACT) is the time interval after which we get a set of 2D DCT coefficients. ACT is the product of the number of clock cycles required for the 2D DCT computation and the duration of a clock cycle. The 2D DCT computation requires 86 cycles, which is comprised of 8 cycles for register inputs, 7 cycles for the first stage 1D DCT, 64 cycles for transpose memory, and 7 cycles for the second stage of 1D DCT.

Finally, we use the energy per output coefficient (EoC) as power metric which amounts to the average of energy

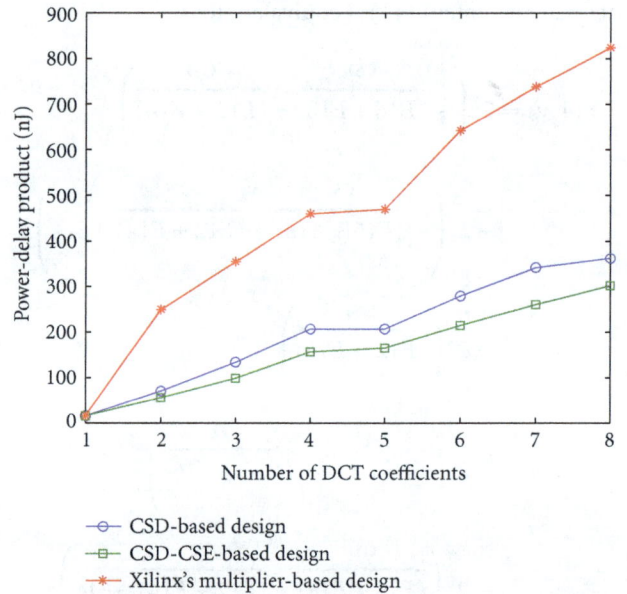

FIGURE 8: Power-delay product estimation with 1.6 V design.

required to compute one value of 2D DCT output. EoC is calculated by multiplying the ACT by the power consumption and dividing the product by 64. It is shown in Table 6 that the design based on Xilinx's multiplier IP involves more than twice EoC compared with all the other designs. Moreover, the proposed 2D DCT with CSD-CSE technique needs less energy compared with CSD-based design and Xilinx's multiplier-based design. It can be further observed that the proposed CSD-CSE and quantization technique has 46% to 65% of EoC compared to CSD-based design.

5.2. FPGA Implementation of Image Compression. In this subsection, we examine the quality of reconstructed image using an FPGA prototype of the proposed DCT-based image

TABLE 6: Performance analysis of the 2D DCT design using Loeffler algorithm.

Constant multiplication design	Slice	MULT18x18SIOs	Power dissipation (mW)	Delay (ns)	ACT (μs)	EoC (nJ)
Xilinx-embedded multipliers (Block)	960	20	176	20.58	1.769	311.344
Xilinx-embedded multipliers (LUT)	1836	0	66.4	24.02	2.065	137.116
CSD	1423	0	74	22.39	1.925	142.45
CSD-CSE	1048	0	76	19.94	1.714	130.264
CSD-CSE and quantization	$[625, 765]$	0	$[48, 64]$	$[15.90, 16.89]$	$[1.367, 1.452]$	$[65.616, 92.928]$

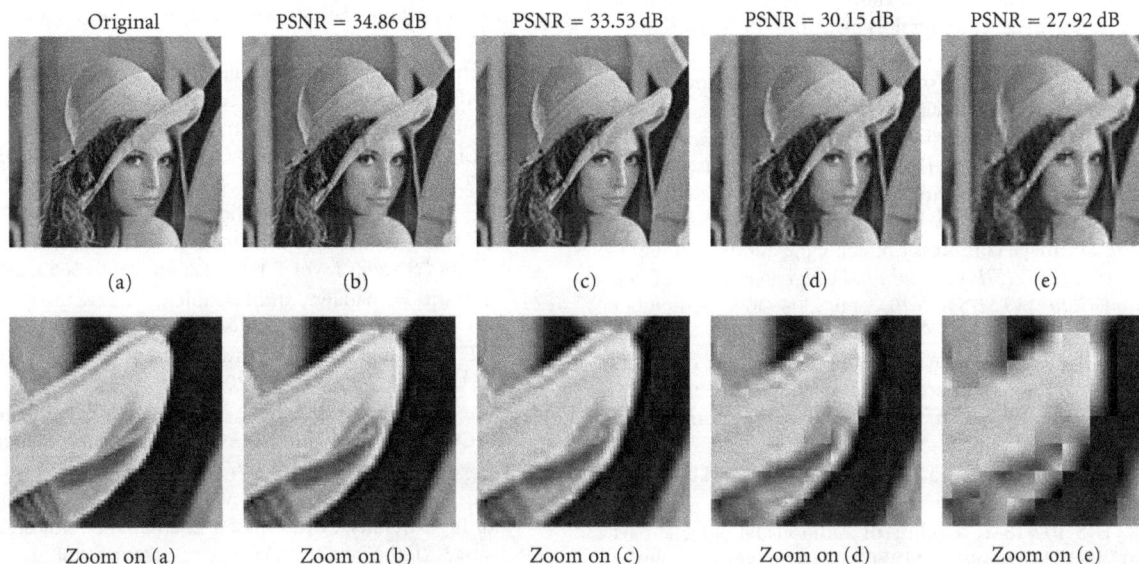

FIGURE 9: Decoded images after joint optimization. (a) (Original), (b) (quantization zone = 4, bpp = 2.87, PSNR = 33.24 dB), (c) (quantization zone = 3, bpp = 2.0, PSNR = 30.26 dB), (d) (quantization zone = 2, bpp =1.26, PSNR = 28.23 dB), and (e) (quantization zone = 1, bpp=0.64, PSNR = 25.38 dB).

TABLE 7: PSNR (dB) versus bpp evaluation.

bpp	0.64	1.26	2.0	2.87
Lena	25.38	28.23	30.26	33.24
Mandrill	20.84	25.27	28.20	29.36
Peppers	27.92	30.15	33.53	34.86
Goldhill	27.51	29.22	32.43	33.12

compression unit. The test images are saved in a ROM in order to avoid the transmission time between the PC and the FPGA. It is important to mention that in the final design we need to use a 2-bit word to indicate 4 available compression ratios. To measure the visual quality of the reconstructed image and to validate the proposed DCT design, we use the Xilinx's integrated logic analyzer (ILA). This module works as a digital oscilloscope and enables to trigger on signals in the hardware design.

To reconstruct back the images, a floating point inverse 2D DCT function of Matlab tool is applied to the FPGA output. PSNR of different 255×255 gray scale images are evaluated and listed in Table 7. The bit per pixel (bpp) depends on the quantization zone selection and varies from 0.64 to 2.87 (The compression is due to DCT only. To increase the compression ratio further the quantized DCT output needs to pass through the entropy coding which we have not performed here.). As shown in Table 7, a good or acceptable image visual qualities can be obtained by joint optimization of the quantization and the DCT. Moreover, to underline the adequacy between PSNR results and the user perception, in Figure 9 we have shown the decoded images for different selection of quantization zones. It is found that the higher the PSNR of reconstructed image, the better the quality is.

6. Conclusion

In this paper, we have presented a low-complexity DCT-based image compression. We presented a novel common subexpression sharing of intermediate signals of the DCT computation based on CSD representation of Loeffler's 8-point DCT algorithm. Finally, we have combined the quantization process with the second stage of DCT computation in order to optimize the bit width of computation of DCT coefficient according to the quantization of different zones.

We would like to point out that a prior detection of zero-quantized coefficients along with the proposed techniques could be used to further reduce the complexity of DCT computations.

References

[1] A. Alkholidi, A. Alfalou, and H. Hamam, "A new approach for optical colored image compression using the JPEG standards," *Signal Processing*, vol. 87, no. 4, pp. 569–583, 2007.

[2] M. Jridi and A. Alfalou, "Joint Optimization of Low-power DCT Architecture and Effcient Quantization Technique for Embedded Image Compression," in *VLSI-SoC: Forward-Looking Trends in IC and System Design*, J. Ayala, A. Alonso, and R. Reis, Eds., pp. 155–181, Springer, Berlin, Germany, 2012.

[3] M. Jridi and A. Alfalou, "A low-power, high-speed DCT architecture for image compression: Principle and implementation," in *18th IEEE/IFIP International Conference on VLSI and System-on-Chip (VLSI-SoC '10)*, pp. 304–309, September 2010.

[4] M. Jridi and A. AlFalou, "A VLSI implementation of a new simultaneous images compression and encryption method," in *IEEE International Conference on Imaging Systems and Techniques (IST '10)*, pp. 75–79, July 2010.

[5] "Video coding for low bit rate communication," (ITU-T Rec. H.263), February 1998.

[6] ISO/IEC DIS 10 918-1, "Coding of audio visual objects: part 2. visual," ISO/IEC 14496-2 (MPEG-4 Part2), January 1999.

[7] W. H. Chen, C. H. Smith, and S. C. Fralick, "A fast computational algorithm for the discrete cosine transform," *IEEE Transactions on Communications*, vol. 25, no. 9, pp. 1004–1009, 1977.

[8] B. G. Lee, "A new algorithm to compute the discrete cosine transform," *IEEE Transactions on Acoustics, Speech, and Signal Processing*, vol. 32, no. 6, pp. 1243–1245, 1984.

[9] M. Vetterli and H. J. Nussbaumer, "Simple FFT and DCT algorithms with reduced number of operations," *Signal Processing*, vol. 6, no. 4, pp. 267–278, 1984.

[10] N. Suehiro and M. Hatori, "Fast algorithms for DFT and other sinusoidal tranforms," *IEEE Transactions on Acoustics, Speech and Signal Processing*, vol. 34, pp. 642–664, 1986.

[11] H. Hou, "A fast recursive algorithm for computing the discrete cosine transform," *IEEE Transactions on Acoustics, Speech and Signal Processing*, vol. 35, pp. 1455–1461, 1987.

[12] C. Loeffler, A. Lightenberg, and G. S. Moschytz, "Practical fast 1-D DCT algorithm with 11 multiplications," in *International Conference on Acoustics, Speech, and Signal Processing (ICASSP '89)*, pp. 988–991, May 1989.

[13] P. Duhamel and H. H'mida, "New 2n DCT algorithm suitable for VLSI implementation," in *International Conference on Acoustics, Speech, and Signal Processing (ICASSP '87)*, pp. 1805–1808, November 1987.

[14] D. Slawecki and W. Li, "DCT/IDCT processor design for high data rate image coding," *IEEE Transactions on Circuits and Systems for Video Technology*, vol. 2, no. 2, pp. 135–146, 1992.

[15] S. A. White, "Applications of distributed arithmetic to digital signal processing: a tutorial review," *IEEE ASSP Magazine*, vol. 6, no. 3, pp. 4–19, 1989.

[16] A. Madisetti and A. N. Willson, "100 MHz 2-D 8 × 8 DCT/IDCT processor for HDTV applications," *IEEE Transactions on Circuits and Systems for Video Technology*, vol. 5, no. 2, pp. 158–165, 1995.

[17] S. Yu and E. E. Swartzlander, "DCT implementation with distributed arithmetic," *IEEE Transactions on Computers*, vol. 50, no. 9, pp. 985–991, 2001.

[18] D. W. Kim, T. W. Kwon, J. M. Seo et al., "A compatible DCT/IDCT architecture using hardwired distributed arithmatic," in *IEEE International Symposium on Circuits and Systems (ISCAS '01)*, pp. 457–460, May 2001.

[19] A. Shams, W. Pan, A. Chidanandan, and M. Bayoumi, "A low power high performance distributed DCT architecture," in *IEEE Computer Society Annual Symposium on VLSI (ISVLSI '02)*, pp. 21–27, 2002.

[20] A. Chidanandan, J. Moder, and M. Bayoumi, "Implementation of NEDA-based DCT architecture using even-odd decomposition of the 8 × 8 DCT matrix," in *49th Midwest Symposium on Circuits and Systems (MWSCAS '06)*, pp. 600–603, August 2007.

[21] P. K. Meher, "Unified systolic-like architecture for DCT and DST using distributed arithmetic," *IEEE Transactions on Circuits and Systems I*, vol. 53, no. 12, pp. 2656–2663, 2006.

[22] M. Alam, W. Badawy, and G. Jullien, "A new time distributed DCT architecture for MPEG-4 hardware reference model," *IEEE Transactions on Circuits and Systems for Video Technology*, vol. 15, no. 5, pp. 726–730, 2005.

[23] S. Yu and E. E. Swartzlander, "A scaled DCT architecture with the CORDIC algorithm," *IEEE Transactions on Signal Processing*, vol. 50, no. 1, pp. 160–167, 2002.

[24] C. C. Sun, S. J. Ruan, B. Heyne, and J. Goetze, "Low-power and high-quality Cordic-based Loeffler DCT for signal processing," *IET Circuits, Devices and Systems*, vol. 1, no. 6, pp. 453–461, 2007.

[25] A. P. Vinod and E. M. K. Lai, "Hardware efficient DCT implementation for portable multimedia terminals using subexpression sharing," in *IEEE Region 10 Annual International Conference (TENCON '04)*, pp. A227–A230, November 2004.

[26] C. Cheng and K. K. Parhi, "A novel systolic array structure for DCT," *IEEE Transactions on Circuits and Systems II*, vol. 52, no. 7, pp. 366–369, 2005.

[27] P. K. Meher, "Systolic designs for DCT using a low-complexity concurrent convolutional formulation," *IEEE Transactions on Circuits and Systems for Video Technology*, vol. 16, no. 9, pp. 1041–1050, 2006.

[28] T. Xanthopoulos and A. P. Chandrakasan, "A low-power dct core using adaptive bitwidth and arithmetic activity exploiting signal correlations and quantization," *IEEE Journal of Solid-State Circuits*, vol. 35, no. 5, pp. 740–750, 2000.

[29] J. Huang and J. Lee, "A self-reconfigurable platform for scalable dct computation using compressed partial bitstreams and blockram prefetching," *IEEE Transactions on Circuits and Systems for Video Technology*, vol. 19, no. 11, pp. 1623–1632, 2009.

[30] J. Huang and J. Lee, "Efficient VLSI architecture for video transcoding," *IEEE Transactions on Consumer Electronics*, vol. 55, no. 3, pp. 1462–1470, 2009.

[31] E. H. Yang and L. Wang, "Joint optimization of run-length coding, Huffman coding, and quantization table with complete baseline JPEG decoder compatibility," *IEEE Transactions on Image Processing*, vol. 18, no. 1, pp. 63–74, 2009.

[32] C. L. Hsu and C. H. Cheng, "Reduction of discrete cosine transform/quantisation/inverse quantisation/inverse discrete cosine transform computational complexity in H.264 video encoding by using an efficient prediction algorithm," *IET Image Processing*, vol. 3, no. 4, pp. 177–187, 2009.

[33] A. Avizienis, "Signed-digit number representations for fast parallel arithmetic," *IRE Transaction on Electronic Computers*, vol. 10, pp. 389–400, 1961.

[34] R. H. Seegal, "The canonical signed digit code structure for FIR filters," *IEEE Transactions on Acoustics, Speech, and Signal Processing*, vol. 28, no. 5, pp. 590–592, 1980.

[35] R. T. Hartley, "Subexpression sharing in filters using canonic signed digit multipliers," *IEEE Transactions on Circuits and Systems II*, vol. 43, no. 10, pp. 677–688, 1996.

[36] C. Y. Pai, W. E. Lynch, and A. J. Al-Khalili, "Low-power data-dependent 8 × 8 DCT/IDCT for video compression," *IEE Proceedings: Vision, Image and Signal Processing*, vol. 150, no. 4, pp. 245–255, 2003.

[37] B. I. Kim and S. G. Ziavras, "Low-power multiplierless DCT for image/video coders," in *13th International Symposium on Consumer Electronics (ISCE '09)*, pp. 133–136, May 2009.

6

Design a Bioamplifier with High CMRR

Yu-Ming Hsiao, Miin-Shyue Shiau, Kuen-Han Li, Jing-Jhong Hou, Heng-Shou Hsu, Hong-Chong Wu, and Don-Gey Liu

Department of Electronic Engineering, Feng Chia University, Taichung 40724, Taiwan

Correspondence should be addressed to Don-Gey Liu; dgliu@fcu.edu.tw

Academic Editor: Yeong-Lin Lai

A CMOS amplifier with differential input and output was designed for very high common-mode rejection ratio (CMRR) and low offset. This design was implemented by the 0.35 μm CMOS technology provided by TSMC. With three stages of amplification and by balanced self-bias, a voltage gain of 80 dB with a CMRR of 130 dB was achieved. The related input offset was as low as 0.6 μV. In addition, the bias circuits were designed to be less sensitive to the power supply. It was expected that the whole amplifier was then more independent of process variations. This fact was confirmed in this study by simulation. With the simulation results, it is promising to exhibit an amplifier with high performances for biomedical applications.

1. Introduction

For biomedical applications, a voltage amplifier with a gain of 80 dB and a high CMRR is required as a building block in front-end subsystems [1, 2]. Since the voltage level of physiologic signals at the front-end subsystem is very weak, processes for analog signals usually include several steps of amplification, filtering, offset adjustment, and electrical conditioning. After suitable processing, the signal will then be large enough and effectively suitable for analog-to-digital conversion at later stages [3–5].

In considering the physiological signals extracted from human bodies, the amplitude of an electrocardiographic (ECG) signal is usually less than 100 μV. Such value is very weak as compared to the noise floor and imperfection of the commonly used operational amplifiers (OPAs). An instrumentation amplifier (IA) is usually employed to achieve the required performances.

In addition to the requirement of high voltage gain in constructing the amplifiers for an IA, another important requirement for the amplifiers is CMRR. According to the recommendations of Association of the Advancement of Medical Instrumentation (AAMI), CMRR is required to be higher than 90 dB with the open-loop voltage gain higher than 80 dB.

In this study, the 0.35 μm CMOS technology of TSMC was employed in designing a high performance amplifier.

In our study, a high-voltage-gain amplifier was tried with a self-biasing technique to have a high CMRR and low input offset and to be less sensitive to process variations. The simulation was performed based on the models supported by Chip Implementation Center (CIC). The related results will be illustrated.

2. Design Details

2.1. Design of the Differential Amplifier. For the purposes of high CMRR and low offset at the input, differential configuration with a symmetrical floor planning in layout will be preferred in the design of an amplifier.

Figure 1 shows the schematic of an amplifier with the differential configuration both at the input and at the output. In this circuit, transistors M_1 and M_2 are the differential pair for amplification. The block with I_T and R_T forms a tail current bias. The resistors R_{D1}, R_{D2}, R_{L1}, and R_{L2} are taken as the loads.

Figure 2 shows an alternative representation of the amplifier in Figure 1. The input and output signals can be decomposed into the common and the differential modes. With this decomposition, the performance of the amplifier in the common mode and the differential mode can be discussed separately.

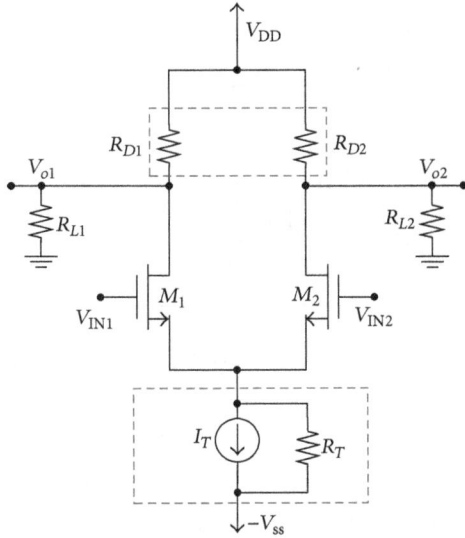

FIGURE 1: Schematic of the differential amplifier.

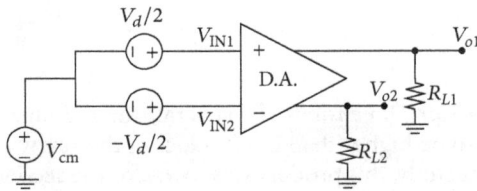

FIGURE 2: Representation of the amplifier in the common and the differential modes.

In the common mode, the two output voltages will be the same if the circuit is ideal in a form of total symmetry. This requires that the branches for I_{D1} and I_{D2} are matched with $R_{L1} = R_{L2}$ and $R_{D1} = R_{D2}$.

For the current bias as the tail, a current mirror with a stable reference current, I_{REF}, can be employed in the integrated circuits to give a high output resistance, R_T.

Other techniques to improve the performance of this amplifier will be discussed in detail in the following.

For practical design, there exist variations in the devices even with the integrated circuit technology. The output voltage will not be zero for the common input condition. For example, the imperfections in the threshold voltage and the transconductance of the MOS transistors and the variation in R_D are uncorrelated. The resulted input offset voltage can be expressed as

$$V_{OS} \equiv \frac{V_{OUT}|_{V_d=0}}{A_d}$$

$$= \frac{V_{OV}}{2}\sqrt{\left(\frac{\Delta R_D}{R_D}\right)^2 + \left(\frac{\Delta (W/L)}{(W/L)}\right)^2 + \left(\frac{\Delta V_{TH}}{(V_{OV})/2}\right)^2}.$$

$$(1)$$

According to the analysis in the common mode and the differential mode, the output voltage can be expressed as the sum of the amplification of the signals of both modes. The relations for the outputs can be written at follows:

$$V_{O1} = \frac{A_d}{2} \times v_d + A_{cm} \times v_{cm}, \qquad (2a)$$

$$V_{O2} = -\frac{A_d}{2} \times v_d + A_{cm} \times v_{cm}, \qquad (2b)$$

where $v_d \equiv v_{in1} - v_{in2}$, $v_{cm} \equiv (v_{in1} + v_{in2})/2$, A_d is the differential-mode voltage gain, and A_{cm} is the common-mode gain. It is similar for the expression for V_{O2}. The common-mode rejection ratio (CMRR) is then defined as

$$CMRR \equiv \left|\frac{A_d}{A_{cm}}\right|. \qquad (3)$$

A good amplifier is required to have a high A_d with a nearly zero A_{cm}. Due to the variations in the fabrication process, it is a big challenge to achieve a high CMRR with a low input offset. In this study, a balanced bias technique was employed to reduce the sensitivity to the process variation. Good properties of this amplifier have been confirmed in the postlayout simulation.

2.2. Tristage Amplifier. In this design, three stages of amplification were employed to achieve the required voltage gain and CMRR at the same time for weak biosignals. Figure 3 shows the detailed circuit in this design. Table 1 gives the specifications for this design.

As seen in Figure 3, the first stage is composed of M_1–M_5. The second stage includes M_6–M_{12}. These two stages can be used as an operational transconductance amplifier (OTA) [6, 7] or a folded cascade amplifier [8]. The third stage comprising M_{13} and M_{14} forms a type A common-source (CS) amplifier to drive loads.

Transistors M_{S1}–M_{S6} provide a bias current for the first-stage amplifier. The source of biasing for the second-stage amplifier comes from the balanced self-bias current mirror, M_8–M_{11}, in Figure 3. In this part, the biasing currents were less sensitive to the level of the power supply. In addition, the complementary arrangement of the loads at the first-stage and the second-stage amplifiers would reduce the variation of the amplification if there are changes in the NMOS and PMOS. The bias voltage for the third stage comes from M_9 in the second stage. Since the bias currents in M_9 and M_{11} were constant, the gate bias for M_{13} would be constant. Therefore, the properties of the whole amplifier would be less affected by the uncertainties in fabrication.

For the design strategy, the first stage was designed to achieve a high CMRR rather than a high voltage gain. The overall voltage gain was boosted at the second and the third stages. Since this amplifier was designed for biomedical applications, the voltage gain was tried to be as high as possible with a moderate small bandwidth around 100 Hz. At the third stage, a clamping circuit can keep dynamic tracking of the output gain such that the voltage gain would be less affected by variations in the transistors.

In addition to the electrical considerations, the layout and circuit for the first and second stages were designed as

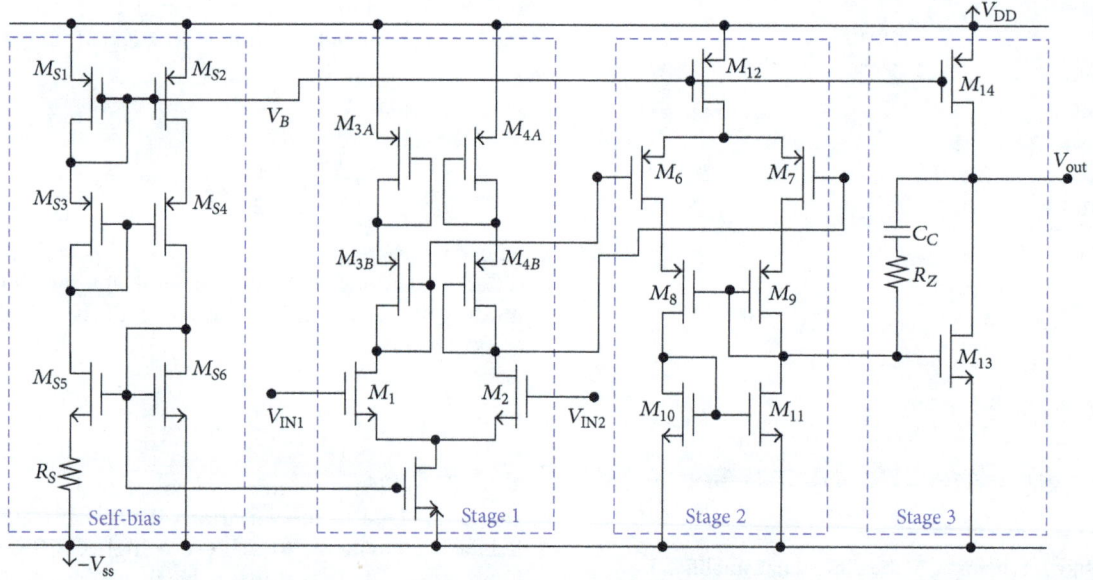

FIGURE 3: Structure of the tristage amplifier.

TABLE 1: Specifications for the bioamplifier.

Parameter	Spec.	Value
V_o/V_i	$\geq$	20 kV/V
A_{VO}	$\geq$	80 dB
PM	$\geq$	60°
UGF	$\leq$	2 MHz
CMRR	$\geq$	90 dB
PD	$\leq$	1 mW

symmetrical as possible. In this way, the common signals would be cancelled out in the differential structure. Therefore, the equivalent input offset would be suppressed effectively.

With the above techniques, an amplifier with high voltage gain, high CMRR, low offset, and low drift voltage can be achieved and confirmed in the simulation.

For our circuit, the level of the power supply was set at 3.3 V by setting V_{DD} at 1.65 V and $-V_{SS}$ at -1.65 V. In the meanwhile, the power dissipation was specified below 1 mW for portable operations. With this constraint, as explained by (4), the total current consumed in this circuit cannot be more than 0.303 mA:

$$I_{\text{Supply}} \leq \frac{P_{D,\text{Spec.}}}{V_{\text{Supply}}} = 0.303 \text{ mA.} \quad (4)$$

As for the stability consideration, the phase margin (PM) of this amplifier was tuned to be 60° in our simulation [9]. In this design, we select $G_{m3} \geq 10G_{m2}$. And the second pole was set as $\omega_{P2} \geq 2.2 \times \omega_T$. Therefore, the Miller compensation capacitor, C_C, was selected to be 18 pF by the following calculation:

$$C_C \geq 2.2 \times \frac{G_{m1}}{G_{m2}} \times C_{O2} = 0.22C_{O2}. \quad (5)$$

2.3. Transistor Dimensions. In general, the tail current was required to be higher than the product of the screw rate and C_C. In this study, this product was 10 μA for the second stage. Since the screw rate for the case of light loads is not required strictly, the tail current at stage 2 was selected as 15 μA.

The bias currents for the two branches through M_6 and M_7 equally divide the tail current into $I_{D6} = I_{D8} = I_{D10} = I_{D7} = I_{D9} = I_{D11} = 7.5$ μA.

With this bias current, the transconductance, g_m, and the gate-to-source voltage, V_{GS}, can be designed by a suitable dimension ratio, W/L, by the following relations:

$$I_D = \frac{1}{2} \cdot \mu_n \cdot C_{ox} \cdot \left(\frac{W}{L}\right) \cdot (V_{OV})^2 \cdot (1 + \lambda V_{DS}), \quad (6)$$

$$g_m = (\mu_n \cdot C_{ox}) \cdot \left(\frac{W}{L}\right) \cdot (V_{OV}) \cdot (1 + \lambda V_{DS})$$

$$= \sqrt{2\mu_n \cdot C_{ox} \cdot \left(\frac{W}{L}\right) \cdot I_D \cdot (1 + \lambda V_{DS})} \quad (7)$$

$$= \frac{2 \times I_D}{V_{OV}},$$

where the overdrive voltage $V_{OV} \equiv V_{GS} - V_{th}$.

For the third stage, its transconductance gain, G_{m3}, is the same as g_{m13} of M_{13}. We chose $G_{m3} \geq 10G_{m2}$; that is, $G_{m3} \geq 590$ μA/V and $g_{m13} = G_{m3} = 600$ μA/V. In addition, the overdrive voltage for M_{13} was selected as $V_{OV13} = 0.2$ V; that is, $V_{GS13} \equiv V_{OV13} + V_{tn} = 0.85$ V. Therefore, $V_{DS13} = V_{DS14} = 1.65$ V. The dimension ratio for $(W/L)_{13}$ can be determined by (6).

In this design, we used $M_{S1} \sim M_{S6}$ and a resistor R_S to form a self-bias circuit with a boot-strapping positive

TABLE 2: Variations of the simulated properties of the bio-amplifier with 5 fabrication corners.

Items	SS	SF	TT	FS	FF
V_o/V_i	−33.1 k	−29.7 k	−29.3 k	−27.7 k	−22.6 k
A_{VO} (dB)	90.4	89.5	89.3	88.9	87.1
PM	88.8°	89.8°	89.7°	89.7°	91.0°
f_{H3db} (Hz)	63.0	63.1	79.4	63.1	82
UGF (Hz)	$1.41E + 06$	$1.47E + 06$	$1.46E + 06$	$1.44E + 06$	$1.52E + 06$
CMRR (dB)	136.4351	136.2238	136.7342	136.5733	135.778
PD (μW)	357	332	328	324	316
V_{os} (μV)	10.8	10.1	5.2	0.033	20.8

feedback. The current controlled by I_{REF} can be expressed as

$$I_{REF} \times R_S = V_{GS_{S6}} - V_{GS_{S5}} = \sqrt{\frac{I_{REF}}{K_{S6}}} - \sqrt{\frac{I_{REF}}{K_{S5}}}. \quad (8)$$

By (8), we selected the dimension ratio of M_{S5} to be 1/4 of that of M_{S6}, that is,

$$\left(\frac{W}{L}\right)_{S5} = \frac{1}{4}\left(\frac{W}{L}\right)_{S6}. \quad (9)$$

The resistance can be obtained as follows:

$$R_S = \frac{1}{I_{REF}} \sqrt{\frac{I_{REF}}{K_{S5}}}. \quad (10)$$

The reference current was set as $I_{REF} = 5\,\mu$A. The dimension ratio for $M_{S1} \sim M_{S4}$ can also be derived by (6).

3. Simulation and Verification before Fabrication

In this study, HSPICE with the device models for $0.35\,\mu$m CMOS technology from TSMC was employed for the simulation and analysis. The performance of the whole circuit was verified first in the prelayout simulation. Then, the physical layout was implemented and the related parameters were extracted. With the obtained information of the physical layout, the postlayout simulation was performed to check the feasibility of our layout. Corner simulations were also performed to check the effect of the process variation on the performance of our amplifier.

Figures 4–7 illustrate the related performances with 5 corner conditions in fabrication. Table 2 lists other performance items with the 5 corners. With these results, we can find that the voltage gain in Figure 6 can be kept higher than 80 dB. And the variation of the obtained gains due to the uncertainty in fabrication can be smaller than 5 dB. It can also be found in Figures 4 and 5 that the phase margin is much larger than 60°. In Figure 5, we can confirm that the variation of phases is insignificant. As shown in Figure 7, the obtained CMRR is as high as 130 dB for frequency up to 10 kHz. The variation of CMRR due to the fabrication is also insignificant. With these results, a bio-amplifier both with a very high CMRR and a high voltage gain at the same time can be expected for the fabricated chips.

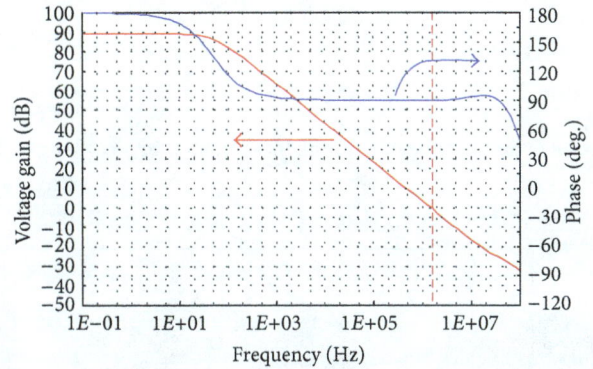

FIGURE 4: Frequency response for the bio-amplifier in the typical fabrication condition (TT).

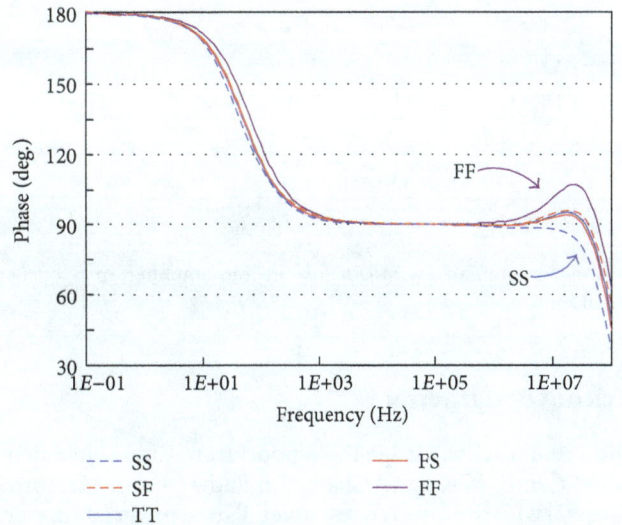

--- SS	—— FS
--- SF	—— FF
—— TT	

FIGURE 5: Simulated phases for the bio-amplifier in 5 corner conditions.

4. Conclusion

A bio-amplifier with high gain and high CMRR was designed and verified in this study. According to the obtained performance properties in Table 2, it is promising that a process independent performance can be obtained for this amplifier.

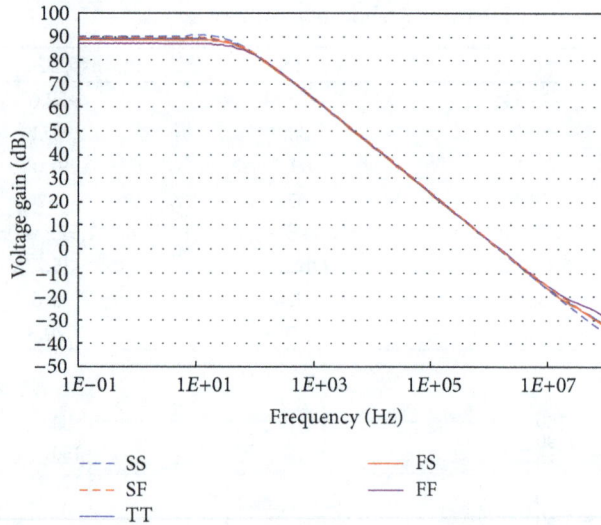

FIGURE 6: Simulated gains for the bio-amplifier in 5 corner conditions.

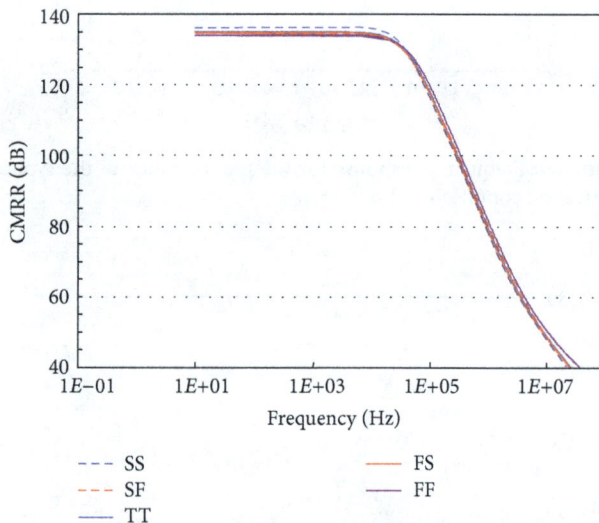

FIGURE 7: Simulated CMRRs for the bio-amplifier in 5 corner conditions.

Acknowledgments

The authors acknowledge the support from Chip Implementation Center (CIC) and Chang Bin Show Chwan Memorial Hospital with the research resources. Partial financial support from National Science Council (NSC), republic of china, is also acknowledged.

References

[1] K. A. Ng and P. K. Chan, "A CMOS analog front-end IC for portable EEG/ECG monitoring applications," *IEEE Transactions on Circuits and Systems I*, vol. 52, no. 11, pp. 2335–2347, 2005.

[2] B. Wang, H. Ji, Z. Huang, and H. Li, "A high-speed data acquisition system for ECT based on the differential sampling method," *IEEE Sensors Journal*, vol. 5, no. 2, pp. 308–311, 2005.

[3] C. H. Chan, J. Wills, J. LaCoss, J. J. Granacki, and J. Choma Jr., "A novel variable-gain micro-power band-pass auto-zeroing CMOS amplifier," in *Proceedings of the IEEE International Symposium on Circuits and Systems (ISCAS '07)*, pp. 337–340, May 2007.

[4] B. Premanode, N. Silawan, and C. Toumazou, "Drift reduction in ion-sensitive FETs using correlated double sampling," *Electronics Letters*, vol. 43, no. 16, pp. 857–859, 2007.

[5] J. Wu, G. K. Fedder, and L. R. Carley, "A low-noise low-offset chopper-stabilized capacitive-readout amplifier for CMOS MEMS accelerometers," in *Proceedings of the IEEE International Solid-State Circuits Conference (ISSCC '02)*, pp. 428–425, February 2002.

[6] G. Nicollini and C. Guardiani, "3.3-V 800-nV rms noise, gain-programmable CMOS microphone preamplifier design using yield modeling technique," *IEEE Journal of Solid-State Circuits*, vol. 28, no. 8, pp. 915–921, 1993.

[7] V. Ivanov, J. Zhou, and I. M. Filanovsky, "A 100-dB CMRR CMOS operational amplifier with single-supply capability," *IEEE Transactions on Circuits and Systems II*, vol. 54, no. 5, pp. 397–401, 2007.

[8] P. C. de Jong, G. C. M. Meijer, and A. H. M. van Roermund, "A 300°C dynamic-feedback instrumentation amplifier," *IEEE Journal of Solid-State Circuits*, vol. 33, no. 12, pp. 1999–2008, 1998.

[9] K. N. Leung and P. K. T. Mok, "Analysis of multistage amplifier-frequency compensation," *IEEE Transactions on Circuits and Systems I*, vol. 48, no. 9, pp. 1041–1056, 2001.

7

Verification of Mixed-Signal Systems with Affine Arithmetic Assertions

Carna Radojicic,[1] Christoph Grimm,[1] Florian Schupfer,[2] and Michael Rathmair[2]

[1] *Design of Cyber-Physical Systems, Kaiserslautern University of Technology, Postfach 3049, 67663 Kaiserslautern, Germany*
[2] *Institute of Computer Technology, Vienna University of Technology, Gushausstraße 27-29, 1040 Vienna, Austria*

Correspondence should be addressed to Carna Radojicic; radojicic@cs.uni-kl.de

Academic Editor: Chang-Ho Lee

Embedded systems include an increasing share of analog/mixed-signal components that are tightly interwoven with functionality of digital HW/SW systems. A challenge for verification is that even small deviations in analog components can lead to significant changes in system properties. In this paper we propose the combination of range-based, semisymbolic simulation with assertion checking. We show that this approach combines advantages, but as well some limitations, of multirun simulations with formal techniques. The efficiency of the proposed method is demonstrated by several examples.

1. Introduction

Analog/mixed-signal (AMS) systems are a crucial part of today's embedded systems. Typical AMS components such as sensors, transceivers, and signal conditioning enable interaction of embedded HW/SW systems with its physical environment. In today's embedded systems, the functionality of the analog components is tightly interwoven with the digital HW/SW system. A particular challenge of AMS systems is that parameters cannot be assumed to be fixed to a deterministic value like in a digital system.

Behavior of AMS systems cannot be assumed to be fixed for the following reasons: *variations* of parameters due to variations in the manufacturing process, but as well during operation (e.g., different temperatures, aging, and supply voltage) introduce deviations compared to an ideal reference. (*Modeling*) *uncertainties* are introduced by the fact that all models represent more or less accurate abstractions of physical reality. No model can be assumed to be absolutely accurate. Furthermore, computation with fixed-point arithmetic in the digital domain can contribute significantly to deviation from expected ideal behavior (rounding errors, quantization). In the following we refer to such deviations of a simulation run from possible real behavior in general as "*deviations*."

A communication system with typical variations and deviations is shown in Figure 1 as an example. Variations of gain, offset, or due to temperature (Figure 1, left) are compensated in software. This is done at lower layers of the software by controlling variable gain amplifier (VGA), voltage controlled oscillator (VCO). Higher layers of the software stack introduce further error correction mechanisms in software. Dependability of the overall system is defined by complex interaction of AMS parts with the software stack. While known statistical methods (e.g., Monte Carlo simulation) allow us computing other statistic properties like Bit or Packet Error Rates (BER, PER), open issues are questions such as

(i) how can we *guarantee* some system properties, for example, for safety relevant systems?

(ii) can we get information from the analysis that assists us in design and debug, such as counter examples?

This paper proposes a new methodology that for the first time combines high verification coverage of formal verification on one hand with the general applicability of simulation-based approaches on the other hand. To achieve this goal, we combine assertion checking with symbolic simulation:

(1) assertions specify required properties of a system;

(2) in an overall system model, deviations and variations are represented by symbols that capture size and correlations of the deviations, variations, respectively;

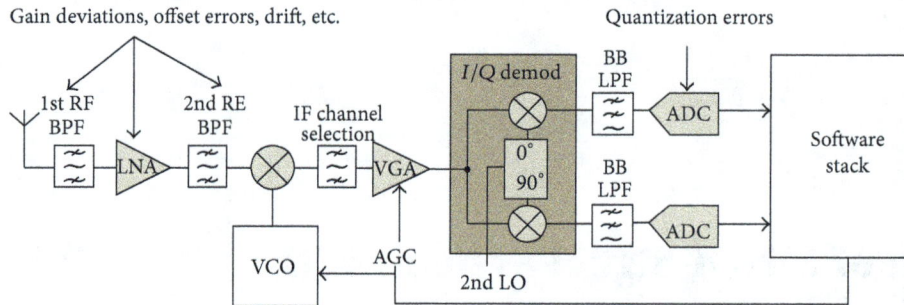

FIGURE 1: System model including parameter deviations.

(3) for verification of "worst case" behavior, a range-based simulation using Affine Arithmetic shows that for given ranges of inputs and deviations the required properties are valid, and that no "forbidden state" is reached.

We mostly focus on level of block diagrams, but the methodology is as well applicable on circuit simulation. We implemented it based on SystemC and its AMS extensions. Section 2 gives a review of state of the art and related work. In Section 3 Affine Arithmetic is described, and some modeling examples are given. Section 4 describes semisymbolic simulation and the verification method proposed in this work. The applicability of the verification technology is demonstrated by examples given in Section 5. Section 6 concludes the paper and identifies future work.

2. State of Art and Related Work

When verifying AMS systems with parameter deviations, application of multirun simulation techniques (Monte-Carlo, worst-case analysis) can be considered as state-of-the art. *Monte Carlo simulation* [1] is a statistical technique. However, statistical techniques do not provide dependable "worst case" results. While the number of simulation runs can be reduced by *importance sampling* [2], the number of simulation runs required may still be prohibitive for analysis of complex systems. *Corner case analysis* [3] is a more appropriate means for finding worst case performances of AMS systems. Unfortunately, the number of simulation runs grows exponentially with the number of parameters considered. However, even if all corner cases are considered, the dependability of the result cannot be guaranteed since corner cases are not necessarily worst cases. *Design of Experiments* [4] allows reducing the number of simulation runs significantly and finding worst case performances more accurately.

Even with high number of simulation runs, there is no guarantee that worst case performances are found. A drawback of multi-run simulation methods is that the dependable operation of AMS systems cannot be guaranteed under all circumstances. For safety-critical systems, for example, in aviation or automotive systems this is a major drawback and motivation for further research.

In order to find counter examples, rapidly exploring random trees [5, 6] and robust test case generation have been proposed in [7–9]. Further, the simulation techniques

proposed in [10, 11] guarantee that a system is "safe" if a set of trajectories lie within certain regions defined by previously found conditions. In contrast to these approaches the techniques proposed in [12, 13] compute an overapproximation of the set of states reached by all trajectories. While these methods support debugging and introduce coverage metrics, they are not able to deal with the increased complexity of systems that with deviations and variations. An approach that enables safety verification of hybrid systems with uncertain parameters is the use of barrier certificates proposed in [14]. Those methods can verify if a set of system trajectories crosses a barrier previously defined by a barrier certificate. Finding a proper barrier certificate is not easy and makes this approach difficult for system verification.

To cope with the drawbacks of simulation-based techniques, formal verification methods were proposed. The idea of formal methods is to use the formal checkers which automatically explore all possible states and transitions in the system model to check if the desired output behaviour is met or not. Hence, in contrast to simulation-based methods which can verify only one behaviour (for only one input stimuli) per operation, the formal methods deal with the set of behaviours at a time. Approaches for formal verification were firstly applied on digital systems [15–17], and due to their efficiency they found a good way in industrial applications. Approaches that also cover analog/mixed-signal or hybrid systems are rare and still in infancy.

In [18–20] hybrid systems with linear and nonlinear dynamics are approximated by timed automata in order to simplify their analysis. Reference [21] describes a model checking tool which requires discrete and (for continuous parts) linear system descriptions. For nonlinear continuous behavior, such approximations are too simple. Linear phase-portrait approximation [22] is a general technique, because its approximation does not depend on the order of the differential equation to be approximated. There is no standard method for partitioning the state space, and therefore it seems to be complicated to find proper discrete models for strongly nonlinear models. In [23–25] focus is on nonlinear analog behavior for which discretized models in the state space are used for verification. The efficiency of these approaches seems to be limited to smaller analog systems. With increasing complexity, the number of states in the discretized model grows which leads to the state explosion problem and high run time of verification algorithms applied on this model.

TABLE 1: Tradeoffs of verification methods for AMS systems with deviations.

Verification techniques	Disadvantages	Advantages
Multirun simulations	Low verification coverage, no "guarantee"	General applicability
Formal methods	State explosion problem, no complex, heterogeneous systems	High verification coverage "guarantee," counter examples
Assertion-based techniques (MSA, AMT tool)	Only nominal behaviour verified, no "guarantee"	Well suited for complex systems, increased coverage

Due to limitation of formal methods for analog/continuous systems to small systems, simulation-based techniques are still the only way for verification of more complex analog/continuous and AMS systems. To formalize verification of AMS systems, assertions that describe typical properties of analog systems are the focus of recent research. In [26] mixed-signal assertions (MSA) were proposed to check properties of mixed-signal systems during simulation. These assertions were implemented in a separate SCAC (SystemC AMS Temporal Checker) library. This library is easily integrated in SystemC AMS simulation environment, due to its C++ based nature. In contrast to this approach, [27] features AMT, an offline tool for monitoring temporal properties of mixed-signal systems for verification. Both verification methodologies simulate and evaluate a nominal system model without taking into account any deviations caused by variations in design process.

A particular challenge in design of complex analog/mixed-signal systems is parameter variations. In any physical system, values are not implemented in an accurate way and change in a partially unpredictable way over time (e.g., due to temperature, aging, etc.). Such deviations form an ideal model change system behavior and can potentially cause malfunctions. For conventional simulation, multi-run methods as described in the first paragraph do not provide the result dependability and require a high number of simulation runs to explore the system behaviour while considering process variations. For formal approaches, such issues are still in infancy, because its applicability simply does not yet allow handling complex and heterogeneous systems such as AMS systems.

A first approach to cover deviation effects in AMS systems and compute the guaranteed worst case results at the same time was introduced in [28–31]. Deviations are modeled as ranges, superimposed on the nominal system model, and modified during system simulation to obtain the formally guaranteed range-based system quantities. For this purpose, Affine Arithmetic was applied. Using this approach the variations in parameter values are represented with deviated symbols which are traced to the system output. Hence, the contribution of all variations in the system is contained in the system response which simplifies analysis of the system robustness.

In [32] it is proposed that how Affine Arithmetic approach can be used to analyze worst case behavior of electrical circuits. Further, in [33] this methodology found its application in sizing of analog circuits. Using Affine Arithmetic the bounds on the worst case circuit behavior are calculated and the global minimum of sizing problem

is determined due to inclusion isotonicity. Beside analog domain, Affine Arithmetic models can also be used in Digital Signal Processing (DSP) applications to represent errors introduced by calculations in floating-point arithmetic [34, 35].

Within this work, semisymbolic simulation based on Affine Arithmetic is combined with the assertion-based technology. Concretely, assertions based on Affine Arithmetic (AAF+A) are introduced to include range-based system quantities and allow specification and automatic verification of typical time and frequency-domain properties of systems considering variations in their parameter values.

Using the proposed verification method system verification is done during simulation.

(i) The desired output behaviour is described with assertion which is embedded into simulation process.

(ii) The assertion is verified automatically. In the case where the design requirement is not met the simulation process is stopped reporting the user about the assertion violation.

Table 1 summarizes the advantages and disadvantages of previously described verification techniques.

The verification method proposed in this work copes with the disadvantages of previous verification methods. Concretely, combining the assertion-based technology with semisymbolic simulation, which generates the dependable guaranteed result in which all output values for the considered parameter set are contained, 100% coverage can be obtained.

3. Affine Arithmetic and Its Use for Modeling Deviations

3.1. Affine Arithmetic. Affine Arithmetic (AA) is a range arithmetic that overcomes the error explosion problem of Interval Arithmetic (IA) [36]. AA keeps track of correlations between quantities represented as ranges. This in particular enables application for simulation of control systems. A feedback loop, for instance, can be simulated keeping the correlation of identical ranges. A subtraction of related ranges therefore results in a reduced range avoiding the overapproximation inherent to Interval Arithmetic [29].

An affine expression $\tilde{x}$ can be represented as

$$\tilde{x} = x_0 + \sum_{i \in \mathcal{N}_{\tilde{x}}} x_i \varepsilon_i, \quad \varepsilon_i \in [-1, 1], \tag{1}$$

where a sum of deviation terms $\sum_{i\in\mathcal{N}_{\tilde{x}}} x_i\varepsilon_i$ models the impact of independent deviations from the ideal system behavior described with the nominal value x_0. The values of deviation symbols ε_i lie in the range $[-1, 1]$ which is scaled by the numerical value x_i. Linear mathematical operations in Affine Arithmetic allow accurate symbolic computations and are defined as follows:

$$\tilde{x} \pm \tilde{y} = (x_0 \pm y_0) + \sum_{i\in\mathcal{N}_{\tilde{x}}} (x_i \pm y_i)\,\varepsilon_i,$$

$$c\tilde{x} = cx_0 + \sum_{i\in\mathcal{N}_{\tilde{x}}} cx_i\varepsilon_i, \tag{2}$$

where $\mathcal{N}_{\tilde{x}}$ defines a set of natural numbers identifying all deviation terms $x_i\varepsilon_i$ in symbol $\tilde{x}$. In contrast to linear operations, nonlinear operations introduce an overapproximation of the exact solution, for example, multiplication as follows:

$$\tilde{x} \cdot \tilde{y} := (x_0 \cdot y_0) + \sum_{i\in\mathcal{N}_{\tilde{x}}} (x_0 y_i + x_i y_0)\,\varepsilon_i$$

$$+ \mathrm{rad}\,(\tilde{x}) \cdot \mathrm{rad}\,(\tilde{y})\,\varepsilon_{\mathcal{N}_{\tilde{x}}+1}, \tag{3}$$

where $\mathrm{rad}(\tilde{x})$ is equal to $\sum_{i\in\mathcal{N}_{\tilde{x}}} |x_i|$ and represents the total deviation of $\tilde{x}$. Although the multiplication operation results in an overapproximation, deviations in the result of this operation are traced to deviations contained in the quantities $\tilde{x}$ and $\tilde{y}$. The overapproximation is contained in residual term.

3.2. Modeling Examples with Affine Arithmetic. In the following we show how to model different deviations. We focus on block-diagram like representations with transfer functions as common in control theory, because this model of computation is generally applicable to a vast set of different domains, including communication systems and electronic circuits. We focus on giving some mathematical background that can be applied in C-language (as in the examples in later sections), but as well, for example, in Matlab/Simulink.

(a) Simple Example—Modeling Gain Variation. To model a gain variation of a block (e.g., a low noise amplifier (LNA)) we assume that the exact gain value is not known but lies in interval $[K_{min}, K_{max}]$. The range for K $[K_{min}, K_{max}]$ can be modeled using Affine Arithmetic as follows:

$$K_{nom} + \varepsilon K_{dev}, \quad \varepsilon \in [-1, 1], \tag{4}$$

where K_{nom} and K_{dev} correspond to the center value of the range and the maximum absolute deviation from the center value, respectively. To model the gain variation in the SystemC AMS (used as a simulation environment in this work) that extended with an abstract data type AAF and some constructors for typical deviations, it is only necessary to call the constructor of the amplifier module with K_{nom} and K_{dev} as the second and the third argument. The first argument is always the name of the module. Therefore, the amplifier can be instantiated by the following line of code:

```
amp amp_("amp_", Knom, Kdev).
```

(b) Modeling (Parameter) Uncertainties. Modeling uncertainties are due to lack of capturing absolute accurate models

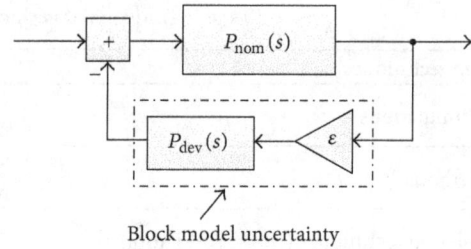

FIGURE 2: Block-diagram level representation of model transfer function with uncertain parameters.

of "real" behavior. In order to describe simple modeling uncertainties, a more general model is assumed in which the "real" behavior can be included by parameter variations. As simple example, the following transfer function of a system block will be supposed:

$$P(s) = \frac{1}{s^2 + as + 1}. \tag{5}$$

Further, it will be supposed that the exact value of the parameter a is not known, but it is known that it lies in interval $[a_{min}, a_{max}]$. The range for parameter a can be modeled using Affine Arithmetic as

$$a_{nom} + \varepsilon a_{dev}, \quad \varepsilon \in [-1, 1], \tag{6}$$

where a_{nom} represents the midpoint of the range and a_{dev} the maximum absolute deviation from the midpoint. Now the transfer function $P(s)$ can be expressed as

$$P(s) = \frac{P_{nom}(s)}{1 + \varepsilon P_{dev}(s) P_{nom}(s)}, \quad \varepsilon \in [-1, 1], \tag{7}$$

where $P_{nom}(s)$ represents the nominal model with the parameter value a_{nom} as follows:

$$P_{nom}(s) = \frac{1}{s^2 + a_{nom}s + 1} \tag{8}$$

and $P_{dev}(s)$ is the deviation function modeled as $P_{dev}(s) = a_{dev}s$. The system block model with parameter deviation can be represented with the block diagram shown in Figure 2.

(c) Variation of Time Delay, Jitter. Time delays are often varying, even in digital systems ("jitter"). A time delay can be modeled by the following transfer function:

$$P(s) = e^{-\tau s} P_{nom}(s), \tag{9}$$

where $P_{nom}(s)$ models an ideal behavior of a block (without time delay) and τ represents a time delay for which it is supposed that its exact value is not known, but it is known that it lies in interval $[0, \tau_{max}]$. The time delay causes deviation of the block from its ideal behavior $P_{nom}(s)$. This deviation will be modeled using Affine Arithmetic. In order to do this the exponential function $e^{-\tau s}$ will be approximated using the first-order Taylor polynomial:

$$e^{-\tau s} := e^{-\tau_{nom}s} + \left.\frac{(e^{-s\tau})'}{1!}\right|_{\tau=\tau_{nom}} * \varepsilon\tau_{dev} \tag{10}$$

$$= e^{-\tau_{nom}s} + (-s) * e^{-\tau_{nom}s} * (\varepsilon\tau_{dev}),$$

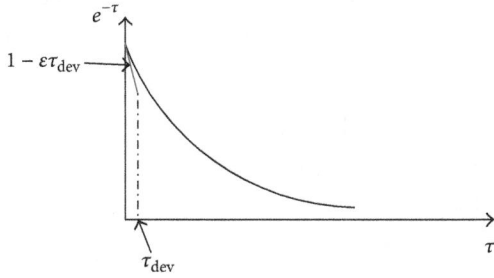

FIGURE 3: Linearization of exponential function $e^{-\tau}$.

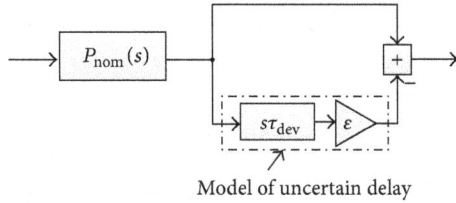

Model of uncertain delay

FIGURE 4: Block-diagram representation of model of uncertain delay respectively jitter.

where τ_{nom} represents time delay in ideal conditions whose value is zero. The symbol τ_{dev} represents the maximum absolute deviation of time delay from its nominal value, and ε is a real number whose value lies in interval $[-1, 1]$. Substituting $\tau_{\text{nom}} = 0$ in the previous equation we get

$$e^{-\tau s} := 1 + (-s) * \varepsilon\tau_{\text{dev}}, \quad \varepsilon \in [-1, 1]. \tag{11}$$

Since the approximation of the exponential function $e^{-\tau s}$ with the first-order Taylor polynomial represents the linearization of $e^{-\tau s}$ around τ_{nom}, this polynomial is actually tangent line on the function at point τ_{nom}, as it can be seen in Figure 3. Replacing $e^{-\tau s}$ in (9) with this approximation the transfer function $P(s)$ can be approximated with

$$P(s) := P_{\text{nom}}(s)\left(1 + (-s) * \varepsilon\tau_{\text{dev}}\right)$$
$$= P_{\text{nom}}(s)\left(1 + \varepsilon P_{\text{dev}}\right), \quad \varepsilon \in [-1, 1], \tag{12}$$

where $P_{\text{nom}}(s)$ models the block ideal behavior and $P_{\text{dev}}(s)$ models the deviation from the ideal behavior. The block diagram corresponding to this model is shown in Figure 4.

3.3. Abstraction of Accurate Model with Affine Arithmetic.
For verification of accurate system models in the presence of parameter deviations, a high number of simulation runs is required to achieve a sufficient verification coverage. To deal with this drawback "semisymbolic" approach based on Affine Arithmetic is introduced. Using this method an abstract system model is created in which the accurate system behavior is included. Concretely, abstraction of system model gets an overapproximation of accurate models and therefore gives a guaranty that if the abstract model satisfies desired specifications, the accurate model will also meet them.

To create the abstract model it will be supposed that there is a small change of the input voltage signal v_d around DC

operating point (I_D, V_D) Δv_d (see Figure 5(b)). This change will be modeled using Affine Arithmetic as follows:

$$v_d = V_D + \varepsilon\Delta v_d, \quad \varepsilon \in [-1, 1]. \tag{13}$$

The accurate model of a diode can be described with the following equation:

$$i_d = I_s\left(e^{v_d/\eta V_T} - 1\right). \tag{14}$$

To get the abstract model of a diode which is more simple for analysis, the accurate model will be approximated linearizing the nonlinear equation around DC operating point (I_D, V_D). This linearization will be performed using the first-order Taylor's series as follows:

$$i_d = I_D + \frac{\partial i_d}{\partial v_d}(V_D)(v_d - V_D) + \text{lin\_error}$$
$$= I_D + I_s e^{V_D/\eta V_T}\frac{1}{\eta V_T}\varepsilon\Delta v_d + \text{lin\_error}, \quad \varepsilon \in [-1, 1], \tag{15}$$

where the symbol lin_error assigns linearization error which is added to enclose the accurate model in the abstracted one. The absolute value of this error represents the maximum absolute value of the Lagrange remainder:

$$|\text{lin\_error}| = \max\left(\frac{1}{2}\left|\frac{\partial^2 i_d}{\partial v_d^2}(\xi)\right|(v_d - V_D)^2\right)$$
$$= \max\left(\frac{1}{2}\left|\frac{\partial^2 i_d}{\partial v_d^2}(\xi)\right|\right)\left(\Delta v_d^2\right), \tag{16}$$

where ξ can take any value from $\xi \in [V_D - \Delta v_d, V_D + \Delta v_d]$. To include the accurate model, linearization error will be represented as

$$\text{lin\_error} = \varepsilon'|\text{lin\_error}|, \quad \varepsilon' \in [-1, 1]. \tag{17}$$

Figure 5(b) shows approximation of nonlinear diode (from Figure 5(a)) at the operating point.

3.4. Time-Domain Properties Modeled with Affine Arithmetic.
To analyze system behavior it is necessary to specify values of its properties. In the following there will be given a list of properties in time, but also in frequency domain whose specified values can be modeled using Affine Arithmetic approach.

3.4.1. Settling Time Property.
This time is defined as the maximum time necessary for the output signal to settle within the error band, usually symmetrical around the value of the output signal asymptote, from the time at which an ideal step input is applied. The specified values for the settling time and the error band can be modeled with Affine Arithmetic as follows:

$$\text{spec}(t_s) = \frac{t_s}{2} + \varepsilon\frac{t_s}{2}, \quad \varepsilon \in [-1, 1], \tag{18}$$

$$\text{error\_band} = y_{\text{asimp}} + \varepsilon\delta, \quad \varepsilon \in [-1, 1],$$

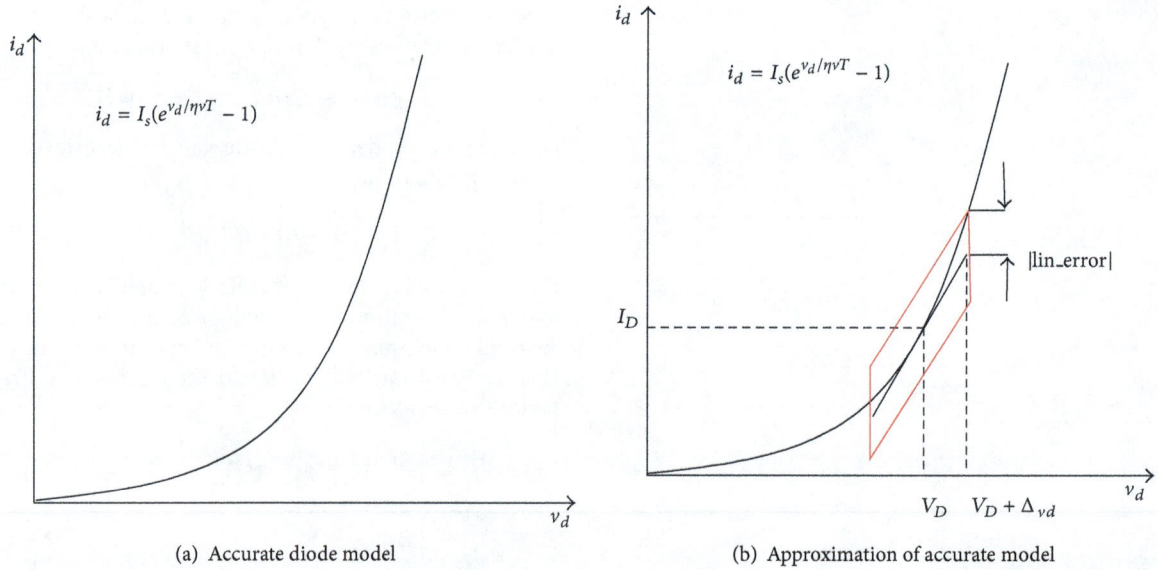

(a) Accurate diode model

(b) Approximation of accurate model

FIGURE 5: Forward diode characteristic.

where δ represents an allowed tolerance from the asymptote y_{asimp}.

(d) Operational Range Specification. This property defines allowed swinging in the output voltage, which does not cause a system make distortions on its output. This range will be represented with Affine Arithmetic as

$$\mathrm{spec}\,(\mathrm{swing}) = SW + \varepsilon\beta \quad \varepsilon \in [-1,1], \tag{19}$$

where SW represents the nominal value of range defining the swing property and β the allowed deviation from the nominal value.

3.5. Frequency-Domain Properties Modeled with Affine Arithmetic. The properties determining system behavior in frequency domain, whose specified values can be modeled with Affine Arithmetic, refer to low pass filter design specifications.

3.5.1. The Allowed Ripple in the Pass Band. This property defines maximum allowed deviation from the DC filter gain $S(0)$. Using Affine Arithmetic this requirement can be represented as

$$S(f) \in S(0) + \varepsilon_p\delta_p, \quad 0 \le f \le f_p, \tag{20}$$

where $\varepsilon_p \in [-1,1]$, $S(0)$ is a DC gain in the pass band, δ_p is the maximum allowed deviation from the DC gain, and f_p is the pass band edge frequency.

3.5.2. The Allowed Ripple in the Stop Band. This design specification defines the minimum allowed attenuation in the stop band and can be modeled in the similar way as

$$S(f) \in \frac{\delta_n}{2} + \varepsilon_n\frac{\delta_n}{2}, \quad f \ge f_n, \tag{21}$$

where $\varepsilon_n \in [-1,1]$, δ_n represents the minimum attenuation, and f_n represents the stop band edge frequency.

4. Semisymbolic Simulation and Assertion-Based Verification

In order to reduce high number of simulation runs required for simulation of systems with parameter variations, semisymbolic simulation is introduced. The idea of this approach is to model parameter deviations using Affine Arithmetic and simulate systems including these deviations. The simulation result is range-based system response which can be obtained in only one simulation run. Semisymbolic simulation can be performed on two levels: system level and circuit level.

4.1. Semisymbolic Simulation on System and Circuit Level. For semisymbolic simulations on system level SystemC AMS environment is used. Affine Arithmetic is implemented in a separate library which is due to SystemC AMS C++ based nature very easily integrated.

Beside system level analog circuits can also be simulated on lower level like transistor level. For this purpose, a semisymbolic circuit simulator has been developed [37]. Circuit simulation is performed in two steps.

(1) A netlist of a simulated system is converted into the according system of differential algebraic equations (DAE) using the Modified-Node-Analysis (MNA):

$$F\left(\underline{x}\,(t),\underline{x}'\,(t),\underline{p}\,(t),t\right) = \underline{0}, \tag{22}$$

where the vector $\underline{x}(t)$ represents the vector of time dependable state variables and $\underline{p}(t)$ represents the vector of time dependable circuit parameters.

TABLE 2: AAF+A operators.

AAF+A operators	Symbols	Operator meaning
Arithmetic operators	$\ominus \in \{+, -, *, /, {}^\wedge\}$	Symbols for the usual arithmetic operations
Relation operators	$\bullet \in \{<, >, \leq, \geq, ==\}$	The usual relation operations
Logic operators	$\oslash \in \{\&\&, \|, \rightarrow\}$!	Logic and, or, implies not
Affine analog operator	IN	Compares two affine terms a and b
Affine analog frequency operator	$AFO \in \{FIN, GFIN\}$ min, max	Compares two affine terms a and b within specified frequency interval, FIN assigns eventually; GIN assigns always Min finds frequency component with minimum amplitude value within specified frequency interval; max finds frequency component with maximum amplitude value within specified frequency interval
Slope operator	DV	Calculates slope of a signal
Temporal operators	$\square \in \{G, F\}$	Always, eventually
Affine analog time interval	Time $\in$ AAF, time $= t_0 + \varepsilon \Delta t, t_0, \Delta t \in \mathbb{R}$	Specified signal time interval of affine analog signal
Verification time interval	$[t], t \in \mathbb{R}$	Specified time within which the assertion is verified

(2) The equation system is passed through a numerical equation solver which performs DC, AC, and Transient Analysis [37]. The solver uses numerical methods as forward, backward Euler or trapezoidal methods to solve the equations.

Since at transistor level analog circuits are usually described with nonlinear differential algebraic equations the numerical solving is followed by linearization of equation system in the operating point with respect to variables and parameters. To deal with affine terms the algorithm [37] with the following steps is applied in the simulator.

In the first step the nominal solution is computed using Newton-Raphson method. In the second step the equation system is linearized in the operating point. Result of linearization is the linear dependency of variables according to the parameter deviations. In the case of linear system with constant parameters and variable inputs the result of linearized equation system is exact affine solution and algorithm ends. As the equation system usually contains nonlinear expressions, the affine solution of the linearized system is usually an underestimation of the exact solution, and therefore it has to be extended to include the exact area. This is done in the third step of algorithm.

4.2. Assertion-Based Verification with Affine Arithmetic. The verification technology proposed in this work is based on assertions which use Affine Arithmetic to model specified values of system properties as ranges. As simulation and verification environment SystemC AMS is used. The assertions representing specifications are verified within simulation run. In the case where the specification is violated the specification violation is reported and simulation run is stopped. To describe specifications with AAF+A, the set of operators defining the syntax of these assertions is used. Table 2 summarizes available operators whose meaning will be briefly described in the following.

TABLE 3: Time and frequency domain formulas (TBF and FBF).

TBF	FBF
$s \bullet d \in$ TBF	$\{GFIN, FIN\}\,(f_1, f_2, FFT\,\langle N\rangle\,(\beta), \gamma) \in$ FBF
s - affine signal, $d \in \mathbb{R}$	$f_1, f_2 \in \mathbb{R}, N \in \mathbb{N}, \beta$-affine signal, $\gamma \in$ AAF
$DV(s) \bullet d \in$ TBF	$IN\,(\zeta, \psi) \in$ FBF
s, affine signal, $d \in \mathbb{R}$	$\zeta \in$ FF, $\psi \in$ AAF
$IN\,\{[time]\}\,(\varphi, \vartheta) \in$ TBF	$c \bullet d \in$ FBF
$\varphi \in \{s, DV(s)\}$, time $\in$ AAF, $\vartheta \in$ AAF	$c \in$ FF $\wedge\, d \in \mathbb{R}$

The label AAF in the table assigns the set of affine terms. The set of assertions based on Affine Arithmetic (AAF+A) is comprised of two sets. The first set defines Boolean formulas checking validation of properties in time-domain TBF and the second one in frequency domain FBF. The operators from Table 2, comprising, respectively, the sets TBF and FBF, are given in Table 3.

In order to simplify the description of the FBF set, the new set of frequency formulas FF is introduced. This set is determined with

$$\{min, max\}\,\{[f_1], [f_1\ \ f_2]\}\,(FFT\,\langle N\rangle\,(\beta))$$
$$\in \text{FF} \wedge \{min, max\}\,(FFT\,\langle N\rangle\,(\beta)) \in \text{FF}, \tag{23}$$

where $\{f_1, f_2\} \in \mathbb{R}$, $f_1 < f_2$, β is an affine signal, and N is an integer number representing the length of Fast Fourier Transform (FFT). Note that for the operators min, max the frequency interval $[f_1, f_2]$ or the frequency f_2 does not need to be specified. In that case default values for f_1 and f_2 are 0 Hz and $f_s/2$ (f_s represents a sampling frequency), respectively.

TABLE 4: The meaning of operators in AAFA.

Operator	Explanation
$IN(s, h)$	Satisfied at time t' in which $(s(t') \leq h))$
$IN[time](s, h)$	Satisfied at time $t'((t' \in time)$ in which $(s(t') \leq h))$
$FIN(f_1, f_2, FFT(s), \beta)$	Satisfied if $\exists f \in [f_1, f_2] \Rightarrow FFT(s)(f) \leq \alpha$
$GFIN(f_1, f_2, FFT(s), \beta)$	Satisfied if $\forall f \in [f_1, f_2] \Rightarrow FFT(s)(f) \leq \alpha$
$G\{[t]\}(h), F\{[t]\}(h),$ $h \in AAFA+A$	Satisfied if the formula h holds always or at least once during simulation, respectively

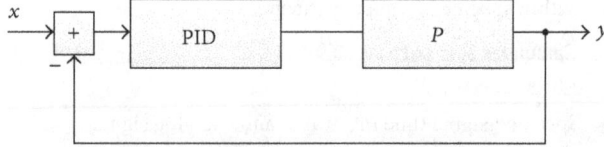

FIGURE 6: Block diagram of a system with PID controller.

The sets TBF and FBF comprise the smallest set of AAF+A as follows:

$$TBF \cup FBF \subset AAF + A,$$

$$(\alpha \in AAF+A \wedge t \in \mathbb{R}) \rightarrow \Box(\alpha) \in AAF+A \wedge \Box[t](\alpha) \in AAF + A,$$

$$(\alpha, \beta \in AAF+A) \rightarrow \alpha \oslash \beta \in AAF+A \wedge !(\alpha) \in AAF+A.$$

The following table (Table 4) gives a brief description of operators given in Table 2. If α and β represent the ranges modeled with Affine Arithmetic, then the operator $\leq$ in Table 4 assigns that the first range α lies in the second range β ($\alpha \subset \beta$). If β is a real value, then this operator assigns that the upper bound of α is lower or equal to the real variable β.

5. Demonstration Examples

Within this work the applicability of the proposed verification method will be shown through several examples. The examples are chosen to demonstrate ability to handle typical challenges for symbolic simulation like feedback, nonlinearities, and discontinuities. Note that complexity itself is not a challenge by itself. As the first example a closed loop control system composed of a PID controller and a plant is chosen. Its block diagram is shown in Figure 6. Further, as the second example also a feedback system, which needs to set a room temperature to a certain value considering variation in the external temperature, is chosen. The third example through which the performance of the method will be illustrated is a PLL (Phase-Locked Loop) circuit containing the loop and nonlinear elements like a phase detector or a voltage controlled oscillator.

5.1. A Control System including a Parameter Uncertainty of a Plant.
For the control system from Figure 6 a plant with the following transfer function will be considered:

$$P(s) = \frac{1}{s^2 + as + 1}, \qquad (24)$$

where for the parameter a it will be supposed that it lies in the interval $[0.4, 0.8]$. This range is modeled using Affine Arithmetic as $a = 0.6 + \varepsilon 0.2$ where $\varepsilon \in [-1, 1]$.

Substituting the affine form of the parameter a, $P(s)$ can be rewritten as

$$P(s) = \frac{P_{nom}(s)}{1 + \varepsilon P_{dev}(s) P_{nom}(s)} \quad \varepsilon \in [-1, 1], \qquad (25)$$

where $P_{nom}(s) = 1/(s^2 + 0.6s + 1)$ and $P_{dev}(s) = 0.2s$. For a PID controller model it is supposed that a noise filter for the derivative term is included, yielding to the following controller structure:

$$C(s) = K_p \left(1 + \frac{1}{T_i s} + \frac{T_d s}{(T_d/20)s + 20} \right), \qquad (26)$$

where the proportional gain K_p is 1.8, the integral time $T_i = 0.38$ s, and the derivative time $T_d = 0.095$ s. Between the integral and derivative times the ratio of 4 ($T_i = 4 * T_d$) is chosen. In [38] it is shown that this ratio is appropriate for many industrial processes.

In order to behave appropriately whether in time or frequency domain, it is required for a control system to satisfy the certain number of specifications. One of the most important specifications on control systems is the stability of the closed loop system. The gain and phase margin of the closed loop system are typical stability criteria. The gain margin is the maximum amount of the gain which is allowed to increase in the loop before a closed loop system becomes unstable, and the phase margin tells how much the phase lag must increase to make the system unstable. Since it is necessary to specify both margins to ensure appropriate behavior of a system, they can be replaced by a single parameter named the stability margin M_s. This parameter is defined as the shortest distance between the Nyqvist curve of the loop transfer function and the critical point -1. This distance is actually the inverse of the maximum sensitivity. The loop transfer function is determined with $L(s) = C(s)P(s)$ where $C(s)$ and $P(s)$ are the controller transfer function and the plant transfer function, respectively. Mathematically, the stability margin can be expressed as

$$M_s = \inf_\omega |-1 - L(j\omega)| = \inf_\omega |1 + L(j\omega)|$$
$$= \left[\sup_\omega \left| \frac{1}{1 + L(j\omega)} \right| \right]^{-1} = \left[\sup_\omega |S(j\omega)| \right]^{-1}, \qquad (27)$$

where $S(j\omega) = 1/(1+L(j\omega))$ is the sensitivity function. In particular, the sensitivity function represents the disturbances amplification at the output of the plant by the closed loop system. Recommended values for the stability margin M_s lie in the range of $[0.5, 0.75]$ [38].

Within this paper it will be verified if the control system meets the stability margin specification. In order to satisfy the stability margin specification the system stability margin must lie in the recommended range. This range representing the specification for M_s will be modeled with Affine Arithmetic:

$$spec(M_s) = M_s' + \varepsilon\delta, \quad \varepsilon \in [-1, 1], \qquad (28)$$

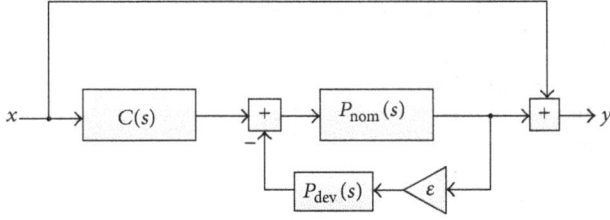

FIGURE 7: M_s calculation block diagram for parameter uncertainty.

where M_s' represents the center value of the specified range and δ represents the maximum absolute distance from the center value. Concretely, since the specified range is [0.5, 0.75], spec(M_s) is equal to

$$\text{spec}(M_s) = 0.625 + \varepsilon 0.125, \quad \varepsilon \in [-1, 1]. \quad (29)$$

Including the parameter uncertainty in the plant, the stability margin M_s can be rewritten as

$$M_s = \inf_\omega \left(\inf_\varepsilon \left| 1 + C(j\omega) P(j\omega) \right| \right)$$

$$= \inf_\omega \left(\inf_\varepsilon \left| 1 + C(j\omega) \frac{P_{\text{nom}}(j\omega)}{1 + \varepsilon P_{\text{dev}}(j\omega) P_{\text{nom}}(j\omega)} \right| \right). \quad (30)$$

This equation expresses that for every frequency it is necessary to determine the shortest distance of the loop transfer function from the critical point −1. It is very easy to be seen that the shortest distance will be obtained for $\varepsilon = 1$, which corresponds to the lower bound of range $1 + C(j\omega)P(j\omega)$.

A control system including uncertainties meets the stability margin specification if its stability margin M_s lies in the range $[0.5, 0.75]$. Since M_s is crucial to verify the control system against the given specification, the proposed verification method will use the system given in Figure 7 to calculate the stability margin M_s.

Using the proposed verification method the specification to be verified will be described with AAF+A assertion which will be verified during simulation run. In order to calculate the stability margin following Expression 4 the minimum value of $|1 + C(s)P(s)|$ with respect to frequency will be determined. This value can be found using AAF+A frequency operator min. One has

$$M_s = \min\left(|1 + C(j\omega) P(j\omega)|\right) = \min\left(|H(j\omega|\right)$$
$$= \min\left(\text{FFT} \langle N \rangle (h(t))\right), \quad (31)$$

where $h(t)$ represents the response of the system with transfer function $1 + C(j\omega)P(j\omega)$ (Figure 7) on Dirac impulse. The operator FFT assigns the Fast Fourier Transform, and N is the number of points for which the Fourier is calculated. In this work N will be set to 2048.

It is important to note that the values calculated by FFT operator are range-based values, because the uncertainty of the plant model is modeled with Affine Arithmetic. The infimum with respect to frequency can be determined as the minimum of all FFT frequency components min(FFT). The

stability margin corresponds to the lower bound of the range representing min(FFT) ($\varepsilon = 1$) (Expression 4). As it is said, in order to meet design specification, it is required from the control system that its M_s lies in the specified range spec(M_s). Using the proposed method this requirement can be written as AAF+A assertion which will be verified during simulation as follows:

$$G\left(\text{IN}\left(\min\left(\text{FFT}\langle 2048\rangle\left(h(t)\right)\right), 0.625 + \varepsilon 0.125\right)\right). \quad (32)$$

This assertion expresses that infimum with respect to frequency determined with min(FFT$\langle 2048\rangle(h(t))$ must always (assigned with operator G) lie in the range modeling the stability margin specification ($0.625 + \varepsilon 0.125$) during simulation.

5.2. *A Room Heating Control System.* As the second demonstration example a system which controls a room temperature is chosen. The block diagram of the system is shown in Figure 8.

For room modeling the model including the thermal resistance of the wall between the room and the ambient R_{ra}, and the thermal capacitance of the room C_r is used. One part of heat, brought into the room, leaves the room through the resistor with R_{ra} and the other part is stored in the capacitor with thermal capacity C_r. This mathematical model can be described with the following equation:

$$q = \frac{\theta_r - \theta_e}{R_{ra}} + C \frac{d\theta_r}{dt}. \quad (33)$$

Using Laplace transformation the equation can be transformed into

$$q = \frac{\theta_r - \theta_e}{R_{ra}} + Cs\theta_r, \quad (34)$$

where q is the heat bringing into the system, θ_r is the temperature of the room, and θ_e is the external temperature. To determine the value of the thermal capacitance C_r, the certain number of factors needs to be considered (the heat capacity of the stuff in the room, the air in the room...). Within this work the value $C_r = 10^7$ (J/K) is chosen (Appendix C in [39]). For the thermal resistance R_{ra} the value of $R_{ra} = 0.0846 * 10^{-6}$ (°C/W).

The controller used for this system is a PID controller with the following coefficients:

$$C(s) = c_1 + c_2 s + \frac{c_3}{s}$$

$$= 0.1 * 10^8 + 0.15 * 10^8 s + \frac{0.15 * 10^8}{s}. \quad (35)$$

According to (34) the room temperature θ_r can be calculated as

$$\theta_r = q \frac{R_{ra}}{1 + sC_r R_{ra}} + \frac{\theta_e}{1 + sC_r R_{ra}}. \quad (36)$$

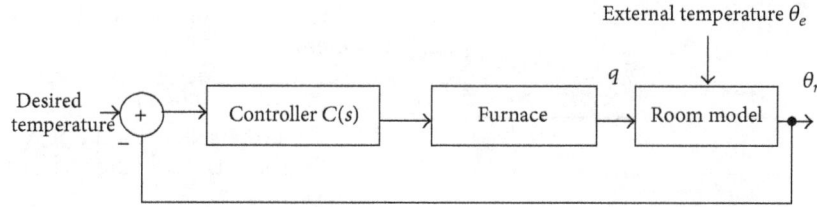

FIGURE 8: The block diagram of a room heating control system.

TABLE 5: Simulation results.

Example	Time without AAFA (s)	Time with AAFA (s)	Overhead (s)
The control system with parameter uncertainty	17.6	19.5	1.9
The room heating control system	9.3	10.8	1.5
PLL circuit	7.3	Simulation run stopped	—

For the external temperature θ_e it will be supposed that its value varies and lies in the range $[-2, 1]$. This range will be modeled using Affine Arithmetic as

$$[-2, 1] = -0.5 + \varepsilon 1.5, \quad \varepsilon \in [-1, 1]. \tag{37}$$

Considering the variation in the external temperature it will be verified

(1) if the room temperature θ_r within 5 seconds reaches the value varying within 3% of the final value,

(2) if the final value is reached.

These two requirements will be described using Affine Arithmetic Assertions (AAF+A) and verified during simulation. The final value of the temperature is supposed to be 30°C. The assertion corresponding to the first requirement can be written as

$$F\left(\text{IN}\left[\frac{t_s}{2} + \varepsilon_1 \frac{t_s}{2}\right](\theta_r, \theta_{\text{final}} + \varepsilon_2 \delta)\right. \\ \left. \longrightarrow G\left(\text{IN}\left(\theta_r, \theta_{\text{final}} + \varepsilon_2 \delta\right)\right)\right), \tag{38}$$

where t_s is 5 s, θ_{final} is 30°C, and $\delta = 0.03 * 30 = 0.9$ C and $\varepsilon_1, \varepsilon_2 \in [-1, 1]$. The second requirement can be described with the following assertion:

$$F\left(\theta_r == 30\right). \tag{39}$$

The simulation results are given in Table 5 in Section 5.4.

5.3. A PLL Circuit including Parameter Uncertainties.
As the second demonstration example a phased-locked loop (PLL) circuit is chosen. Due to high number of applications phased-locked loops found their place in analog-mixed-signal (AMS) systems. Some of these applications are listed in the following:

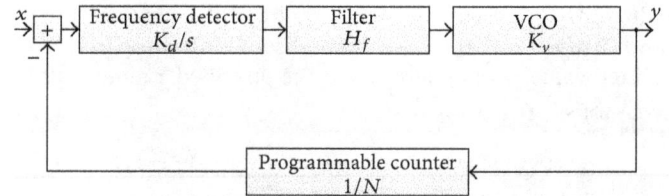

FIGURE 9: PLL circuit on block level.

(i) in radio transmitters it is used to synthesize frequencies, which are a multiple of a reference frequency;

(ii) clock generators which multiply a low-frequency reference clock to higher operating frequencies of microprocessors;

(iii) in communication systems for coherent demodulation and demodulation of frequency and phase modulated signals.

In this paper a PLL circuit as a frequency synthesizer is considered. Its block diagram is shown in Figure 9. This PLL model models the behavior of the system in ideal conditions. In reality certain numbers of deviations influence the behavior of the PLL causing design specifications to violate from their values previously defined by design. Since the PLL is often the part of more complex AMS systems (e.g., in the role as a frequency synthesizer it is embedded into communication systems to generate carrier frequencies) it is of a great importance to verify if the desired output behaviour in the presence of parameter deviations is still met. Within this work a time delay of the filter from Figure 9 will be considered and added to the PLL model.

As it can be seen from the figure the PLL model is comprised of a frequency detector, a filter, and a voltage controlled oscillator (VCO). The filter with the following transfer function is used:

$$H_f(s) = K_f \frac{(bs + 1)}{as} e^{-\tau s}, \tag{40}$$

where K_f is the filter gain, and within this work it will be supposed that its value is one. For parameters a and b the values $2e^{-3}$ and $680 * 0.5 * 10^{-6}$ are supposed, respectively. The parameter τ represents time delay of the filter, and for this example it will be supposed that its value lies in the range $[0, 100\ \mu s]$. In Section 3.2 it is shown that time delay causes

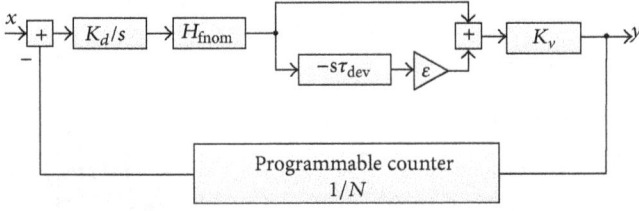

FIGURE 10: PLL model with the filter time delay.

deviation from the nominal filter behavior approximating the filter transfer function $H_f(s)$ with

$$
\begin{aligned}
H_f(s) &:= H_{f,\text{nom}}(s)\left(1 + (-s) * \varepsilon\tau_{\text{dev}}\right) \\
&= H_{f,\text{nom}}(s)\left(1 + \varepsilon H_{f,\text{dev}}\right) \quad \varepsilon \in [-1, 1],
\end{aligned} \tag{41}
$$

where $H_{f,\text{nom}}$ represents the filter transfer function in ideal conditions (time delay τ is zero). Substituting the values for parameters K_f, a, and b, $H_{f,\text{nom}}$ can be rewritten to

$$
H_{f,\text{nom}}(s) = \frac{\left(680 * 0.5 * 10^{-6}s + 1\right)}{2 * 10^{-3}s}. \tag{42}
$$

The parameter τ_{dev} represents the maximum value of time delay which is for this example $100\,\mu s$. The block diagram of the PLL model considering the filter time delay is shown in Figure 10.

For this PLL model it will be verified if the lock time of the PLL to switch from one frequency to within 5 kHz of another frequency is not greater than 1 ms. The PLL parameters such as f_{step}, the minimum obtained value of the output frequency $f_{o,\text{min}}$ and the maximum obtained value of the output frequency $f_{o,\text{max}}$, are supposed to be 100 KHz, 2 MHz, and 3 MHz, respectively. The detector gain K_d and the gain of the voltage controlled oscillator K_v are supposed to be

$$
K_d = 0.111\frac{V}{\text{rad}}, \tag{43}
$$

$$
K_v = 11.2 * 10^6\frac{\text{rad}}{Vs}.
$$

The loop gain of the system from Figure 10 is equal to

$$
L(s) = \frac{K_d}{s} * H_f(s) * K_v * \frac{1}{N}. \tag{44}
$$

From this formula it can be seen that for the maximum ratio of a programmable counter the loop gain will have the minimum value causing maximum lock time. Thus, only this ratio value will be considered. For considered PLL circuit this value is equal to

$$
N_{\text{max}} = \frac{f_{o,\text{max}}}{f_{\text{step}}} = \frac{3\,\text{MHz}}{100\,\text{kHz}} = 30. \tag{45}
$$

Using the verification method proposed in this work the specification to be verified will be described with AAF+A

assertion. This assertion will be verified during simulation as follows:

$$
\begin{aligned}
F&\left(\text{IN}\left[\frac{t_s}{2} + \varepsilon_1\frac{t_s}{2}\right]\left(f_o, f_{\text{steady}} + \varepsilon_2\delta\right)\right) \\
&\longrightarrow G\left(\text{IN}\left(f_o, f_{\text{steady}} + \varepsilon_2\delta\right)\right),
\end{aligned} \tag{46}
$$

where the values of ε_1 and ε_2 lie in the range $[-1, 1]$. The operator F in the assertion assigns that the output frequency f_o must eventually within the settling time t_s enter the error band around the value of the steady state $f_{\text{steady}} + \varepsilon_2\delta$ and stay there (assigned with operator G).

We consider only the maximum ratio of the programmable counter, and therefore the frequency value in the steady state is $f_{\text{steady}} = N * f_{\text{step}} = 30 * 100\,\text{KHz} = 3\,\text{MHz}$. The other parameters are according to desired specification equal to

$$
\begin{aligned}
t_s &= 1\,\text{ms}, \\
\delta &= 5\,\text{kHz}.
\end{aligned} \tag{47}
$$

Substituting these values in the previous assertion we have

$$
\begin{aligned}
F&\left(\text{IN}\left[0.5 * 10^{-3} + \varepsilon_1 0.5 * 10^{-3}\right]\left(f_o, 3 * 10^6 + \varepsilon_2 5 * 10^3\right)\right) \\
&\longrightarrow G\left(\text{IN}\left(f_o, 3 * 10^6 + \varepsilon_2 5 * 10^3\right)\right).
\end{aligned} \tag{48}
$$

5.4. Experimental Results. The SystemC AMS is used as simulation and verification environment. The control system considering the filter parameter uncertainty was simulated and verified for 10^6 s with the sampling rate $T_s = 1$ s. The specification (Assertion 6) passed and the stability margin calculated for the control system is shown in Figure 11. The symbol k in the figure assigns the k frequency component of Fast Fourier Transform which was used to determine the stability margin with respect to frequency.

Using the sampling period $T_s = 0.5$ s the heating control system was simulated for $2 * 10^5$ s. The system met desired requirements and both assertions (Assertions 8 and 9) passed. The signal representing the room temperature θ_r is shown in Figure 11.

The PLL circuit was simulated and verified for 20 s with the sampling period $T_s = 0.1$ ms. It was verified if the lock time of the PLL to switch from 2.9 MHz to within 5 kHz of 3 MHz is less than 1 ms. The Assertion 10 failed, and simulation run was stopped reporting the information about the specification violation. The fact that the PLL output frequency did not (within 1 ms) set to the desired value within the specified tolerance does not imply that it will not do so after some time. To prove this the PLL simulation result is given in Figure 12.

Since the assertion failed and the simulation run was stopped, the PLL response from Figure 12 is the result of simulation in the case where the assertion is omitted. From the figure it can be concluded that the output frequency converges to its final value and deviated terms converge

(a)

(b)

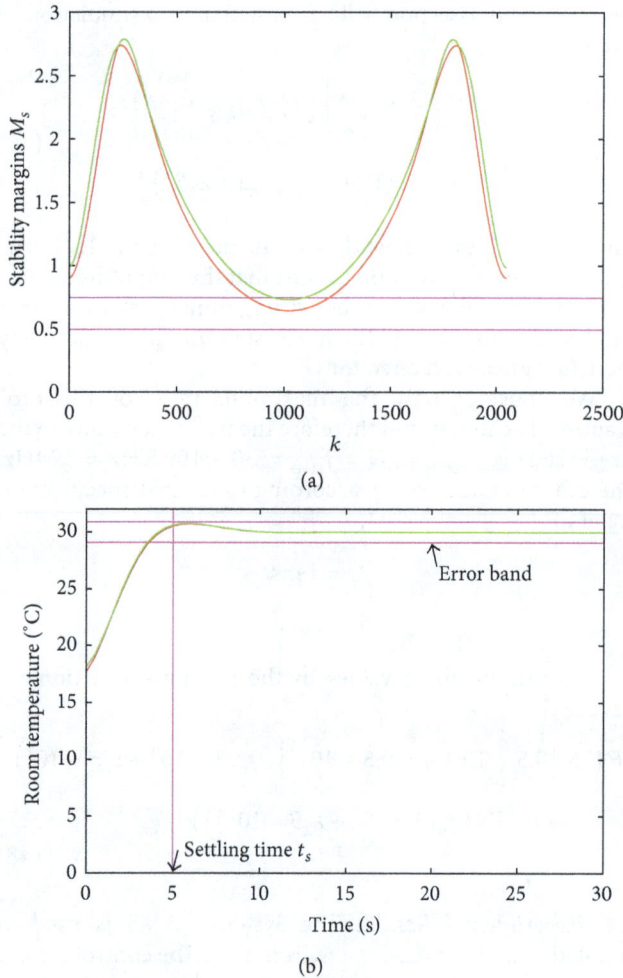

FIGURE 11: Simulation results. The stability margin of the control system M_s and the room temperature θ_r.

f_{min}

f_{max}

FIGURE 12: The PLL output frequency.

to zero. This fact is one of the main advantages of the Affine Arithmetic approach. Its ability to identify the correlation between system quantities reaches its maximum in the systems with feedback loops like the case with our demonstration examples.

Hence, even in the case where the loop contains the nonlinear elements the additional terms (which are the result of overapproximation introduced by nonlinear operations) will have negligible values, and the output will converge to its finite value. To free the memory of unnecessary variables, the authors in [28] propose the (cleanup) method. Concretely, all terms under some user specified value are replaced with only two symbols, the one representing the sum of all terms with a positive and the other with a negative sign. In this way the safe inclusion of the result is kept, and the number of terms is drastically decreased.

Table 5 summarizes the simulation times necessary for all designs in the case where AAF+A assertions were included into design simulation and when they were omitted. It can be noted that in the case where the assertions were satisfied, the proposed verification method generated additional simulation time, but the overhead was not high.

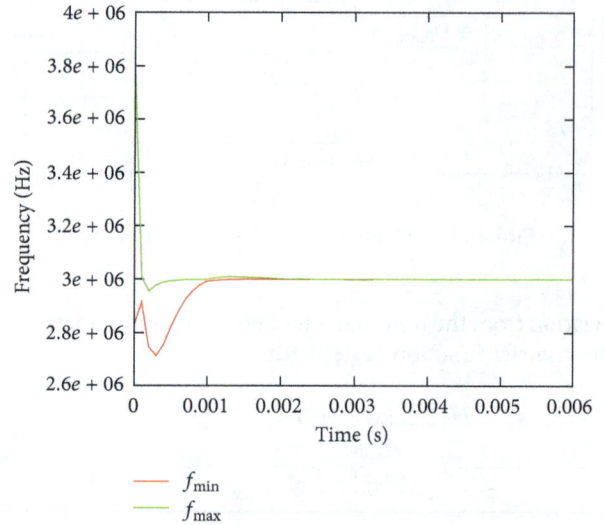

The simulation time of the PLL in the case where the assertion was embedded into simulation was omitted. The reason lies in the fact that the assertion failed stopping the simulation process, and hence the time required for simulation was much lower than for the case in which the assertion was not included.

6. Conclusion and Future Work

This work introduces a methodology that enables verification of analog/mixed-signal systems including deviations. The verification method is combined with symbolic simulation which generates the worst case dependable response adding deviations to a system model and modeling them as ranges. Since the generated output behaviour contains all possible traces for the considered parameter set the proposed assertion-based technology can provide formal verification result using simulation-based techniques. The assertions use Affine Arithmetic to model allowed or forbidden areas of typical system properties as ranges. The specified ranges are further combined with Boolean logic, frequency operators, and temporal logic which allows us to verify the system behaviour in time, but also in the frequency domain. The assertions are embedded into simulation, and as soon as the assertion violation is detected, the simulation run is stopped, and information about the assertion harm is reported.

Overapproximation is a challenge that can become a problem for strongly nonlinear systems. The first step to deal with this problem was proposed in [28] and found its applicability in the systems containing the loops. The further step towards the problem solution is to modify Affine Arithmetic in the way that we keep the second order terms in symbolic representation and in this way that we reduce overapproximation.

Furthermore, the AAF+A assertions (Table 2) will be extended from rather formal operators to libraries of application-specific properties that are close to requirement specifications found in various application domains. Also, the method up to now verifies the system against the specifications for which time and frequency requirements must be known in advanced. One interesting direction in the future would be to extend the method to extract the information about lower and upper bounds of time or frequency for which the system behaviour is still desirable under the considered set of deviated parameters.

Acknowledgment

This work has been funded by the Vienna Science and Technology Fund (WWTF) through Project ICT08_012.

References

[1] R. Y. Rubinstein, *Simulation and the Monte Carlo Method*, John Wiley & Sons, New York, NY, USA, 1981.

[2] D. E. Hocevar, M. R. Lightner, and T. N. Trick, "Study of variance reduction techniques for estimating circuit yields," *IEEE Transactions on Computer-Aided Design of Integrated Circuits and Systems*, vol. 2, no. 3, pp. 180–192, 1983.

[3] K. Antreich, H. Gräb, and C. Wieser, "Practical methods for worst-case and yield analysis of analog integrated circuits," *International Journal of High Speed Electronics and Systems*, vol. 4, no. 3, pp. 261–282, 1993.

[4] M. Rafaila, C. Grimm, C. Decker, and G. Pelz, "Sequential design of experiments for effective model-based validation of electronic control units," *e&i Elektrotechnik Und Informationstechnik*, vol. 127, pp. 164–170, 2010.

[5] M. S. Branicky, M. M. Curtiss, J. A. Levine, and S. B. Morgan, "RRTs for nonlinear, discrete, and hybrid planning and control," in *Proceedings of the 42nd IEEE Conference on Decision and Control*, pp. 657–663, December 2003.

[6] A. Bhatia and E. Frazzoli, "Incremental search methods for reachability analysis of continuous and hybrid systems," in *Proceedings of the International Workshop Hybrid Systems: Computation and Control (HSCC '04)*, vol. 2993 of *Lecture Notes in Computer Science*, pp. 142–156, Springer, 2004.

[7] A. Julius, G. Fainekos, M. Anand, I. Lee, and G. Pappas, "Robust test generation and coverage for hybrid systems," in *Proceedings of the International Workshop Hybrid Systems: Computation and Control (HSCC '07)*, vol. 4416 of *Lecture Notes in Computer Science*, pp. 329–342, Springer, 2007.

[8] J. M. Esposito, "Randomized test case generation for hybrid systems: metric selection," in *Proceedings of the 36th Southeastern Symposium on System Theory*, pp. 236–240, 2004.

[9] Q. Zhao, B. H. Krogh, and P. Hubbard, "Generating test inputs for embedded control systems," *IEEE Control Systems Magazine*, vol. 23, no. 4, pp. 49–57, 2003.

[10] J. Kapinski, B. H. Krogh, O. Maler, and O. Stursberg, "On Systematic simulation of open continuous systems," in *Proceedings of the International Workshop Hybrid Systems: Computation and Control (HSCC '03)*, vol. 2623 of *Lecture Notes in Computer Science*, pp. 283–297, Springer, 2003.

[11] A. Girard and G. Pappas, "Verification using simulation," in *Proceedings of the International Workshop Hybrid Systems: Computation and Control (HSCC '06)*, vol. 3927 of *Lecture Notes in Computer Science*, pp. 272–286, Springer, 2006.

[12] A. Chutinan and B. H. Krogh, "Computing polyhedral approximations to flow pipes for dynamic systems," in *Proceedings of the 37th IEEE Conference on Decision and Control (CDC '98)*, pp. 2089–2094, December 1998.

[13] A. B. Kurzhanski and P. Varaiya, "Ellipsoidal techniques for reachability analysis," in *Proceedings of the International Workshop Hybrid Systems: Computation and Control (HSCC '00)*, vol. 1790 of *Lecture Notes in Computer Science*, pp. 202–214, Springer, 2000.

[14] S. Prajna and A. Jadbabaie, "Safety verification of hybrid systems using barrier certificates," in *Proceedings of the International Workshop Hybrid Systems: Computation and Control (HSCC '04)*, vol. 2993 of *Lecture Notes in Computer Science*, pp. 477–492, Springer, 2004.

[15] E. M. Clarke, O. Grumberg, and D. A. Peled, *Model Checking*, The MIT Press, 1999.

[16] J. R. Burch, E. M. Clarke, K. L. McMillan, D. L. Dill, and L. J. Hwang, "Symbolic model checking: 1020 states and beyond," *Information and Computation*, vol. 98, no. 2, pp. 142–170, 1992.

[17] E. Clarke, O. Grumberg, and D. Long, "Verification tools for finite-state concurrent systems," in *Proceedings of the Decade of Concurrency, Reflections and Perspectives*, vol. 803 of *Lecture Notes in Computer Science*, pp. 124–175, Springer, 1994.

[18] O. Maler and G. Batt, "Approximating continuous systems by timed automata," in *Proceedings of the 1st international workshop on Formal Methods in Systems Biology (FMSB '08)*, vol. 5054 of *Lecture Notes in Computer Science*, Springer, 2008.

[19] O. Stursberg, S. Kowalewski, and S. Engell, "On the generation of timed discrete approximations for continuous systems," *Mathematical and Computer Modelling of Dynamical Systems*, vol. 6, no. 1, pp. 51–70, 2000.

[20] O. Stursberg, S. Kowalewski, and S. Engell, "Timed approximations of hybrid processes for controller verification," in *Proceedings of the 1st IFAC Conference on Analysis and Design of Hybrid Systems*, pp. 289–295, 2003.

[21] R. Alur, T. A. Henzinger, G. Lafferriere, and G. J. Pappas, "Discrete abstractions of hybrid systems," *Proceedings of the IEEE*, vol. 88, no. 7, pp. 971–984, 2000.

[22] T. Henzinger and H. Wong-Toi, "Linear phase-portrait approximations for nonlinear hybrid systems," in *Proceedings of the Hybrid Systems III: Verification and Control*, vol. 1066 of *Lecture Notes in Computer Science*, pp. 377–388, Springer, 1996.

[23] W. Hartong, L. Hedrich, and E. Barke, "Model checking algorithms for analog verification," in *Proceedings of the 39th Annual Design Automation Conference (DAC '02)*, pp. 542–547, June 2002.

[24] D. Grabowski, D. Platte, L. Hedrich, and E. Barke, "Time constrained verification of analog circuits using model-checking algorithms," *Electronic Notes in Theoretical Computer Science*, vol. 153, no. 3, pp. 37–52, 2006.

[25] W. Hartong, L. Hedrich, and E. Barke, "On discrete modelling and model checking of nonlinear analog systems," in *Proceedings of the 14th International Conference on Computer Aided Verification (CAV '02)*, pp. 401–413, 2002.

[26] S. Lämmermann, A. Jesser, M. Rathgeber et al., "Checking heterogeneous signal characteristics applying assertion-based verification," in *Proceedings of the Frontiers in Analog Circuit Verification (FAC '09)*, Grenoble, France, 2009.

[27] A. Pnueli and O. Maler, "Extending PSL for analog circuits," Tech. Rep., 2005, PROSYD Deliverable D 1.3/1.

[28] W. Heupke, C. Grimm, and K. Waldschmidt, "Semi-symbolic simulation of nonlinear systems," in *Proceedings of the Forum on Specification and Design Languages (FDL '05)*, ECSI, Lausanne, Switzerland, September 2005.

[29] C. Grimm, W. Heupke, and K. Waldschmidt, "Refinement of mixed-signal systems with affine arithmetic," in *Proceedings of the Design, Automation and Test in Europe Conference and Exhibition (DATE '04)*, pp. 372–377, IEEE Press, February 2004.

[30] C. Grimm, W. Heupke, and K. Waldschmidt, "Analysis of mixed-signal systems with affine arithmetic," *IEEE Transactions on Computer-Aided Design of Integrated Circuits and Systems*, vol. 24, no. 1, pp. 118–123, 2005.

[31] D. Grabowski, M. Olbrich, C. Grimm, and E. Barke, "Range arithmetics to speed up reachability analysis of analog systems," in *Proceedings of the Forum on Specification, Verification and Design Languages (FDL '07)*, Barcelona, Spain, 2007.

[32] N. Femia and G. Spagnuolo, "True worst-case circuit tolerance analysis using genetic algorithms and affine arithmetic," *IEEE Transactions on Circuits and Systems*, vol. 47, no. 9, pp. 1285–1296, 2000.

[33] A. Lemke, L. Hedrich, and E. Barke, "Analog circuit sizing based on formal methods using affine arithmetic," in *Proceedings of the IEEE/ACM International Conference on Computer Aided Design (ICCAD '02)*, pp. 486–489, November 2002.

[34] C. F. Fang, R. A. Rutenbar, M. Püschel, and T. Chen, "Toward efficient static analysis of finite-precision effects in DSP applications via affine arithmetic modeling," in *Proceedings of the 40th Design Automation Conference (DAC '03)*, pp. 496–501, June 2003.

[35] C. F. Fang, T. Chen, and R. A. Rutenbar, "Floating-point error analysis based on affine arithmetic," in *Proceedings of IEEE International Conference on Accoustics, Speech, and Signal Processing (ICASSP '03)*, vol. 2, pp. 561–564, April 2003.

[36] R. E. Moore, *Interval Analysis*, Prentice-Hall, Englewood Cliffs, NJ, USA, 1966.

[37] D. Grabowski, M. Olbrich, and E. Barke, "Analog circuit simulation using range arithmetics," in *Proceedings of the Asia and South Pacific Design Automation Conference (ASP-DAC '08)*, pp. 762–767, IEEE Computer Society Press, Seoul, Korea, March 2008.

[38] K. J. Åström and T. Hägglund, *PID Controllers: Theory, Design and Tuning*, Instrument Society of America, 2nd edition, 1995.

[39] D. Ryder-Cook, "Thermal modelling of buildings," Tech. Rep., University of Cambridge, 2009.

8

Low Cost Design of a Hybrid Architecture of Integer Inverse DCT for H.264, VC-1, AVS, and HEVC

Muhammad Martuza and Khan A. Wahid

Department of Electrical and Computer Engineering, University of Saskatchewan, Saskatoon, SK, Canada S7N 5A9

Correspondence should be addressed to Khan A. Wahid, khan.wahid@usask.ca

Academic Editor: Maurizio Martina

The paper presents a unified hybrid architecture to compute the 8 × 8 integer inverse discrete cosine transform (IDCT) of multiple modern video codecs—AVS, H.264/AVC, VC-1, and HEVC (under development). Based on the symmetric structure of the matrices and the similarity in matrix operation, we develop a generalized "decompose and share" algorithm to compute the 8 × 8 IDCT. The algorithm is later applied to four video standards. The hardware-share approach ensures the maximum circuit reuse during the computation. The architecture is designed with only adders and shifters to reduce the hardware cost significantly. The design is implemented on FPGA and later synthesized in CMOS 0.18 um technology. The results meet the requirements of advanced video coding applications.

1. Introduction

In recent years, different video applications use different video standards, such as H.264/AVC [1], VC-1 [2], and AVS [3]. To improve the coding efficiency further, recently a joint collaboration team on video coding (JCT-VC) is drafting a next generation video coding standards, known tentatively as high efficient video coding (HEVC or H.265) [4]. The target bit rate is half of that of H.264/AVC. Besides, several other effective techniques are proposed in the draft to reduce the complexity of the encoder such as improved intrapicture coding, and simpler VLC coefficients [5]. As a result of these new features, experts predict that the HEVC will dominate the future multimedia market.

In order to meet up the present and future demands of different multimedia applications, it becomes necessary to develop a unified video decoder that can support all popular video standards on a single platform. In recent years, there is a growing interest to develop multistandard inverse transform architectures for advanced multimedia applications. However, most of them do not support AVS, the video

codec developed by Chinese government that became the core technology of China Mobile Multimedia Broadcasting (CMMB) [6]. None of the existing works supports the HEVC; thought it is not finalized yet, considering the future prospective of the HEVC [7], it is important to start exploring possible implementation in hardware of the transform unit discussed in the draft.

In this paper, we present a new generalized algorithm and its hardwire implementation of an 8 × 8 IDCT architecture. The scheme is based on matrix decomposition with sparse matrices and offset computations. These sparse matrices are derived in a way that can be reused maximum number of times during decoding different inverse matrices. All multipliers in the design are replaced by adders and shifters. In the scheme, we first split the 8 × 8 transformation matrix into two small 4 × 4 matrices by applying permutation techniques. Then we concurrently perform separate operations on these two matrices to compute the output. It enables parallel operation and yields high throughput, which eventually helps meet the coding requirement of the high resolution video.

The proposed generalized algorithm is later applied to compute the 8×8 integer IDCT of AVS. Then we identify the submatrices of AVS and reuse them to compute the IDCT of VC-1. We follow the same principle to compute the other two IDCTs of H.264 and HEVC. For HEVC, we have used the draft matrix discussed in the recent meeting [7]; since it is not yet finalized, we have developed the generalized architecture in such a way that can be easily adjusted to accommodate any changes to the final HEVC format.

2. Previous Works

In recent years, some multistandard inverse transform architectures have been proposed for video applications. Lee's work in [8] presents a 8×8 multistandard IDCT architecture based on delta coefficient matrices which can support VC-1, MPEG4, and H.264. It can process up to 21.9 fps for full HD video. Kim's work in [9] describes a design following similar approach of [8] to unify the IDCT and inverse quantization (IQ) operations for those three codecs. However, the design cannot support full HD video format. Qi's work in [10] shows an efficient integrated architecture designed for multistandard inverse transforms of MPEG-2/4, H.264, and VC-1 using factor share (FS) and adder share (AS) strategies for saving circuit resource. The work achieves 100 MHz working frequency for full HD video resolution, but does not support AVS. In another interesting design [11], the authors devise a common architecture by sharing adders and multipliers to perform transform and quantization of H.264, MPEG-4, and VC-1. The common shortcoming of all these designs discussed in [8–11] is that none of them supports the Chinese standard, AVS, nor the HEVC.

In our previous work [12], we have developed a resource shared design using delta coefficient matrices which can compute the 8×8 IDCT of VC-1, JPEG, MPEG4, H.264/AVC, and AVS. But due to complex data scheduling and the integration of JPEG (which is an image codec), the decoding capability is limited. The design supports both HD formats, but fails to comply with super resolution (WQXGA). Liu [13] introduces another design to support multiple standards where the design throughput is low (110.8 MHz) and cannot decode HD and WQXGA video. Fan's works in [14, 15] are based on another efficient matrix decomposition algorithm to compute multiple transforms; however, the work is limited to only H.264 and VC-1. There are similar works in [16–18], which are also limited to these two codecs (H.264 and VC-1).

In this paper, we present a generalized low-cost algorithm and its single chip implementation to compute all four modern video standards (AVS, H.264, VC-1, and HEVC). The design meets the requirement of high performance video coding as it can process the HD video at 145 fps, the full HD video at 62 fps, and the WQXGA video at 32 fps. The proposed scheme can be applied to both forward and inverse transformation; however, here we only show the implementation for the inverse process (targeted for decoders).

TABLE 1: Matrix coefficients of 8-point IDCT.

	AVS [3]	VC-1 [2]	H.264 [1]	HEVC [7]
a	8	12	8	64
b	10	16	12	89
c	9	15	10	75
d	6	9	6	50
e	2	4	3	18
f	10	16	8	83
g	4	6	4	36

3. Proposed Generalized Algorithm for 8×8 IDCT

In a video compression system, the transform coding usually employs an 8-point II-type DCT. Since, the forward DCT uses the same basis coefficients and is the transpose of the IDCT matrix, the proposed IDCT scheme is easily applicable to it without any added cost or complexity. The 8-point 1D forward and inverse DCT coefficient matrices are expressed in general form as F and I respectively (below in (1), where, $a, b, c \ldots, g$ denote seven different transform coefficients):

$$
F = \begin{bmatrix}
a & a & a & a & a & a & a & a \\
b & c & d & e & -e & -d & -c & -b \\
f & g & -g & -f & -f & -g & g & f \\
c & -e & -b & -d & d & b & e & -c \\
a & -a & -a & a & a & -a & -a & a \\
d & -b & e & c & -c & -e & b & -d \\
g & -f & f & -g & -g & f & -f & g \\
e & -d & c & -b & b & -c & d & -e
\end{bmatrix},
$$

$$
I = \begin{bmatrix}
a & b & f & c & a & d & g & e \\
a & c & g & -e & -a & -b & -f & -d \\
a & d & -g & -b & -a & e & f & c \\
a & e & -f & -d & a & c & -g & -b \\
a & -e & -f & d & a & -c & -g & b \\
a & -d & -g & b & -a & -e & f & -c \\
a & -c & g & e & -a & b & -f & d \\
a & -b & f & -c & a & -d & g & -e
\end{bmatrix}.
$$

$$(1)$$

In this paper, we have denoted the 8×8 IDCT transform matrices for AVS, VC-1, H.264/AVC, and HEVC by the letters A, V, H and HV respectively. These seven coefficients $(a, b, c \ldots, g)$ for each of the transforms are different, but integer in nature (as shown in Table 1).

3.1. Development of a Generalized "Decompose and Share" Algorithm. First of all, we derive a generalized matrix decomposition scheme by utilizing the symmetric structure of the matrices and factoring the 8×8 matrix into two 4×4 sub-matrices as shown below:

$$I = P_0 \cdot I_0, \tag{2}$$

where

$$
I_0 = \begin{bmatrix}
a & 0 & f & 0 & a & 0 & g & 0 \\
a & 0 & g & 0 & -a & 0 & -f & 0 \\
a & 0 & -g & 0 & -a & 0 & f & 0 \\
a & 0 & -f & 0 & a & 0 & -g & 0 \\
0 & -e & 0 & d & 0 & -c & 0 & b \\
0 & -d & 0 & b & 0 & -e & 0 & -c \\
0 & -c & 0 & e & 0 & b & 0 & d \\
0 & -b & 0 & -c & 0 & -d & 0 & -e
\end{bmatrix},
$$

(3)

$$
P_0 = \begin{bmatrix}
1 & 0 & 0 & 0 & 0 & 0 & 0 & -1 \\
0 & 1 & 0 & 0 & 0 & 0 & -1 & 0 \\
0 & 0 & 1 & 0 & 0 & -1 & 0 & 0 \\
0 & 0 & 0 & 1 & -1 & 0 & 0 & 0 \\
0 & 0 & 0 & 1 & 1 & 0 & 0 & 0 \\
0 & 0 & 1 & 0 & 0 & 1 & 0 & 0 \\
0 & 1 & 0 & 0 & 0 & 0 & 1 & 0 \\
1 & 0 & 0 & 0 & 0 & 0 & 0 & 1
\end{bmatrix}.
$$

The computational complexity of P_0 is only 8 additions. To reduce the complexity of I_0, we use permutation techniques by performing the operations: $I_0 = \tilde{I} \cdot P_C$.
Where

$$
P_C = \begin{bmatrix}
1 & 0 & 0 & 0 & 0 & 0 & 0 & 0 \\
0 & 0 & 1 & 0 & 0 & 0 & 0 & 0 \\
0 & 0 & 0 & 0 & 1 & 0 & 0 & 0 \\
0 & 0 & 0 & 0 & 0 & 0 & 1 & 0 \\
0 & 1 & 0 & 0 & 0 & 0 & 0 & 0 \\
0 & 0 & 0 & 1 & 0 & 0 & 0 & 0 \\
0 & 0 & 0 & 0 & 0 & 1 & 0 & 0 \\
0 & 0 & 0 & 0 & 0 & 0 & 0 & 1
\end{bmatrix},
$$

(4)

$$
\tilde{I} = \begin{bmatrix}
a & f & a & g & 0 & 0 & 0 & 0 \\
a & g & -a & -f & 0 & 0 & 0 & 0 \\
a & -g & -a & f & 0 & 0 & 0 & 0 \\
a & -f & a & -g & 0 & 0 & 0 & 0 \\
0 & 0 & 0 & 0 & -e & d & -c & b \\
0 & 0 & 0 & 0 & -d & b & -e & -c \\
0 & 0 & 0 & 0 & -c & e & b & d \\
0 & 0 & 0 & 0 & -b & -c & -d & -e
\end{bmatrix}.
$$

There is no computational cost for P_C as it only permutes the input data set (just needs rewiring). $\tilde{I}$ can be further decomposed into two 4×4 submatrices, $\tilde{I}_{00}$ and $\tilde{I}_{11}$, by the direct sum operation ("$\oplus$") as shown below:

$$
\tilde{I} = \tilde{I}_{00} \oplus \tilde{I}_{11}.
$$

(5)

Thus,

$$
I = P_0 \cdot \left(\tilde{I}_{00} \oplus \tilde{I}_{11} \right) \cdot P_C,
$$

(6)

where

$$
\tilde{I}_{00} = \begin{bmatrix}
a & f & a & g \\
a & g & -a & -f \\
a & -g & -a & f \\
a & -f & a & -g
\end{bmatrix},
$$

$$
\tilde{I}_{11} = \begin{bmatrix}
-e & d & -c & b \\
-d & b & -e & -c \\
-c & e & b & d \\
-b & -c & -d & -e
\end{bmatrix}.
$$

(7)

Equation (6) forms the general expression of (1). We will use $\tilde{I}_{00}$ and $\tilde{I}_{11}$ as the basic building blocks to compute other 8×8 IDCTs. Since, the coefficients in $\tilde{I}_{00}$ and $\tilde{I}_{11}$ are fixed, they can be independently implemented, enabling fast computation.

In the following section, we show how (6) can be applied to different IDCT matrices. Another new feature of the proposed scheme is that we take the advantage of the similarity in matrix operation to further optimize the implementation. First of all, we apply (6) to efficiently implement the transformation matrix of AVS. Based on it and the generalized structure, we develop the matrix of VC-1 so that we can share as many units (from AVS) as possible. Next, we develop the IDCT matrix of H.264 based on the same principle (decompose and share from AVS and VC-1). In this stage, we are able to achieve the maximum sharing as it will be shown later (in Section 3.4) that the implementation of H.264 does not cost any extra hardware. Finally, we develop the IDCT of HEVC by further decomposing and reusing the units already implemented (with a minimum addition of extra units).

3.2. Matrix Decomposition for AVS.

Let us now construct A (from (1) and Table 1) and apply (6) to compute the 4×4 submatrices, $\tilde{A}_{00}$ and $\tilde{A}_{11}$. We then right shift $\tilde{A}_{00}$ by three bits and decompose it as follows:

$$
\frac{\tilde{A}_{00}}{8} = \begin{bmatrix}
1 & \dfrac{5}{4} & 1 & \dfrac{1}{2} \\
1 & \dfrac{1}{2} & -1 & -\dfrac{5}{4} \\
1 & -\dfrac{1}{2} & -1 & \dfrac{5}{4} \\
1 & -\dfrac{5}{4} & 1 & -\dfrac{1}{2}
\end{bmatrix} = A_1 \cdot A_2,
$$

(8)

where

$$
A_1 = \begin{bmatrix}
1 & 0 & 1 & 0 \\
0 & 1 & 0 & 1 \\
0 & 1 & 0 & -1 \\
1 & 0 & -1 & 0
\end{bmatrix}, \quad
A_2 = \begin{bmatrix}
1 & 0 & 1 & 0 \\
1 & 0 & -1 & 0 \\
0 & \dfrac{5}{4} & 0 & \dfrac{1}{2} \\
0 & \dfrac{1}{2} & 0 & -\dfrac{5}{4}
\end{bmatrix}.
$$

(9)

Like P_0, the computational cost of A_1 is only 4 additions. For A_2, we implement $(5/4) \cdot x$ as $(1 + 1/4) \cdot x$—that is right shift x (arbitrary data) by two bits and then add with x. So, the cost is 6 add and 6 shift operations. Thus in (8), the total computational cost is 10 addition and 6 shift operations. In similar way, we can decompose $\tilde{A}_{11}$ as shown below:

$$\frac{\tilde{A}_{11}}{4} = \begin{bmatrix} -\frac{1}{2} & \frac{3}{2} & -\frac{9}{4} & \frac{5}{2} \\ -\frac{3}{2} & \frac{5}{2} & -\frac{1}{2} & -\frac{9}{4} \\ -\frac{9}{4} & \frac{1}{2} & \frac{5}{2} & \frac{3}{2} \\ -\frac{5}{2} & -\frac{9}{4} & -\frac{3}{2} & -\frac{1}{2} \end{bmatrix} = A_3 \cdot A_4, \quad (10)$$

where

$$A_3 = \begin{bmatrix} -1 & 0 & \frac{3}{2} & 1 \\ 0 & 1 & 1 & -\frac{3}{2} \\ -\frac{3}{2} & 1 & -1 & 0 \\ -1 & -\frac{3}{2} & 0 & -1 \end{bmatrix}, \quad A_4 = \begin{bmatrix} \frac{3}{2} & 0 & 0 & -1 \\ 0 & \frac{3}{2} & 1 & 0 \\ 0 & 1 & -\frac{3}{2} & 0 \\ 1 & 0 & 0 & \frac{3}{2} \end{bmatrix}. \quad (11)$$

For both A_3 and A_4, the coefficient (3/2) can be shared and the cost is: 12 additions and 4 shift operations for A_3; 8 additions and 4 shift operations for A_4. From (8)–(10), we can summarize the final expression of the 8×8 IDCT for AVS as:

$$A = 4 \cdot P_0 \cdot [(A_1 \cdot 2A_2) \oplus (A_3 \cdot A_4)] \cdot P_C. \quad (12)$$

Thus, the total computational cost to implement A is 38 additions and 26 shift operations. In the next section, we will apply (6) to VC-1 and subsequently decompose the matrix in a way so that we can reuse the units already developed for the AVS (from (12)).

3.3. Matrix Decomposition for VC-1.

We follow the same principles, as discussed in (8) and (10), to decompose the 8×8 IDCT for the VC-1:

$$V = P_0 \cdot \tilde{V} \cdot P_C, \quad (13)$$

where

$$\tilde{V} = \tilde{V}_{00} \oplus \tilde{V}_{11}. \quad (14)$$

Now considering the symmetric property and the coefficient distribution patterns between $\tilde{A}_{00}/8$ (in (8)) and $\tilde{V}_{00}/8$, we decompose $\tilde{V}_{00}/8$ as:

$$\frac{\tilde{V}_{00}}{8} = \begin{bmatrix} \frac{3}{2} & 2 & \frac{3}{2} & \frac{3}{4} \\ \frac{3}{2} & \frac{3}{4} & -\frac{3}{2} & -2 \\ \frac{3}{2} & -\frac{3}{4} & -\frac{3}{2} & 2 \\ \frac{3}{2} & -2 & \frac{3}{2} & -\frac{3}{4} \end{bmatrix} = A_1 \cdot V_2, \quad (15)$$

where

$$V_2 = \begin{bmatrix} \frac{3}{2} & 0 & \frac{3}{2} & 0 \\ \frac{3}{2} & 0 & -\frac{3}{2} & 0 \\ 0 & 2 & 0 & \frac{3}{4} \\ 0 & \frac{3}{4} & 0 & -2 \end{bmatrix} = 2A_2 - V_3, \quad (16)$$

$$V_3 = \begin{bmatrix} \frac{1}{2} & 0 & \frac{1}{2} & 0 \\ \frac{1}{2} & 0 & -\frac{1}{2} & 0 \\ 0 & \frac{1}{2} & 0 & \frac{1}{4} \\ 0 & \frac{1}{4} & 0 & -\frac{1}{2} \end{bmatrix}.$$

From (16), (15) can be reexpressed as:

$$\frac{\tilde{V}_{00}}{8} = A_1 \cdot (2A_2 - V_3). \quad (17)$$

Now it can be seen how the implementation of AVS matrix (from (12)) can be reused in (17). This matrix decomposition enables hardware sharing and results in significant saving in implementation resources. From (17), the total cost of V_3 and $\tilde{V}_{00}/8$ is 8 additions and 6 shift operations.

Next based on our careful observation between the computational similarities between $\tilde{A}_{11}/4$ (in (10)) and $\tilde{V}_{11}/8$, we devise the decomposition scheme of $\tilde{V}_{11}/8$ as:

$$\frac{\tilde{V}_{11}}{8} = \begin{bmatrix} -\frac{1}{2} & \frac{9}{8} & -\frac{15}{8} & 2 \\ -\frac{9}{8} & 2 & -\frac{1}{2} & -\frac{15}{8} \\ -\frac{15}{8} & \frac{1}{2} & 2 & \frac{9}{8} \\ -2 & -\frac{15}{8} & -\frac{9}{8} & -\frac{1}{2} \end{bmatrix} = V_4 \cdot A_{4v}, \quad (18)$$

where $V_4 = A_3 + A_{3V}$,

$$A_{4v} = \begin{bmatrix} \frac{1}{4} & 0 & 0 & -1 \\ 0 & \frac{1}{4} & 1 & 0 \\ 0 & 1 & -\frac{1}{4} & 0 \\ 1 & 0 & 0 & \frac{1}{4} \end{bmatrix}, \quad A_{3v} = \begin{bmatrix} -1 & -\frac{3}{2} & 0 & -1 \\ \frac{3}{2} & -1 & 1 & 0 \\ 0 & 1 & 1 & -\frac{3}{2} \\ 1 & 0 & -\frac{3}{2} & -1 \end{bmatrix}. \quad (19)$$

By substituting (19) in (18), $\tilde{V}_{11}/8$ is expressed as:

$$\frac{\tilde{V}_{11}}{8} = (A_3 + A_{3V}) \cdot A_{4v}. \quad (20)$$

Note that A_{4v} in (19) is structurally similar to A_4 in (10) except the change in the diagonal coefficients. So we only need to implement it; the rest is shared from the architecture of A_4. We do so by adding 4 multiplexers at the output of the four left diagonal elements of A_4 matrix. Then according to (19), we reuse A_3 to compute V_4. As the new matrix A_{3v} can be derived from A_3 by rearranging the rows and changing the polarity of some input bits, we share it from the design of A_3 by adding 4 multiplexers only. Finally, the expression of $\tilde{V}_{00}$ and $\tilde{V}_{11}$ from (17) and (20) are substituted in (13) to get the final expression of the IDCT for VC-1:

$$V = 8 \cdot P_0 \cdot \{[A_1 \cdot (2A_2 - V_3)] \oplus [(A_3 + A_{3V}) \cdot A_{4v}]\} \cdot P_C. \tag{21}$$

It is seen from (21) that to implement V, the only new unit that is required is V_3; the rest is shared from the implementation of AVS (from (12)). So, the total computational cost for VC-1 is 12 additions and 10 shift operations.

3.4. Matrix Decomposition for H.264/AVC.

Following similar procedure illustrated in the two previous sections, we can simplify the 8×8 transformation matrix for H.264/AVC as shown below:

$$H = P_0 \cdot \tilde{H} \cdot P_C, \tag{22}$$

where

$$\tilde{H} = \tilde{H}_{00} \oplus \tilde{H}_{11}. \tag{23}$$

In order to ensure the maximum unit sharing, we decompose $\tilde{H}_{00}/8$ as below:

$$\frac{\tilde{H}_{00}}{8} = \begin{bmatrix} 1 & 1 & 1 & \frac{1}{2} \\ 1 & \frac{1}{2} & -1 & -1 \\ 1 & -\frac{1}{2} & -1 & 1 \\ 1 & -1 & 1 & -\frac{1}{2} \end{bmatrix} = A_1 \cdot A_{2h}, \tag{24}$$

where

$$A_{2h} = \begin{bmatrix} 1 & 0 & 1 & 0 \\ 1 & 0 & -1 & 0 \\ 0 & 1 & 0 & \frac{1}{2} \\ 0 & \frac{1}{2} & 0 & -1 \end{bmatrix}. \tag{25}$$

In (24), A_1 is directly reused from (12). To share A_{2h} from the architecture of A_2 we simply add two multiplexer units.

So there is no additional cost in terms of adders and shifters to compute $\tilde{H}_{00}/8$. Similarly, we can decompose $\tilde{H}_{11}/8$ as:

$$\frac{\tilde{H}_{11}}{8} = \begin{bmatrix} -\frac{3}{8} & \frac{3}{4} & -\frac{5}{4} & \frac{3}{2} \\ -\frac{3}{4} & \frac{3}{2} & -\frac{3}{8} & -\frac{5}{4} \\ \frac{5}{4} & \frac{3}{8} & \frac{3}{2} & \frac{3}{4} \\ -\frac{3}{2} & -\frac{5}{4} & -\frac{3}{4} & -\frac{3}{8} \end{bmatrix} = A_{3h} \cdot A_{4v}, \tag{26}$$

where

$$A_{3h} = \begin{bmatrix} -\frac{3}{2} & -1 & 1 & 0 \\ 1 & 0 & \frac{3}{2} & -1 \\ -1 & \frac{3}{2} & 0 & -1 \\ 0 & -1 & -1 & -\frac{3}{2} \end{bmatrix}. \tag{27}$$

Here A_{4v} is directly reused from (21) and we share A_{3h} from the architecture of A_3. In this sharing we do not even need to use any multiplexers, because we have already done so while sharing A_{3v} from A_3 in Section 3.3. The final expression of the 8×8 IDCT for H.264 (with all shared units) can be summarized as follows:

$$H = 8 \cdot P_0 \cdot \{[A_1 \cdot A_{2h}] \oplus [A_{3h} \cdot A_{4v}]\} \cdot P_C. \tag{28}$$

It is interesting to note that all terms in (28) are implemented from the terms of (12) and (21); thus, in the proposed scheme, there is no additional cost to implement the IDCT for H.264 which results in significant hardware savings.

3.5. Matrix Decomposition for HEVC.

In this section, we develop the transformation matrix for the HEVC based on the principles described before. The 8×8 matrix can be decomposed as:

$$HV = P_0 \cdot \widetilde{HV} \cdot P_C, \tag{29}$$

where

$$\widetilde{HV} = \widetilde{HV_{00}} \oplus \widetilde{HV_{11}}, \qquad (30)$$

$$\frac{\widetilde{HV_{00}}}{4} = \begin{bmatrix} 16 & \dfrac{83}{4} & 16 & 9 \\[2mm] 16 & 9 & -16 & -\dfrac{83}{4} \\[2mm] 16 & -9 & -16 & \dfrac{83}{4} \\[2mm] 16 & -\dfrac{83}{4} & 16 & -9 \end{bmatrix} = A_1 \cdot (16A_2 + HV_1),$$

$$HV_1 = \begin{bmatrix} 0 & 0 & 0 & 0 \\[1mm] 0 & 0 & 0 & 0 \\[1mm] 0 & \dfrac{3}{4} & 0 & 1 \\[2mm] 0 & 1 & 0 & -\dfrac{3}{4} \end{bmatrix}.$$

$$(31)$$

The computational cost of HV_1 is 4 additions and 4 shift operations. Here the coefficient $(3/4) \cdot x$ is factorized as $(x - x/4)$. So the cost of $\widetilde{HV_{00}}$ in (30) is 4 additions and 8 shift operations. Similarly, we decompose $\widetilde{HV_{11}}/4$ as:

$$\frac{\widetilde{HV_{11}}}{4} = \begin{bmatrix} -\dfrac{9}{2} & \dfrac{25}{2} & -\dfrac{75}{4} & \dfrac{89}{4} \\[2mm] -\dfrac{25}{2} & \dfrac{89}{4} & -\dfrac{9}{2} & -\dfrac{75}{4} \\[2mm] -\dfrac{75}{4} & \dfrac{9}{2} & \dfrac{89}{4} & \dfrac{25}{2} \\[2mm] -\dfrac{89}{4} & -\dfrac{75}{4} & -\dfrac{25}{2} & -\dfrac{9}{2} \end{bmatrix} = (8A_3 + HV_2) \cdot A_{4HV},$$

$$(32)$$

where

$$HV_2 = \begin{bmatrix} \dfrac{7}{4} & \dfrac{5}{4} & -2 & 0 \\[2mm] -\dfrac{5}{4} & 0 & -\dfrac{7}{4} & 2 \\[2mm] 2 & -\dfrac{7}{4} & 0 & \dfrac{5}{4} \\[2mm] 0 & 2 & \dfrac{5}{4} & \dfrac{7}{4} \end{bmatrix}, \qquad A_{4HV} = \begin{bmatrix} 2 & 0 & 0 & -1 \\ 0 & 2 & 1 & 0 \\ 0 & 1 & -2 & 0 \\ 1 & 0 & 0 & 2 \end{bmatrix}.$$

$$(33)$$

Combining (29)–(32), we compute the proposed 8×8 IDCT for HEVC as given below:

$$HV = 4 \cdot P_0 \cdot \{[A_1 \cdot (16A_2 + HV_1)]$$
$$\oplus [(8A_3 + HV_2) \cdot A_{4HV}]\} \cdot P_C. \qquad (34)$$

In (34), only the new matrices, HV_1 and HV_2, will be implemented and the rest will be shared from (12). So the

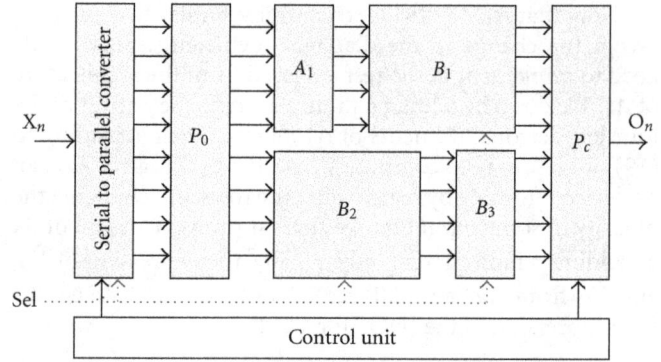

FIGURE 1: Block diagram of the proposed architecture.

total computational cost to implement HV in the proposed design is 24 additions and 28 shift operations. It is important to note that we have carefully decomposed HV so that if there is any change in the final standard, all one needs to do is to update (30) and (32) with new parameters without interrupting the entire design. In summery, the proposed unified design costs 74 additions and 64 shift operations to perform the inverse transformation of four defined video standards.

4. Hardware Implementation of the Shared Architecture

In the implementation of the multistandard architectures on a single platform, we have shared the entire hardware unit of the 4×4 matrices, instead of sharing individual adders, shifters, or other factors (as done in [10]). It ensures maximum reduction of hardware cost in our design. The overall block diagram of our proposed scheme is shown in Figure 1. We can see from Figure 1 that the P_0 block splits the 8-point decomposition process to two independent 4-point processes; since these two processes work concurrently, the design throughput is highly increased. The blocks B_1, B_2, and B_3 perform different operations (shared) as shown in Table 2.

Figure 2(a) shows the design of the serial to parallel converter (S2P) block. It performs left shift and then stores the input one by one into eight registers in 8 clock cycles, and at the 9th cycle, all stored input samples are sent to next block, P_0. Here the S2P block apparently functions like a temporary memory buffer as it stores the rows of the input matrix inside eight registers. As a result, the proposed design does not require additional memory architecture. The wrapper architecture (P_C) is shown in Figure 2(b). In this multicodec system, only one IDCT and its associated computational units are activated at a time by the control unit and the select pin (Sel); the rest is disabled. The other blocks are shown in Figure 3. In different stages of the design, several multiplexers are used to ensure proper computation of the IDCT in operation. Finally, the P_C block combines two different set of data and generates one output. In Figure 3, In_0, In_1, ..., In_3 represent the inputs coming from the previous block and Out_0, Out_1, ..., Out_3 represent the outputs going to the next block. As an example, in

TABLE 2: Controlling selection of subblocks.

Select (Sel)	IDCT	B_1 $(16/2) \cdot (A_2/A_{2h}) - (V_3/HV_1)$	B_2 $(8/1) \cdot [A_3/A_{3h}/(A_3 + A_{3v})] + HV_2$	B_3 $A_4/A_{4v}/A_{4HV}$
00	AVS	$2A_2$	A_3	A_4
01	VC-1	$2A_2 - V_3$	$(A_3 + A_{3v})$	A_{4v}
10	H.264	A_{2h}	A_{3h}	A_{4v}
11	HEVC	$16A_2 - HV_1$	$8A_3 + HV_2$	A_{4HV}

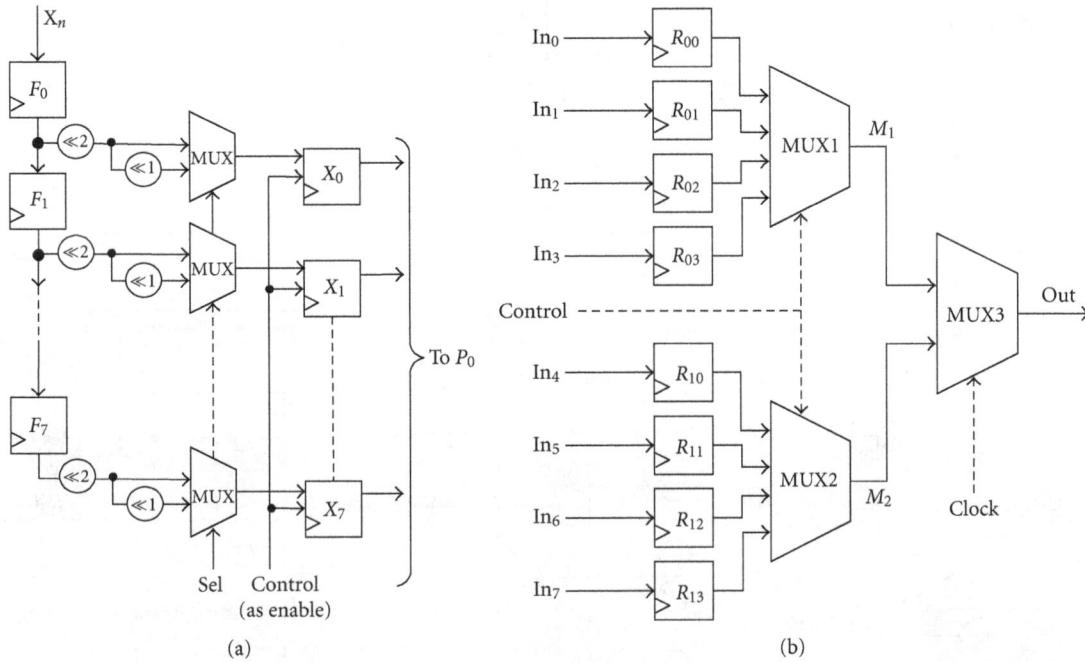

FIGURE 2: (a) Serial to parallel converter with shifting; (b) output wrapper (P_C block).

TABLE 3: Comparison of the cost of adders and shifters.

Codecs	Number of adders	Number of shifters
JPEG + MPEG-2/4 + H.264 + VC-1 in [8]	112	—
JPEG + MPEG-2/4 + H.264 + VC-1 in [10]	70	—
H.264 + VC-1 in [14]	76	28
MPEG-2/4 + H.264 + VC-1 + AVS in [13]	76	—
JPEG + MPEG-2/4 + H.264 + VC-1 + AVS in [12]	58	31
Proposed—H.264 + VC-1 + AVS + HEVC	74	64

Figure 3(c) for the shared design of V_3/HV_1, the inputs are coming from A_2/A_{2h} subblock and the outputs are going to P_C block.

The state diagram of the control unit is shown in Figure 4. Here, "r" is reset and "c" is a 3-bit internal counter run by the system clock. There are one reset and four active states. The states of the control signals are also shown in the diagram; for example, in state 1 (S1), S2P is storing the input vector while the output wrapper (P_C block) enables R_{00} from MUX1 and R_{10} from MUX2. Table 2 shows the units that are active depending on the status of the select pin. For example, the select signal will be "00" when the user wants to perform the IDCT of AVS codec. In that case, B_1, B_2, and B_3 will function as $2 \cdot A_2$, A_3, and A_4, respectively (the rest is inactive as found in (12)).

5. Performance Analysis and Comparisons

The proposed design is implemented in Verilog and its operation is verified using Xilinx Vertex4 LX60 FPGA. The total number of LUTs needed for this proposed architecture is 2,242. The design is later synthesized using 0.18 μm CMOS technology. The architecture costs 39.3 K gates and 12.15 K standard cells with a maximum operating frequency of 200.8 MHz. The estimated power consumption is 29.9 mW with 3 V supply.

In order to demonstrate the sharing efficiency, we have compared the adder count of our design with the 8-point standalone IDCT matrices of three standards: AVS, VC-1, and H.264/AVC (as presented in [12]). The results are shown in Figure 5. As of today, there is no implementation of the 8 × 8 IDCT of HEVC; thus, we have implemented it separately for the sake of better comparison. Now, we

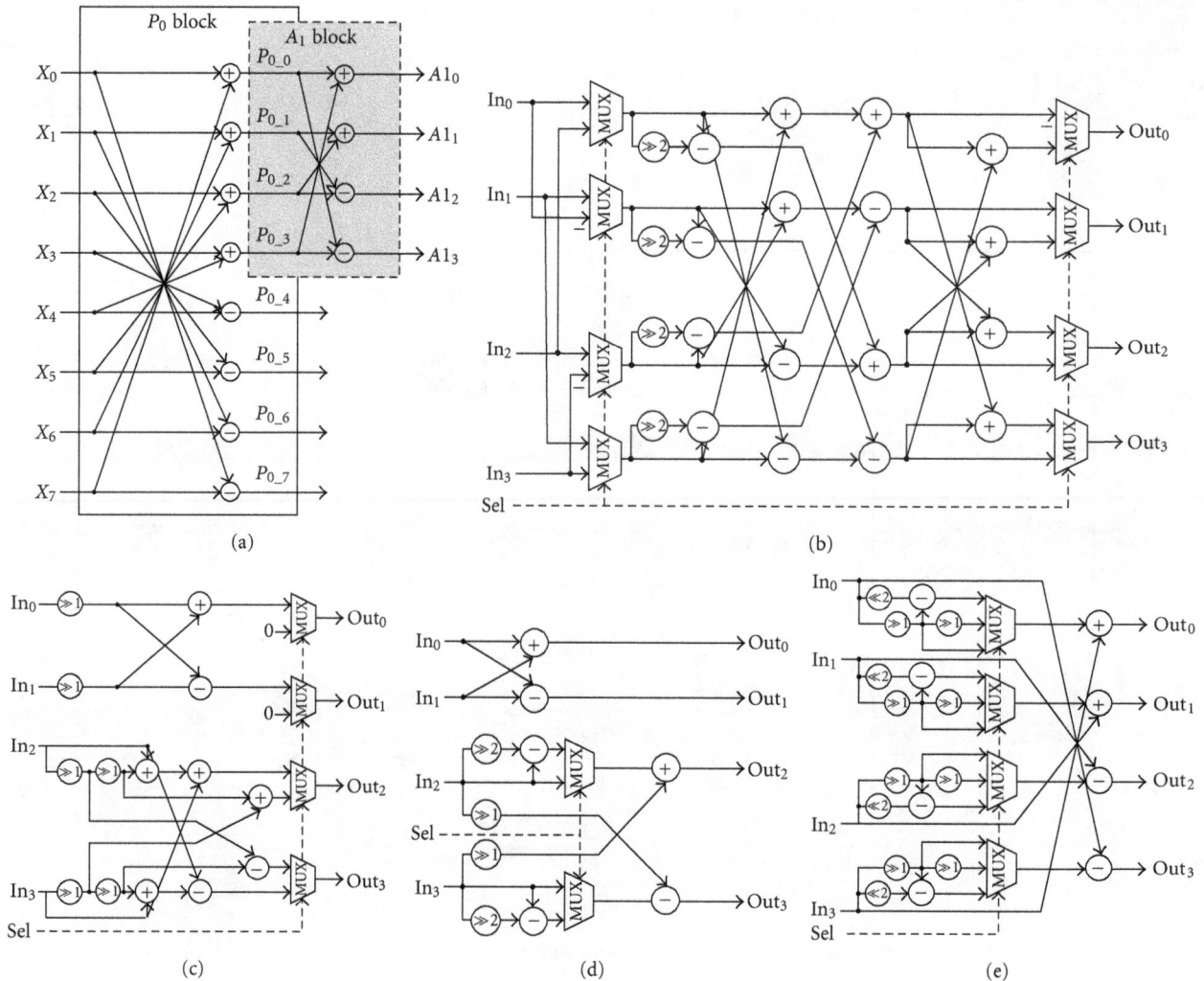

FIGURE 3: Shared architecture of (a) P_0 and A_1; (b) A_3, A_{3h} and $(A_3 + A_{3v})$; (c) V_3 and HV_1; (d) A_2 and A_{2h}; (e) A_4, A_{4v} and A_{4HV}.

TABLE 4: Comparison of the resource-shared 8-point 1-D IDCT architecture.

Scheme	Tech.	Gate count	Freq. (MHz)	Full HD support?	Super Resolution support?	Supporting standards			
						H.264	VC-1	AVS	HEVC
Lee's [8]	0.13 μm	19.1 K	136	Y	o	Y	Y	o	o
Kim's [9]	0.13 μm	30.9 K	151	o	o	Y	Y	o	o
Qi's [10]	0.13 μm	18 K	100	Y	o	Y	Y	o	o
Lee's [11]	0.13 μm	10.5 K	123	o	o	Y	Y	o	o
Wahid's [12]	0.18 μm	19.8 K	194.7	Y	o	Y	Y	Y	o
Liu's [13]	0.13 μm	16.5 K	110.8	—	—	Y	Y	Y	o
Fan's [14]	0.18 μm	7.14 K	100	—	—	Y	Y	o	o
Li's [19]	0.18 μm	13.7 K	200	Y	o	Y	o	o	o
Proposed	0.18 μm	39.3 K	200.8	Y	Y	Y	Y	Y	Y

"Y": yes; "o": No; "—": no information.

can see from Figure 5 that a total of 104 adders is required to implement these four transforms without sharing. The proposed shared design can compute all of them with 28.9% less adders. Moreover, the savings achieved in individual standards due to the sharing are also marked on the figure.

It is important to note that, though the proposed design costs 38 adders to implement AVS, it does not cost any additional adder units to implement H.264. Hence, AVS and H.264 combined together cost only 38 adders (compared to 48 for standalone implementations). The cost

$r = 0$

S0 (reset)
S2P: initializing registers with 0;
P_c: initializing registers with 0;
$c = 0$

$r = 0$ $r = 1$ and $c = 0$ $r = 0$

$r = 1$ and $c = 7$ **S4**
S2P: store input, generate output;
P_c: $M_1 = R_{03}, M_2 = R_{13}$

$r = 1$ and $c \neq 0$ **S1**
S2P: store input;
P_c: $M_1 = R_{00}, M_2 = R_{10}$ $r = 1$ and $c = 1$

$r = 0$ $r = 1$ and $c = 6$ $r = 0$ $r = 0$ $r = 1$ and $c = 2$

$r = 1$ and $c = 5$ **S3**
S2P: store input;
P_c: $M_1 = R_{02}, M_2 = R_{13}$ $r = 1$ and $c = 4$ **S2**
S2P: store input;
P_c: $M_1 = R_{01}, M_2 = R_{11}$ $r = 1$ and $c = 3$

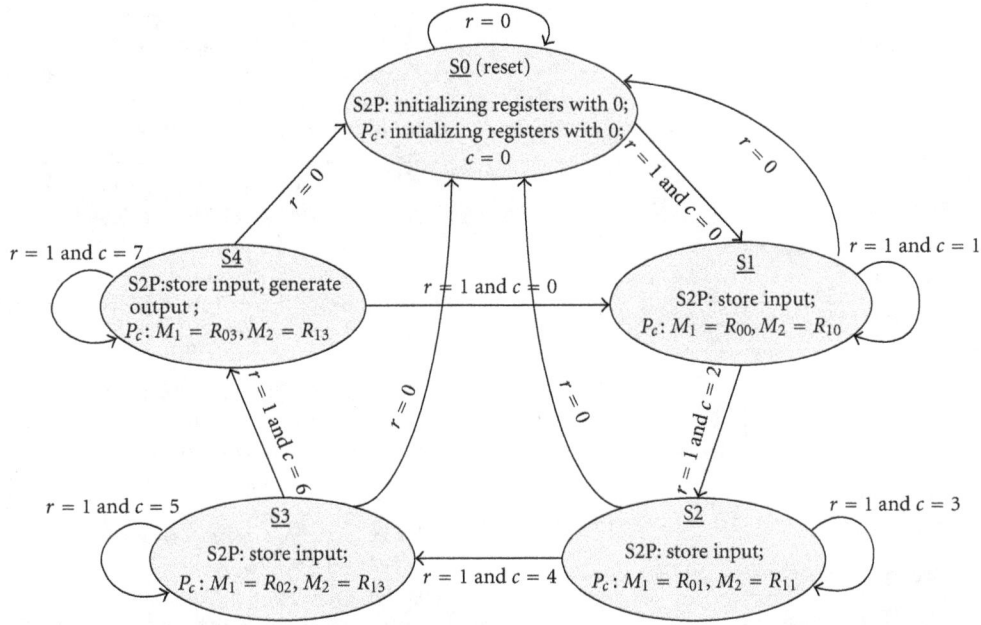

FIGURE 4: State diagram of control unit.

TABLE 5: Comparison of decoding capability (minimum support of three codecs).

| | HD resolution | | | | Super resolution | |
| Scheme | 1920×1080 | | 1280×720 | | 2560×1600 | |
	Time to transmit 1 frame (msec)	Frame per second (fps)	Time to transmit 1 frame (msec)	Frame per second (fps)	Time to transmit 1 frame (msec)	Frame per second (fps)
Lee's [8]	22.8	44	10.2	98	x	x
Kim's[9]	x	x	10.2	98	x	x
Qi's [10]	31.1	32	13.8	72	x	x
Wahid's [12]	16.7	60	7.1	140	x	x
Proposed	15.5	64	6.9	145	30.6	32

"—": no information; "x": not supported by the hardware.

of implementing shift operation is considered insignificant in the computation. In Table 3, we compare the cost of the proposed scheme with available existing designs in the literature. None of the designs in this table supports HEVC (which is computationally expensive due to large matrix parameters as shown in Table 1). Although, the designs in [10, 12] cost fewer adders, it is shown later that the proposed scheme outperforms it in decoding capacity. Considering the fact that, the proposed architecture can decode the IDCT for four video codecs, it consumes the least number of adders compared to others.

In Table 4, we have summarized the performance in terms of gate count, maximum working frequency, and standard support with other designs. Only the design in [19] has frequency closer to us, but it supports only H.264. Similarly, designs in [11, 14] support only two codecs and accordingly cost lesser hardware than ours. Among other designs [8–10, 12, 13] are comparable to our design as they support as many as three codecs. While working at

maximum capacity, the proposed design can process 200.8 million pixels/sec.

In order to have a better assessment among comparable designs (e.g., minimum support of three codecs), in Table 5 we compare the decoding capability (using $4:2:0$ luma-chroma sampling) of the proposed approach with that of [8–10, 12]. In our work, the maximum achieved frame rate of a 1080 p video is $= 200.8 \times 10^6/(1920 \times 1080 + 2 \times 960 \times 540) = 64.56 \approx 64$ fps, which is the highest compared to all other designs in Table 5. Considering the current trends to use super resolution monitors, in this table we have also compared the decoding capabilities for the Wide Quad eXtended Graphics Array (WQXGA, with resolution of 2560 $\times$ 1600 pixels). Thus, it can be seen that the proposed design cannot only decode AVS, H.264/AVC, VC-1, and HEVC videos, but also can maintain relatively higher operational frequency to meet the requirements of real time transmission (the target fps to transmit HD, full HD, and QWXGA video are 120, 60, and 30, resp.). From the performance analysis,

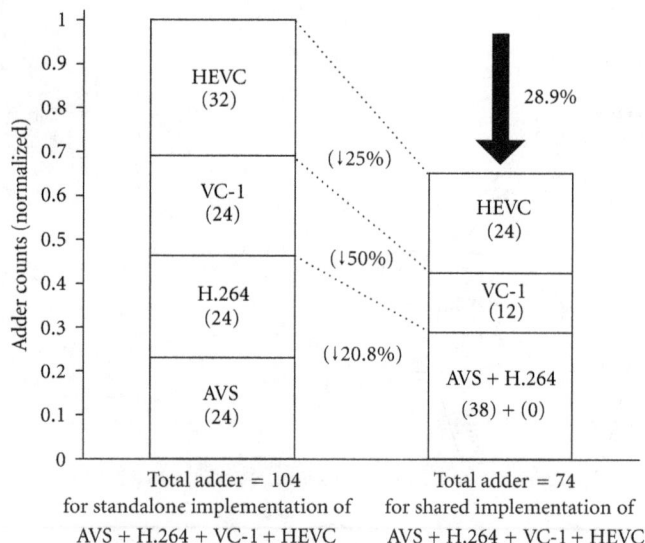

FIGURE 5: Cost of the proposed scheme—standalone versus cost-shared.

the scheme is found to be competitive as it can transmit the highest number of frames per seconds and, hence, takes the least time to transmit one frame at a given resolution.

6. Conclusion

In this paper, we present a generalized algorithm and a hardware-shared architecture by using the symmetric property of the integer matrices and the matrix decomposition to compute the 8-point 1-D IDCT for four modern video codecs: H.264/AVC, VC-1, AVS, and HEVC (draft in stage). The architecture is designed in such a way that can accommodate any change in the final release of the HEVC. We first apply the generalized scheme to AVS-based transform unit, and then gradually build the rest of the transform units on top of another to maximize the sharing. The performance analysis shows that the proposed design satisfies the requirement of all four codecs and achieves the highest decoding capability. Overall, the architecture is suitable for low-cost implementation in modern multicodec systems.

Acknowledgment

The authors would like to acknowledge the Natural Science and Engineering Research Council of Canada (NSERC) for its support to this research paper.

References

[1] ITU-T Rec, "H.264/ISO/IEC 14496-10 AVC," 2003.

[2] "Standard for Television: VC-1 Compressed Video Bitstream Format and Decoding Process," SMPTE 421M, 2006.

[3] GB/T 20090.1, "Information technology - Advanced coding of audio and video – Part 1: System," Chinese AVS standard.

[4] G. J. Sullivan and J.-R. Ohm, "Recent developments in standardization of high efficiency video coding (HEVC)," in

Applications of Digital Image Processing XXXIII, vol. 7798 of *Proceedings of SPIE*, August 2010.

[5] K. Ugur, K. Andersson, A. Fuldseth et al., "High performance, low complexity video coding and the emerging hevc standard," *IEEE Transactions on Circuits and Systems for Video Technology*, vol. 20, no. 12, pp. 1688–1697, 2010.

[6] C. C. Ju, Y. C. Chang, C. Y. Cheng et al., "A full-HD 60fps AVS/H.264/VC-1/MPEG-2 video decoder for digital home applications," in *International Symposium on VLSI Design, Automation and Test (VLSI-DAT '11)*, pp. 117–120, April 2011.

[7] Joint Collaborative Team – Video Coding, *CE10: Core transform design for HEVC*, JCTVC-G495, Geneva, Switzerland, 2011.

[8] S. Lee and K. Cho, "Architecture of transform circuit for video decoder supporting multiple standards," *Electronics Letters*, vol. 44, no. 4, pp. 274–276, 2008.

[9] S. Kim, H. Chang, S. Lee, and K. Cho, "VLSI design to unify IDCT and IQ circuit for multistandard video decoder," in *12th International Symposium on Integrated Circuits (ISIC '09)*, pp. 328–331, December 2009.

[10] H. Qi, Q. Huang, and W. Gao, "A low-cost very large scale integration architecture for multistandard inverse transform," *IEEE Transactions on Circuits and Systems II*, vol. 57, no. 7, pp. 551–555, 2010.

[11] S. Lee and K. Cho, "Circuit implementation for transform and quantization operations of H.264/MPEG-4/VC-1 video decoder," in *International Conference on Design and Technology of Integrated Systems in Nanoscale Era (DTIS '07)*, pp. 102–107, September 2007.

[12] K. A. Wahid, M. Martuza, M. Das, and C. McCrosky, "Efficient hardware implementation of 8 × 8 integer cosine transforms for multiple video codecs," *Journal of Real-Time Image Processing*. In press.

[13] G. Liu, "An area-efficient IDCT architecture for multiple video standards," in *2nd International Conference on Information Science and Engineering (ICISE '10)*, pp. 3518–3522, December 2010.

[14] C. P. Fan and G. A. Su, "Efficient low-cost sharing design of fast 1-D inverse integer transform algorithms for H.264/AVC and VC-1," *IEEE Signal Processing Letters*, vol. 15, pp. 926–929, 2008.

[15] C. Fan and G. Su, "Fast algorithm and low-cost hardware-sharing design of multiple integer transforms for VC-1," *IEEE Transactions on Circuits and Systems II*, vol. 56, pp. 788–792, 2009.

[16] D. Zhou, Z. You, J. Zhu et al., "A 1080p@60fps multi-standard video decoder chip designed for power and cost efficiency in a system perspective," in *Symposium on VLSI Circuits*, pp. 262–263, June 2009.

[17] C. P. Fan and Y. L. Lin, "Implementations of low-cost hardware sharing architectures for fast 8 × 8 and 4 × 4 integer transforms in H.264/AVC," *IEICE Transactions on Fundamentals of Electronics, Communications and Computer Sciences*, vol. 90, no. 2, pp. 511–516, 2007.

[18] Y. C. Chao, S. T. Wei, C. H. Kao, B. D. Liu, and J. F. Yang, "An efficient architecture of multiple 8×8 transforms for H.264/AVC and VC-1 decoders," in *1st International Conference on Green Circuits and Systems (ICGCS '10)*, pp. 595–598, June 2010.

[19] Y. Li, Y. He, and S. Mei, "A highly parallel joint VLSI architecture for transforms in H.264/AVC," *Journal of Signal Processing Systems*, vol. 50, no. 1, pp. 19–32, 2008.

9T Full Adder Design in Subthreshold Region

Shiwani Singh,[1] Tripti Sharma,[2] K. G. Sharma,[2] and B. P. Singh[1]

[1] *Faculty of Engineering & Technology, MITS (Deemed University), Lakshmangarh 332311, India*
[2] *Department of Electronics & Communication, Suresh Gyan Vihar University, Jaipur, India*

Correspondence should be addressed to Tripti Sharma, tripsha@gmail.com

Academic Editor: Jose Silva-Martinez

This paper presents prelayout simulations of two existing 9T and new proposed 9T full adder circuit in subthreshold region to employ in ultralow-power applications. The proposed circuit consists of a new logic which is used to implement Sum module. The proposed design remarkably reduces power-delay product (PDP) and improves temperature sustainability when compared with existing 9T adders. Therefore, in a nut shell proposed adder cell outperforms the existing adders in subthreshold region and proves to be a viable option for ultralow-power and energy-efficient applications. All simulations are performed on 45 nm standard model on Tanner EDA tool version 13.0.

1. Introduction

Advances in CMOS technology have led to a renewed interest in the design of basic functional units for digital systems. The use of integrated circuits in high-performance computing, telecommunications, and consumer electronics has been growing at a very fast pace. This trend is expected to continue, with very important implications for power-efficient VLSI and systems designs.

Digital integrated circuits commonly use CMOS circuits as building blocks. The continuing decrease in feature size of CMOS circuits and corresponding increase in chip density and operating frequency have made power consumption a major concern in VLSI design. Excessive power dissipation in integrated circuits not only discourages their use in portable environment but also causes overheating which reduces chip life and degrades performance.

Computations in these devices need to be performed using low-power, area efficient circuits operating at greater speed. The design of high-speed and low-power VLSI architectures needs efficient arithmetic processing units, which are optimized for the performance parameters, namely, speed and power consumption [1, 2].

Addition is one of the widely used fundamental arithmetic operations. In addition to its main task, which is adding two binary numbers, it is the nucleus of many other useful operations such as subtraction, multiplication, and division. Full adder is an essential component for designing all types of processors, namely, digital signal processors (DSP), microprocessors, and so forth.

In most of the digital systems adder lies in the critical path that affects the overall speed of the system.

It is very important to choose the adder topology that would yield the desired performance. So enhancing the performance of the 1-bit full adder cell is the main design aspect.

One way to achieve ultralow is by running digital circuits in subthreshold mode [1, 2]. Subthreshold current of an MOSFET transistor occurs when the gate-to-source voltage (V_{GS}) of a transistor is lower than its threshold voltage (V_{TH}). When V_{GS} is larger than V_{TH}, majority carriers are repelled from the gate area of the transistor and a minority carrier channel is created. This is known as strong-inversion, as more minority carriers are present in the channel than majority carriers. When V_{GS} is lower than V_{TH}, there are less minority carriers in the channel, but their presence comprises a current and the state is known as weak-inversion. In standard CMOS design, this current is a subthreshold parasitic leakage, but if the supply voltage (V_{DD}) is lowered below V_{TH}, the circuit can be operated using the subthreshold current with ultralow-power consumption.

The proposed circuit operates efficiently in subthreshold region to achieve ultralow power. Results show improvement in temperature sustainability and PDP over the other adders with comparable performance. The rest of the paper is organized as follows. Section 2 briefly describes the previous work reported in the literature. Proposed 9T Adder cell is described in Section 3. Section 4, presents the simulation results and conclusions are drawn in Section 5.

2. Prior Work

The full adder operation can be stated as follows. Given the three 1-bit inputs A, B, and Cin, it is desired to calculate the two 1-bit outputs Sum and Cout, where

$$\text{Sum} = A \oplus B \oplus \text{Cin}, \tag{1}$$

$$\text{Cout} = A \cdot B + \text{Cin}(A \oplus B). \tag{2}$$

The circuit shown in Figure 1 [3–5] is the schematic of modified 8T full adder cell [6–9] using an extra transistor M9 to improve the performance of the 8T full adder cell. The Sum output is basically obtained by a cascaded exclusive ORing of the three inputs in addition to an extra transistor M9. Cout is implemented using 2T multiplexer. 8T full adder is confronted with problems for certain input vectors. This problem is eliminated in the design of Figure 1 by adding an extra transistor M9. Although it has area overhead of one transistor, but still its power consumption is reduced than the 8T adder circuit. The outputs have good logic level only for certain input vectors. For the remaining input vectors, there is a major degradation in output voltage that may lead to functional failure as well as increased power consumption at higher voltages. An extra added transistor M9 remains ON for 010 and 100 input combinations also but does not contribute to produce Sum output and hence results into excess power consumption.

Figure 2 shows schematic of another 9T full adder design [10] reported in literature. In this circuit, a three-transistor XOR gate [6–9] and a multiplexer are used to implement Sum and one multiplexer to implement the Cout. The selector circuit of the output multiplexers is output of first-stage XOR. This circuit shows nominal improvements in power when compared with adder of Figure 1 in subthreshold region. The reason for less power consumption than the circuit of Figure 1 is shown in Table 2.

3. Proposed 9T Full Adder Design

The schematic of proposed 9T full adder cell is shown in Figure 3 and its truth table is stated in Table 1. The operating principle of proposed circuit is different from traditional circuits. For generating the Sum output in the proposed design, the truth table has been divided into two parts, one for input A = "0" and another for A = "1" rather than implementing the conventional Sum module of (1). From the truth table shown in Table 1 it is evident that when A = "0", Sum can be produced by XORing inputs B and Cin. Similarly, when A = "1", Sum is showing the XNORing between inputs B and Cin. Therefore, the operation of Sum module is based on implementing XOR operation and XNOR operation between inputs B and Cin which is

TABLE 1: Truth table of 1-bit full adder.

A	B	Cin	Sum	Cout
0	0	0	0	0
0	0	1	1	0
0	1	0	1	0
0	1	1	0	1
1	0	0	1	0
1	0	1	0	1
1	1	0	0	1
1	1	1	1	1

indicated in (3) and (4). The logic for Cout output is stated in (5) and (6).

When $A = 0$,

$$\text{Sum} = B \oplus \text{Cin}. \tag{3}$$

When $A = 1$,

$$\text{Sum} = B \odot \text{Cin}. \tag{4}$$

For Cout, when $B \oplus \text{Cin} = 0$,

$$\text{Cout} = \text{Cin}. \tag{5}$$

When $B \odot \text{Cin} = 1$,

$$\text{Cout} = A. \tag{6}$$

An inverter is connected at the output of first-stage XOR gate to generate XNOR function. Finally the Sum is implemented by transferring these output levels through 2T multiplexer. Input to the PMOS (M6) of 2T multiplexer is XOR of B and Cin while to NMOS (M7) is XNOR of B and Cin. This 2T multiplexer is controlled by input A. Cout is implemented by using another 2T multiplexer which is controlled by output of first-stage XOR gate and passes either A or Cin accordingly. This circuit reduces the overall PDP at varying input voltages and operating frequencies and also improves the temperature sustainability while operating in subthreshold region. The most demanding design constraint for developing compact systems, that is, area, remains constant for all three designs.

Total power consumption in MOS logic circuits is expressed as sum of three components [2] as shown in (7):

$$P_{\text{Total}} = P_{\text{Switching}} + P_{\text{Sub}} + P_{\text{Short circuit}}, \tag{7}$$

where $P_{\text{switching}}$ denotes the average switching power consumption and is given by (8)

$$P_{\text{Switching}} = \alpha_T C_{\text{load}} V_{\text{DD}}^2 f. \tag{8}$$

Since no external load capacitance is connected in the circuit, hence total capacitance is only due to parasitic present in the design. Therefore at constant frequency and supply voltage, α_T (switching activity factor) will be the dominant factor which determines the total switching power of the circuit.

P_{Sub} denotes subthreshold power consumption and is given by (9)

$$P_{\text{Sub}} = V_{\text{DD}} \times I_{\text{Sub}}, \tag{9}$$

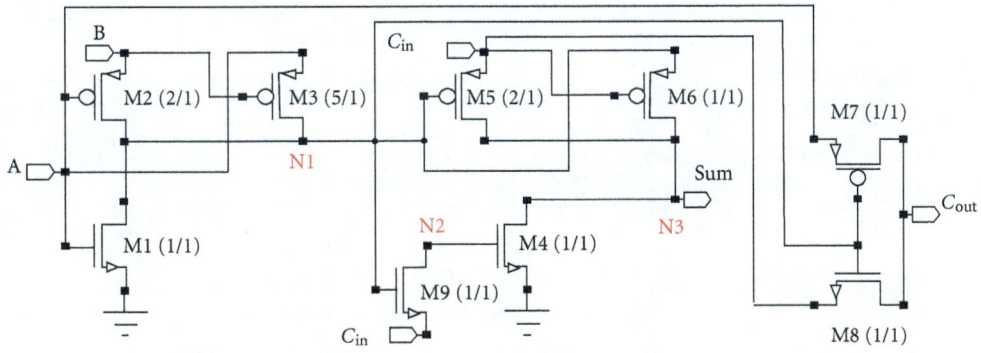

FIGURE 1: Existing 1-bit 9T full adder.

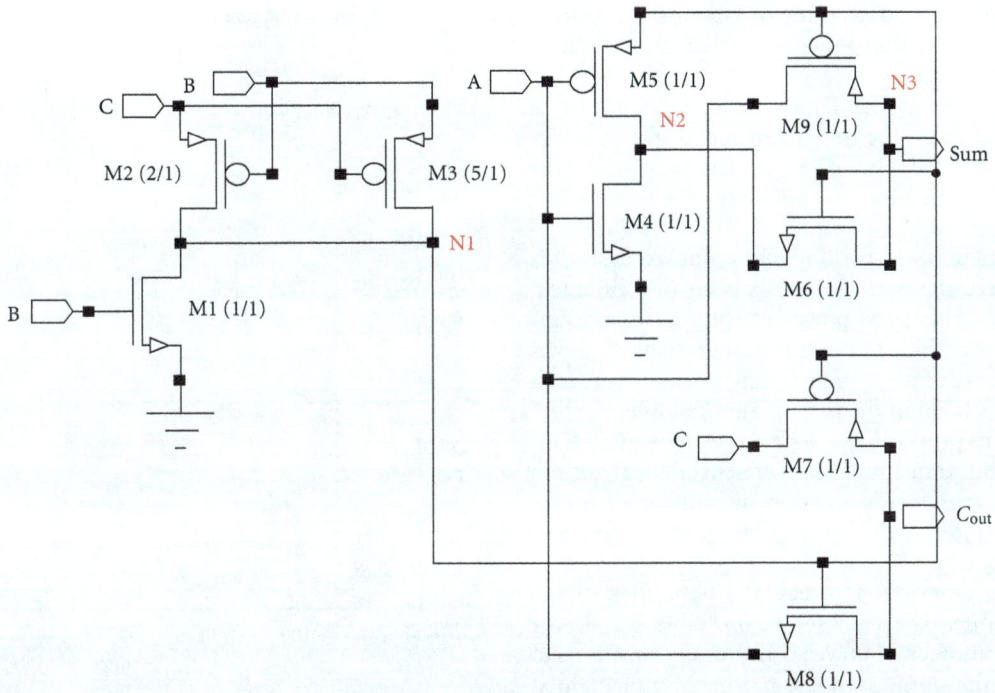

FIGURE 2: Existing 1-bit 9T full adder.

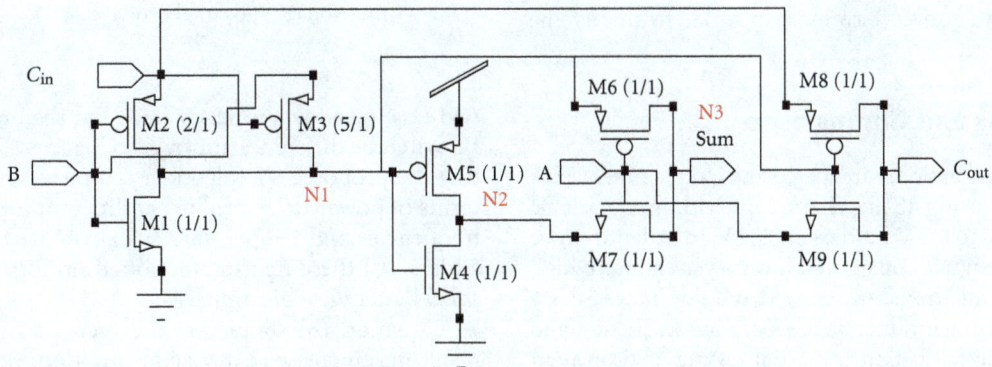

FIGURE 3: Proposed 9T full adder cell.

TABLE 2: No. of power consuming transitions at different internal nodes.

	Figure 1			Figure 2			Proposed 9T (Figure 3)	
N1	N2	N3	N1	N2	N3	N1	N2	N3
1	4	3	1	3	3	2	2	3
Total = 8			Total = 7			Total = 7		

where

$$I_{\text{Sub}} = K \times \exp\left[\frac{(V_{\text{GS}} - V_t)q}{\eta KT}\right],$$

$$\times \left[1 - \exp\left(V_{\text{DS}}\frac{q}{KT}\right)\right]. \qquad (10)$$

As shown in (10), the reduced value of V_{GS} in subthreshold mode decreases current exponentially and thus reduces subthreshold power.

$P_{\text{Short circuit}}$ is due to the large rise and fall times of input voltage. But in all the designs included in this paper the default rise and fall time of 1 nsec has been taken for simulation which results into negligible short circuit power consumption.

The difference between existing and proposed designs is at Node N2 where the main logic has been implemented. In the proposed design, the power consuming transitions at node N2 are less than the existing ones. Table 2 shows the effective number of power consuming transitions at each internal node which lies in the path of Sum module.

As $P_{\text{Switching}}$ is proportional to α_T and the number of power consuming transitions at different nodes decides the value of α_T, therefore, reduction in number of power consuming transitions will result in reduced power consumption.

The previous description reveals the basic idea of the proposed technique. The data shown in Table 2 makes it evident that the number of power consuming transitions at node N2 where the proposed logic has been implemented results into low-power full adder cell.

In a nutshell, the proposed 9T full adder proves itself to be a better option for low-power compact systems. All the substrate terminals in Figures 1, 2 and 3 are connected to their respective source terminals in order to nullify the substrate-bias effect.

4. Simulations and Comparison

All schematic simulations are performed on Tanner EDA tool version 13.0 using 45 nm technology with input voltage ranges from 0.2 V to 0.3 V in steps of 0.02 V. In order to prove that proposed design is consuming low power and have high performance, simulations are carried out for power-delay product at increasing input voltage, operating frequency, and temperature. To establish an impartial testing environment all circuits have been tested on the same input patterns which covers each and every combinations of the input stream. From Figures 4, 5, and 6 it is evident that the performance of

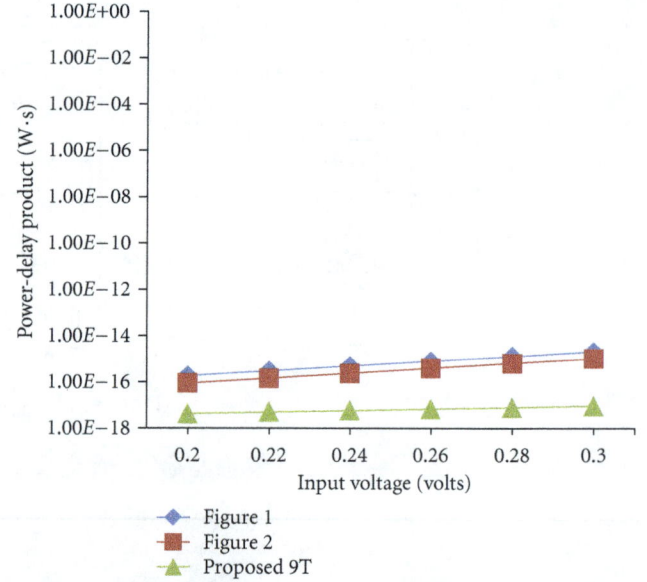

FIGURE 4: Power-delay product with increasing input voltage.

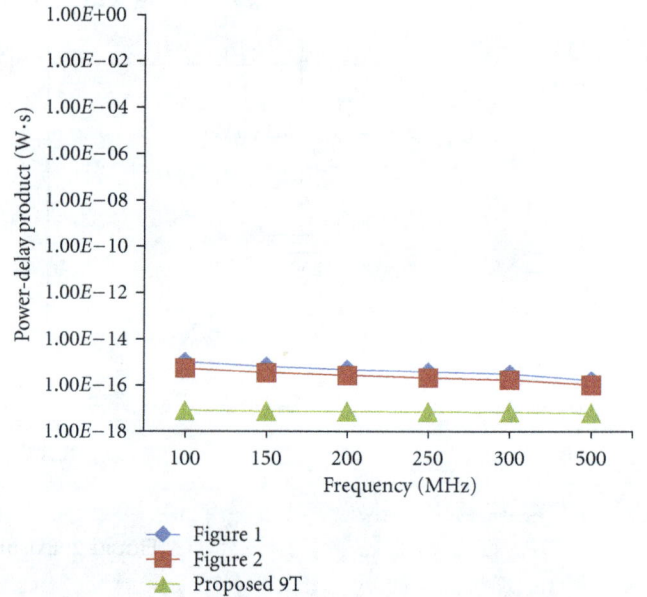

FIGURE 5: Power-delay product with increasing operating frequency at 0.3 V input voltage and supply voltage.

two existing adders is nearly same but the proposed adder is showing the difference at par. Also Figures 4, 5, and 6 reveals that the proposed 9T full adder cell proves its superiority in terms of power-delay product at different input voltages and frequencies and temperature sustainability over existing 9T adders. All these figures are plotted on logarithmic scale to show better view of comparison.

Equation (8) states that the increase in frequency will result into increased power consumption. But as power and delay are inversely proportional to each other, hence, the resulting PDP curve in Figure 5 is showing decreasing slope with increase in frequency.

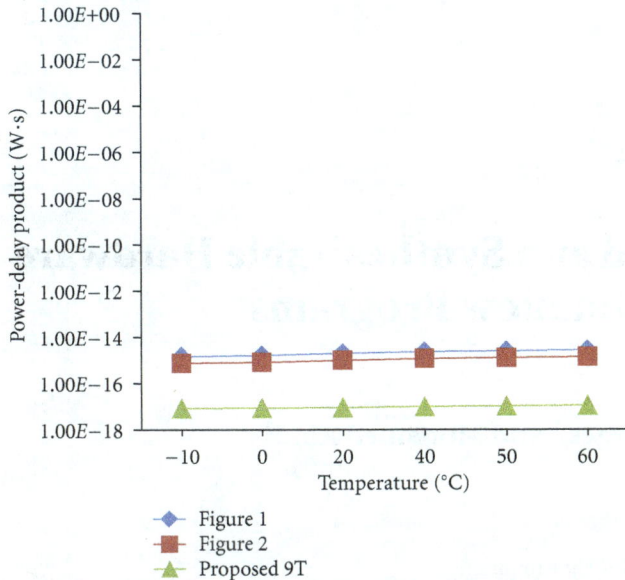

FIGURE 6: Power-delay product with varying temperature at 0.3 V input voltage and supply voltage.

The increase in temperature results into increase carrier mobility, and due to this random motion of electrons and holes will increase and hence more number of electron hole pairs will be formed in the interfacial region of the MOSFET thereby leading to delay in channel formation and increment in threshold voltage. This increment in threshold voltage will give rise to the power consumption of the device.

Also in a MOSFET, a pn junction is formed between the Drain/Source and Bulk of the transistor. The current equation of a pn-junction is expressed as

$$I = I_0 \left(e^{qV/KT} - 1 \right). \tag{11}$$

This states that with increase in temperature, current increases exponentially and results into more power consumption. This is the cause of slight rise in the curve of Figure 6 with increasing temperature.

5. Conclusion

Based on the subthreshold conduction region, the designing of a 1-bit full adder has been done. The proposed 9T 1-bit full adder is found to give better performance than both existing 9T full adders in subthreshold region. The proposed circuit has been tested to have better temperature sustainability and significantly less power delay product to achieve high performance. The proposed 9T adder has been designed and studied using 45 nm technology and can be a viable option for low-power complex circuit design.

References

[1] N. Weste and K. Eshraghian, *Principles of CMOS Digital Design, A System Perspective*, Addisonn Wesley, Massachusetts, Mass, USA, 1993.

[2] S. Kang and Y. Leblebici, *CMOS Digital Integrated Circuit Analysis and Design*, McGraw-Hill, 3rd edition, 2005.

[3] D. Sinha, T. Sharma, K. G. Sharma, and B. P. Singh, "Ultra low power 1-bit full adder," *International Journal of Computer Applications*, vol. 1, article 3, 2011.

[4] T. Sharma, K.G. Sharma, and B. P. Singh, "Low power and high performance 1-bit 9T full adder," in *Proceedings of the i- COST Electronics & Communication Conference*, pp. 5.2.1–5.2.3, B. S. Deore College of Engineering, Dhule, January 2011.

[5] D. Sinha, T. Sharma, K. G. Sharma, and B. P. Singh, "Energy efficient design and analysis of 1-bit full adder cell," *International Journal of Recent Trends in Engineering & Technology*, vol. 5, no. 2, pp. 49–52, 2011.

[6] S. R. Chowdhury, A. Banerjee, A. Roy, and H. Saha, "A high speed 8-transistor full adder design using novel 3 transistor XOR gates," *International Journal of Electronics, Circuits and Systems*, vol. 2, no. 4, pp. 217–223, 2008.

[7] S. Veeramachaneni and M. B. Srinivas, "New improved 1-bit full adder cells," in *Proceedings of the CCECE/ CCGEI*, pp. 735–738, Niagara Falls, Canada, May 2008.

[8] D. Wang, M. Yang, W. Cheng, X. Guan, Z. Zhu, and Y. Yang, "Novel low power full adder cells in 180 nm CMOS technology," in *Proceedings of the 4th IEEE Conference on Industrial Electronics and Applications (ICIEA '09)*, pp. 430–433, May 2009.

[9] T. Sharma, K. G. Sharma, B. P. Singh, and N. Arora, "A novel CMOS 1-bit 8T full adder cell," *WSEAS Transactions on Systems*, vol. 9, no. 3, pp. 317–326, 2010.

[10] M. Hosseinghadiry, M. Mohammadi, and M. Nadisenejani, "Two new low power high performance full adders with minimum gates," *International Journal of Electrical and Computer Engineering*, vol. 4, pp. 671–678, 2009.

10

Automatic Generation of Optimized and Synthesizable Hardware Implementation from High-Level Dataflow Programs

Khaled Jerbi,[1,2] **Mickaël Raulet,**[1] **Olivier Déforges,**[1] **and Mohamed Abid**[2]

[1] *IETR/INSA. UMR CNRS 6164, 35043 Rennes, France*
[2] *CES Laboratory, National Engineering School of Sfax, 3038 Sfax, Tunisia*

Correspondence should be addressed to Khaled Jerbi, khaled.jerbi@insa-rennes.fr

Academic Editor: Maurizio Martina

In this paper, we introduce the Reconfigurable Video Coding (RVC) standard based on the idea that video processing algorithms can be defined as a library of components that can be updated and standardized separately. MPEG RVC framework aims at providing a unified high-level specification of current MPEG coding technologies using a dataflow language called Cal Actor Language (CAL). CAL is associated with a set of tools to design dataflow applications and to generate hardware and software implementations. Before this work, the existing CAL hardware compilers did not support high-level features of the CAL. After presenting the main notions of the RVC standard, this paper introduces an automatic transformation process that analyses the non-compliant features and makes the required changes in the intermediate representation of the compiler while keeping the same behavior. Finally, the implementation results of the transformation on video and still image decoders are summarized. We show that the obtained results can largely satisfy the real time constraints for an embedded design on FPGA as we obtain a throughput of 73 FPS for MPEG 4 decoder and 34 FPS for coding and decoding process of the LAR coder using a video of CIF image size. This work resolves the main limitation of hardware generation from CAL designs.

1. Introduction

User requirements of high quality video are growing which causes a noteworthy increase in the complexity of the algorithms of video codecs. These algorithms have to be implemented on a target architecture that can be hardware or software. In 2007, the notion of Electronic System Level Design (ESLD) has been introduced in [1] as a solution to decrease the time to market using high-level synthesis which is an automatic compilation of high-level description into a low-level one called register transfer level (RTL). The high-level description is governed by models of computation which are the rules defining the way data is transferred and processed. Many solutions were developed to automate the hardware generation of complex algorithms using ESLD. Synopsys developed a C to gate compiler called synphony [2]. Mentor Graphics also created a C to HDL compiler called Catapult C [3, 4]. For their NIOS II, Altera introduces C2H as a converter from C to HDL [5, 6]. To extend Matlab for hardware generation from functional blocks, Mathworks created

a hardware generator for FPGA design [7]. In the university research field, STICC laboratory in France developed a high-level synthesis tool called GAUT that extracts parallelism and generates VHDL code from a pure C description [8, 9]. The common point between all previously quoted tools is the fact that they are application-specific generators which means that they are not always efficient on an entire multi-component system description.

In this context, CAL [10] was introduced in the Ptolemy II project [11] as a general-use dataflow target agnostic language based on the dataflow Process Network (DPN) Model of Computation [12] related to the Kahn Process Network (KPN) [13]. The MPEG community standardized the RVC-CAL language in the MPEG RVC (Reconfigurable Video Coding) standard [14]. This standard provides a framework to describe the different functions of a codec as a network of functional blocks developed in RVC-CAL and called actors. Some hardware compilers of RVC-CAL were developed but their limitation is the fact that they cannot compile high-level structures of the language so these structures have to be manually transformed.

In [15], we presented an original functional method to quicken the HDL generation using a software platform for rapid design and validation of a high complexity dataflow architecture but going from high to low-level representation used to be manual. Therefore, we proposed to add automatic transformations to make any RVC-CAL design synthesizable.

This paper extends a preliminary work presented in [16] by introducing efficient optimizations and their impact on the area and time consumption of the design. The transformation tool analyzes the RVC-CAL code and performs the required transformations to obtain synthesizable code whatever the complexity of the considered actor. In Section 2, we explain the main advantages of using MPEG RVC standard for signal processing algorithms and the key notions of the RVC-CAL language and its behavioral structures and mechanisms. The proposed transformation process is detailed in Section 4 and finally hardware implementation results of MPEG4 Part2 decoder and LAR codec are presented in Sections 5 and 6.

2. Background

Since the beginning of ISO/IEC/WG11 (MPEG) in 1988 with the appearance of MPEG-1, many video codecs have been developed (MPEG-4 part2, MPEG SVC, MPEG AVC, HEVC, etc.) with an increasing complexity and so they take longer time to be produced. In addition, every standard has a set of profiles depending on the implementation target or the user specifications. Consequently, it became a tough task for standard communities to develop, test, and standardize a decoder at any given time. Moreover, the standards specification is monolithic which makes it harder to reuse or update some existing algorithms. This ascertainment originated a new conception methodology standard called Reconfigurable Video Coding introduced by MPEG.

In the following, we present an overview of MPEG RVC standard and associated tools and frameworks, we also present the main features of CAL actor language and the limitations that motivated this work.

2.1. MPEG RVC. RVC presents a modular library of elementary components (actors). The most important and attractive features of RVC are reconfigurability and flexibility. An RVC design is a dataflow directed graph with actors as vertices and unidirectional FIFO channels as edges. An example of a graph is shown in Figure 1.

Actually, defining video processing algorithms using elementary components is very easy and rapid with RVC since every actor is completely independent from the rest of the other actors of the network. Every actor has its own scheduler, variables, and behavior. The only communication of an actor are its input ports connected to the FIFO channels to check the presence of tokens and as explained later an internal scheduler is going to allow or not the execution of elementary functions called actions depending on their corresponding firing rules (see Section 3). Thus, RVC insures concurrency, modularity, reuse, scalable parallelism, and

encapsulation. In [17], Janneck et al. show that, for hardware designs, *RVC standard allows a gain of 75% of development time* and considerably reduces the number of lines compared with the manual HDL code. To manage all the presented concepts of the standard, RVC presents a framework based on the use of the following.

(i) A subset of the CAL actor language called RVC-CAL that describes the behavior of the actors (see details in Section 2.2).

(ii) A language describing the network called FNL (Functional unit Network Language) that lists the actors, the connections and the parameters of the network. FNL is an XML dialect that allows a multilevel description of actors hierarchy which means that a functional unit can be a composition of other functional units connected in another network.

(iii) Bitstream syntax Description Language (BSDL) [18, 19] to describe the structure of the bitstream.

(iv) An important Video Tool Library (VTL) of actors containing all MPEG standards. This VTL is under development and it already contains 3 profiles of MPEG 4 decoders (Simple Profile, Progressive High Profile and Constrained Baseline Profile).

(v) Tools for edition, simulation, validation and automatic generation of implementations:

(a) open DF framework [20] is an interpreter infrastructure that allows the simulation of hierarchical actors network. Xilinx contributed to the project by developing a hardware compiler called OpenForge (available at http://openforge.sourceforge.net/) [21] to generate HDL implementations from RVC-CAL designs.

(b) open RVC-CAL Compiler (Orcc) (available at http://orcc.sourceforge.net/) [19] is an RVC-CAL compiler under development. It compiles a network of actors and generates code for both hardware and software targets. Orcc is based on works on actors and actions analysis and synthesis [22, 23]. In the front-end of Orcc, a graph network and its associated CAL actors are parsed into an abstract syntax tree (AST) and then transformed into an intermediate representation that undergoes typing, semantic checks and several transformations in the middle-end and in the back-end. Finally, pretty printing is applied on the resulting IR to generate a chosen implementation language (C, Java, Xlim, LLVM, etc.).

At this level, the question is that *why RVC-CAL and not C*? Actually, a C description involves not only the specification of the algorithms but also the way inherently parallel computations are sequenced, the way data is exchanged throw inputs and outputs, and the way computations are mapped. Recovering the original intrinsic properties of the algorithms by analyzing the software program is impossible.

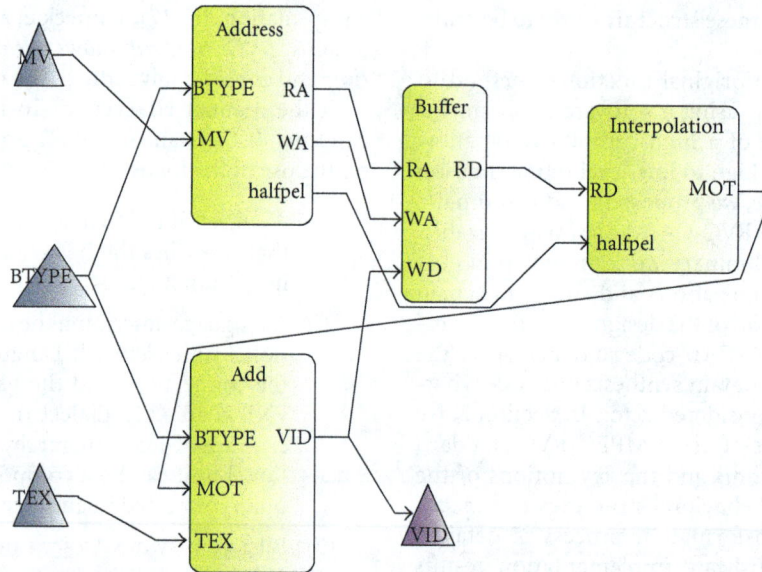

FIGURE 1: Graph example. Block diagram of the motion compensation of an MPEG 4 part 2 decoder.

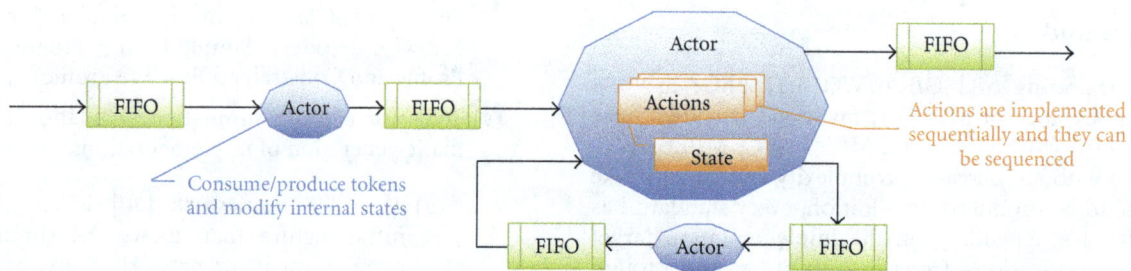

FIGURE 2: CAL actor model.

In addition, the opportunities for restructuring transformations on imperative sequential code are very limited compared to the parallelization potential available on multi-core platforms. For these reasons, RVC adopted the CAL language for actors specification. The main notions of this language are presented below.

2.2. CAL Actor Language. The execution of an RVC-CAL code is based on the exchange of data tokens between computational entities (actors). Each actor is independent from the others since it has its own parameters and finite state machine if needed. Actors are connected to form an application or a design, this connection is insured by FIFO channels. Executing an actor is based on *firing* elementary functions called *actions*. This action firing may change the state of the actor in case of an FSM. An RVC-CAL dataflow model is shown in the network of Figure 2.

Figure 3 presents an example of a CAL actor realizing the sum between two tokens read from its two input ports.

Like in VHDL, an actor definition begins by defining the I/O ports and their types then actions are later listed. An action begins also by defining the I/O ports it uses from the list of ports of the actor and this definition includes the number of tokens this action have to find in the FIFO to be fireable. In the "sum" actor, the internal scheduler allows action "add" only when there is at least one token in the FIFO of port "INPUT1" and one token in the FIFO of port "INPUT2" and this property explains how an actor can be totally independent and can neither read nor modify the state of any other actor. Of course, an actor may contain any number of actions that can be governed by an internal finite state machine. At a specific time two or more actions may have the required conditions to be fired so the notion of priority was introduced (see details in Section 3).

For the same behavior, an actor may be defined in different ways. Let us consider the "sum-5" actor of Figure 4 that reads 5 tokens in a port "IN," computes their sum and produces the result in a port "OUT."

In Figure 4(a), the required algorithm is defined in only one action. The condition of 5 required tokens is expressed by the instruction "repeat 5." Action "add" fires by consuming the 5 tokens from the FIFO into an internal buffer "I." After data storage, the algorithm of the action is applied. Finally the action firing finishes by writing the result in the port "OUT".

Such description is very fast to develop and implement on software targets but for hardware implementations a multitoken read is not appropriate. This is the reason of

```
actor sum ()
(int size=8) INPUT1, (int size=8) INPUT2 ==> int(size=8) OUTPUT:

 add: action INPUT1:[ i1 ], INPUT2[i2] ==> OUTPUT:[s]
 var
   int s
 do
   s:= i1 + i2 ;
 end
end
```

FIGURE 3: Example of sum actor.

```
actor sum-5 () int (size=8) IN
==> int(size=8) OUT:

List (type: int (size=8), size = 5) data;
int counter :=0 ;

read: action IN:[ i ] ==>
do
  data[counter] := i ;
  counter := counter + 1 ;
end

read_done: action ==>
guard
  counter = 5
do
  counter := 0 ;
end

process: action ==> OUT:[ s ]
var
 int s := 0
do
  foreach int k in 0 .. 4 do
    s := s + data[k] ;
  end
end

schedule fsm state0:
  state0 (read)      --> state0;
  state0 (read_done) --> state1;
  state1 (process)   --> state0;
end

priority
  read_done > read;
end

end
```

```
actor sum-5 () int (size=8) IN
==> int(size=8) OUT:

add: action IN:[ i ] repeat 5
==> OUT:[ s ]
var
  int s := 0
do
  foreach int k in 0 .. 4 do
    s := s + i[k] ;
  end
end
end
```

(a) SW-oriented definition   (b) HW-oriented definition

FIGURE 4: Two-way definition example of sum-5 actor behavior.

developing the equivalent monotoken code of Figure 4(b). In this description, we use a finite state machine to lock the actor in the state "state0." While counter¡5, only the action "read" can be fired to store tokens one per one in "data" buffer. Once the condition of action "read_done" (counter = 5) is true, both of "read" and "read_done" actions are fireable. This is why the priority "read_done > read" is important to keep the determinism of the actor. Finally, the firing of "read_done" action involves an FSM update to "state1" where only "process" action can be fired and the actor is back to the initial state.

3. Actor Behavior Formalism

Actor execution is governed by a set of conditions called firing rules. Moreover, during this firing many internal features of the actor are updated (state, state variables, etc.). All these concepts and behavior evolutions are detailed below. The actor execution, so called firing, is based on the dataflow Process Network (DPN) principle [12] derived from the Kahn Process Network (KPN) [13]. Let Ω be the universe of all tokens values exchanged by the actors and $\mathbb{S} = \Omega^*$, the set of all finite sequences in Ω. We denote the length of a sequence $s \in \mathbb{S}^k$ by $|s|$ and the empty sequence by λ. Considering an actor with m inputs and n outputs, $\mathbb{S}^m$ and $\mathbb{S}^n$ are the set of m-tuples and n-tuples consumed and produced. For example, $s_0 = [\lambda[t_0, t_1, t_2]]$ and $s_1 = [[t_0], [t_1]]$ are sequences of tokens that belong to $\mathbb{S}^2$ and we have $|s_0| = [0, 3]$ and $|s_1| = [1, 1]$.

3.1. Actor Firing. A dataflow actor is defined with a pair $\langle f, R \rangle$ such as:

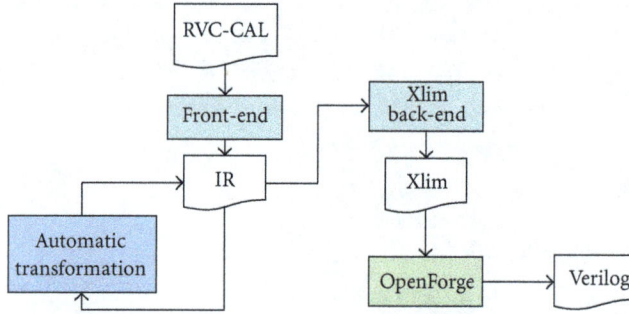

FIGURE 5: Automatic transformation localization in Orcc compiling process.

(i) $f : \mathbb{S}^m \to \mathbb{S}^n$ is the firing function;

(ii) $R \subset \mathbb{S}^m$ are the firing rules;

(iii) for all $r \in R$, $f(r)$ is finite.

An actor may have N firing rules which are finite sequences of m patterns (one for each input port). A pattern is an acceptable sequence of tokens for an input port. It defines the nature and the number of tokens necessary for the execution of at least one action. RVC-CAL also introduces the notion of *guard* as additional conditions on tokens values. An example of firing rule r_j in $\mathbb{S}^2$ is

$$g_{j,k} : [x] \mid x > 0$$
$$r_j = \left[t_0 \in g_{j,k}, [t_1, t_2, t_3] \right], \tag{1}$$

Equation (1) means that if there is a positive token in the FIFO of the first input port and 3 tokens in the FIFO of the second input port then the actor will select and execute a fireable action. An action is fireable or schedulable iff:

(i) the execution is possible in the current state of the FSM (if an FSM exists);

(ii) there are enough tokens in the input FIFO;

(iii) a guard condition returns true.

An action may be included in a finite state machine or untagged making it higher priority than FSM actions.

3.2. Actor Transition. The FSM transition system of an actor is defined with $\langle \sigma_0, \Sigma, \tau, \prec \rangle$ where Σ is the set of all the states of the actor, σ_0 is the initial state, $\prec$ is a priority relation and $\tau \subseteq \Sigma \times \mathbb{S}^m \times \mathbb{S}^n \times \Sigma$ is the set of all possible transitions. A transition from a state σ to a state σ' with a consumption of sequence $s \in \mathbb{S}^m$ and a produced sequence $s' \in \mathbb{S}^n$ is defined with (σ, s, s', σ') and denoted.

$$\sigma \xrightarrow[\tau]{s \mapsto s'} \sigma'. \tag{2}$$

To solve the problem of the existence of more than one possible transition in the same state, RVC-CAL introduced the notion of priority relation such as for the transitions

$t_0, t_1 \in \tau$, t_0 a higher priority than t_1 is written $t_0 \succ t_1$. As explained in [24] a transition $\sigma \xrightarrow[\tau]{s \mapsto s'} \sigma'$ is enabled iff:

$$\neg \exists \sigma \xrightarrow[\tau]{p \mapsto q} \sigma'' \in \tau : p \in \mathbb{S} \wedge \sigma \xrightarrow[\tau]{s \mapsto s'} \sigma'' \succ \sigma \xrightarrow[\tau]{s \mapsto s'} \sigma'. \tag{3}$$

This section presented and explained the main RVC-CAL principles. In the next section we present an automatic transformation as a solution to avoid these limitations without changing the overall macrobehavior of the actor.

3.3. Hardware Generation Problematic. A firing rule is called *multitoken* iff : $\exists e \in |s| : e > 1$ otherwise it is called a *mono-token* rule. The limitation of OpenForge is the fact that it does not support multitoken rules which are omnipresent in most actors. The observation of Figure 4 shows the incontestable complexity difference between the multitoken (a) and the monotoken (b) code. Moreover, manually changing a CAL code from high-level to low-level by creating the new actions, variables and state machine is contradictory to the main purpose of RVC standard which is the fact that CAL is a target agnostic language so we have to write in CAL the same way for hardware of software implementation. Our work consists in automatically transforming the data read/write processes from multitoken to monotoken while preserving the same actor behavior. All the required actions, variables and finite state machines are created and optimized directly in the Intermediate representation of Orcc compiler. The following section explains the achieved transformation mechanism.

4. Methodology for Hardware Code Generation

As shown in Figure 5, our transformation acts on the IR of Orcc. The HDL implementation is later generated using OpenForge.

4.1. Actor Transformation Principle. Let us consider an actor with a multitoken firing rule $r \in \mathbb{S}^k$ such as $|r| = [r_0, r_1, \ldots, r_{k-1}]$, this rule fires a multitoken action a realizing the transition $source \xrightarrow[\tau]{a} target$ and $\mathbb{I}$ the set of all input ports. The transformation creates for every input port an internal buffer with read-and-write indexes and clips r into a set $\mathbb{R}$ of k firing rules so that:

$$\forall i \in \mathbb{I}, \exists ! \rho \in \mathbb{R} : \begin{cases} \rho : \mathbb{S}^1 \longrightarrow \mathbb{S}^0 \\ |r| = 1 \\ g_\rho : IdxWrite_i - IdxRead_i \leq sz_i, \end{cases} \tag{4}$$

with ρ a monotoken firing rule of an untagged action $untagged_i$, g_ρ is the guard of ρ, and sz_i the size of the associated internal buffer defined as the closest power of 2 of r_i. This guard checks that the buffer contains an empty place for the token to read. The multitoken action is consequently removed, and new *read actions* that read one token from the internal buffers are created. While reading tokens another firing rule may be validated and causes the firing of an unwanted action. To avoid the nondeterminism of such a case, we use an FSM to put the actor in a reading loop so it can only read tokens. The loop is entered using a *transition*

```
actor A () int IN1, int IN2, int IN3 ==> int OUT1, int OUT2:
  a: action
  in1:[in1] repeat 2, IN2:[in2] repeat 3, IN3:[in3] ==>
  OUT1:[out1], OUT2:[out2] repeat 2
  do
      {treatment}
  end
end
```

FIGURE 6: RVC-CAL code of actor A.

FIGURE 7: Created FSM macroblock.

action realizing the FSM passage $source \xrightarrow[\tau]{transition} read$ and has the same priority order of the deleted multitoken action but has no process. The read actions loop in the read state with the transition $t = read \xrightarrow[\tau]{read} read$. Then the loop is exited when all necessary tokens are read using a *read done* action and a transition to the process state $t' = read \xrightarrow[\tau]{read\ Done} process \succ t$. The treatment of the multitoken action is put in a *process action* with a transition $process \xrightarrow[\tau]{process} write$. The multitoken outputs are also transformed into a writing loop with *write actions* that store data directly in the output FIFO associated with a transition $w = write \xrightarrow[\tau]{write} write$ and a *write done* action that insures the FSM transition $w' = write \xrightarrow[\tau]{write\ Done} target \succ w$.

For example, the actor A of Figure 6 is defined with $f : \mathbb{S}^3 \rightarrow \mathbb{S}^2$ with a multitoken firing rule:
$r \in \mathbb{S}^3 : r = [[t_0, t_1], [t_2, t_3, t_4], [t_5]]$.

Consequently, $|r| = [2, 3, 1]$ which means that there is an action in A that fires if 2 tokens are present in $IN1$ port, 3 tokens are present in $IN2$ and one token is present in $IN3$. The transformation creates the FSM macroblock of Figure 7.

4.2. FSM Creation Cases. We consider an example of an actor defined as $f : \mathbb{S}^3 \rightarrow \mathbb{S}^2$ containing the actions $a1 \cdots a5$ such as $a3$ is the only action applying a multitoken firing rule $r \in \mathbb{S}^3$.

Creating an FSM only for action $a3$ is not appropriate because $a1$, $a2$, $a4$, $a5$ will be a higher priority which may not be true. The solution is to create an initial state containing all the actions and add the created FSM macroblock of $a3$ (previously presented in Figure 7). The resulting FSM is presented in Figure 8.

We now suppose the same actor scheduled with an initial FSM as shown in Figure 9.

The transition $t = S1 \xrightarrow[\tau]{s \mapsto s'} S2$ is substituted with the macroblock of $a3$ as shown in Figure 10.

4.3. Optimizations. To improve the transformation, some optimization solutions were added. In the previously presented transformation method we used the untagged actions to store data in the internal buffers, then we used read actions to peek the required tokens from the internal buffers using R/W indexes and masks. To preserve the schedulability, the action is split into a transition action that contains the firing rule and a process action that applies the algorithm. The proposed optimization consists in making the action reading directly from the internal buffers. The firing rule of the action is transformed as presented in (4) to detect the presence of enough data in the internal buffers. Let us reconsider the basic example of the "sum-5" actor of Figure 4 of Section 2.2. The transformation explained above and the optimized transformation of this actor are presented in Figure 11. This actor is transformed this way. First an internal buffer and an untagged action are created to store data inside the actor. The input pattern of the *read* action is transformed into a connection to the internal buffer. Every read or write from the internal buffer must be masked to make the modulo of th buffer size since it is circular.

5. RVC Case of Study: MPEG 4 SP Intradecoder

To assess the performance of the previously presented transformation, we applied it on the whole MPEG 4 simple profile intradecoder. This choice is explained by the fact that there exists a stable design in the VTL and also because this decoder includes various image processing algorithms with more or less complexity. In the following we present an overview of this codec architecture and basic actors. We also present the implementation results and a comparison with an academic high-level synthesis tool called GAUT.

5.1. Concept. MPEG codecs have all a common design. It begins with a parser that extracts motion compensation and texture reconstruction data. The parser is then followed by reconstruction blocs for texture and motion and a merger as presented in Figure 12. This decoder is a full example of coding techniques that encapsulates predictions, scan, quantization, IDCT transform, buffering, interpolation, merging and especially the very complex step of parsing.

Table 1 gives an idea about the complexity of parsers in MPEG 4 Simple Profile and MPEG Advanced Video Coding (AVC).

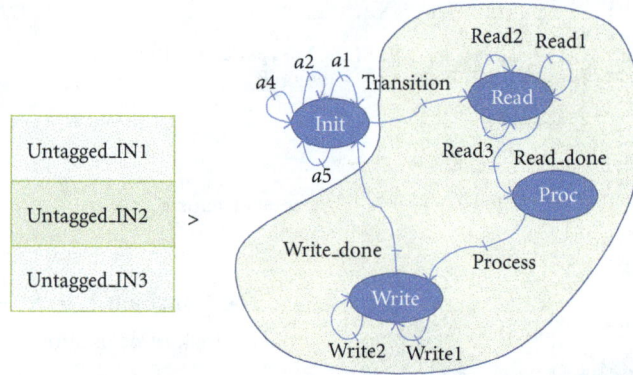

Figure 8: FSM with created initial state.

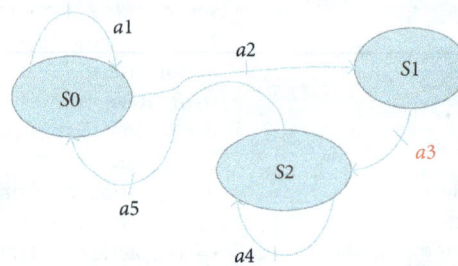

Figure 9: Initial FSM of an actor.

Table 1: Composition of MPEG-4 simple profile and MPEG-4 advanced video coding RVC-CAL description.

	Actors	Levels	Parser size kSLOC	Decoder size kSLOC
MPEG-4 SP	27	3	9.6	2.9
MPEG-4 AVC	45	6	19.8	3.9

Table 2: MPEG4 decoder area consumption.

Criterion	Transformed design	Optimized design
Slice flip flops	21,624/135,168 (15%)	13,575/135,168 (10%)
Occupied slices	45,574/67,584 (67%)	18,178/67,584 (26%)
4 input LUTs	68,962/135,168 (51%)	34,333/135,168 (25%)
FIFO16/RAMB16s	14/288 (4%)	14/288 (4%)
Bonded IOBs	107/768 (13%)	107/768 (13%)

Table 3: MPEG4 decoder timing results.

Criterion	Transformed design	Optimized design
Maximum frequency (MHz)	26.4	26.67
Latency (μs)	381.8	306.4
Cadency (MHz)	1.9	2.33
Processing time (ms/image)	13.55	11.01
Throughput frequency (MHz)	1.8	2.2
Global image processing (FPS)	73.8	90.82

Actors of Figure 12 are the main functional units some of them are hierarchical composition of actor networks. An actor may be instantiated more than one time so for 27 FU there are 42 actor instantiations.

5.2. *Implementation and Results.* The achieved automatic transformation was applied on MPEG4 SP intradecoder (see design in Orcc Applications (available at http://orcc .sourceforge.net/)) which contains 29 actors. We omitted the inter decoder part because it is very memory consuming. The HDL generated code was implemented on a virtex4 (xc4vlx160-12ff1148) and the area consumption results we obtained are presented in Table 2. The removal of read

actions buffers and process actions had an important impact on the area consumption since it has decreased about 50%.

After the synthesis of the design, we applied a simulation stream of compressed videos. Table 3 below presents the timing results of a CIF (352×288) image size video.

We notice that timing results were partially improved. This is due to the presence of division operations in some actors. In our transformation we replaced divisions by an Euclidean division which is very costly and time consuming. The impact is noticeable since these divisions reduced the maximum frequency by 60%. Therefore, we applied the transformation on the inverse discrete cosine 2D transform (IDCT2D). We chose this actor because it contains very complex algorithm, functions and procedures. We tried to compare with an optimal low-level architecture designed by Xilinx experts and also with an existing implementation study of a direct VHDL written algorithm in [25]. For a significant comparison, we used the same implementation target

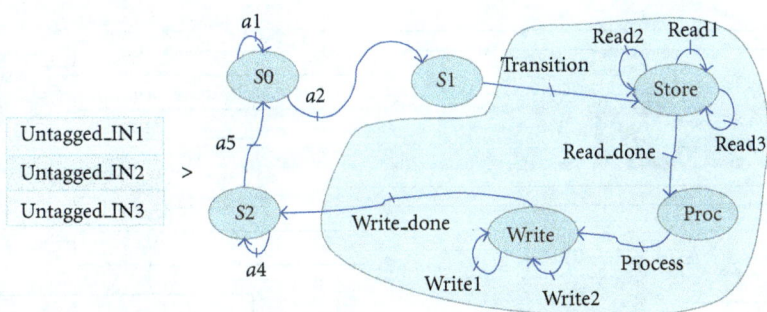

FIGURE 10: Resulting FSM transformation.

(a) Transformed equivalent CAL

```
actor sum-5 () int (size=8) IN
==> int(size=8) OUT:

List (type: int (size=8), size = 8) buffer;
// closest power of 2 for circular buffer
List (type: int (size=8), size = 5) data;
int readIdx := 0;
int writeIdx := 0;
int counter :=0 ;

action IN:[ i ] ==> // untagged action
guard
  readIdx - writeIdx < 8
  // condition that the buffer is not full
do
  buffer[readIdx & 7] := i ;
  // masked read index
  readIdx := readIdx + 1 ;
end

read: action ==>
do
  data[counter] := buffer[writeIdx & 7] ;
  // masked write index
  counter := counter + 1 ;
end

read_done: action ==>
guard
  counter = 5
do
  counter := 0 ;
end

process: action ==> OUT:[ s ]
var
 int s := 0
do
  foreach int k in 0 .. 4 do
    s := s + data[k] ;
  end
  writeIdx := writeIdx + 5; // update writeIdx
end

schedule fsm state0:
  state0 (read)      --> state0;
  state0 (read_done) --> state1;
  state1 (process)   --> state0;
end

priority
  read_done > read;
end

end
```

(b) Optimized equivalent CAL

```
actor sum-5 () int (size=8) IN
==> int(size=8) OUT:

List (type: int (size=8), size = 8) buffer;
int readIdx := 0;
int writeIdx := 0;

action IN:[ i ] ==>
guard
  readIdx - writeIdx < 8
do
  buffer[readIdx & 7] := i ;
  readIdx := readIdx + 1 ;
end

process: action ==> OUT:[ s ]
guard
        readIdx - writeIdx > 5
var
 int s := 0
do
  foreach int k in 0 .. 4 do
    s := s + buffer[k + (writeIdx&7)] ;
  end
  writeIdx := writeIdx + 5; // update writeIdx
end

end
```

FIGURE 11: Transformed and optimized sum-5 actor.

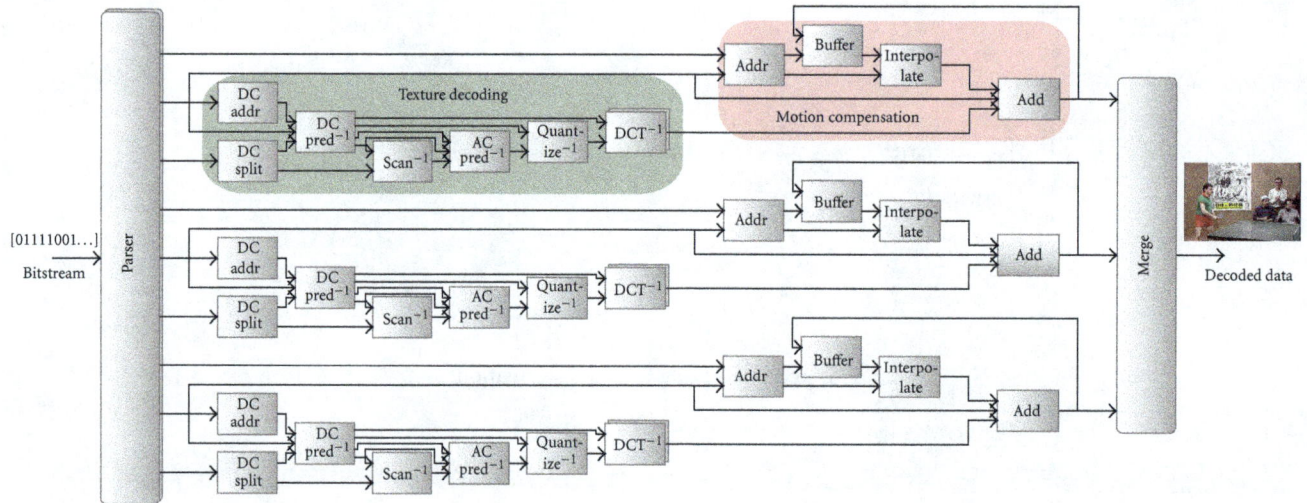

FIGURE 12: MPEG 4 SP architecture.

TABLE 4: IDCT2D timing results.

Image size	Xilinx design	Transformed design	Optimized design	VHDL design
Maximum frequency (MHz)	37	37	43	41
Latency (μs)	11.52	82.7	28.4	*
Cadency (MHz)	30	18.49	21.7	71
Processing time (μs/64 Tokens)	1.99	3.4	2.8	0.89
Throughput frequency (MHz)	26.62	0.72	2.43	62.4
Global image processing (FPS)	1064	31	101	2518

*Not mentioned in the literature.

of the study which is the Xilinx Spartan 3 XC3S4000. Timing and area consumption results comparison are presented in Tables 4 and 5.

Obviously, Table 5 reveals that area results for the optimized design are very close to those of the Xilinx low-level design. This property is noted for all actors containing more computing algorithms then data control and management algorithms. Concerning the area consumption of the VHDL design, it is expectable to find results nearby the optimal design and clearly worse than the Xilinx design and this is due to the synthesis constraints indicated in [25] that favor treatment speed in spite of the surface. This is what explains also the very high FPS rate of the design presented in Table 4. Timing results of the other designs show that the optimized design performances are far from the optimal Xilinx design. This is due to the low level architecture made by Xilinx experts which is completely different and oriented for hardware generation. This architecture is a pipelined set of actors realizing the IDCT2D (rowsort, fairmerge, IDCT1D, separate, transpose, retranspose, and clip) which is a relatively complex design compared with the high-level IDCT2D code used for the transformation.

After comparing with the Xilinx design and a VHDL directly written design, we compared our results with existing generation tools and we considered GAUT hardware generator. This tool is an academic high-level synthesizer from

C to VHDL. It extracts the parallelism and creates a scheduled dependency graph made of elementary operators. Potentially, GAUT synthesizes a pipe-lined design with memory unit, communication interface and a processing unit. However, like most existing hardware generators, GAUT is not able to manage a system level design with very high complexity and a variety of processing algorithms. Moreover, there are so many restrictions on the C input code to have a functioning design. As it was impossible to test the whole MPEG 4 decoder we chose the IDCT2D algorithm to have a comparison with previously presented results.

The IDCT2D is so generated with GAUT and we obtained the results of Table 6 below.

Results show that the optimized transformation generates a better design even for the specific case of study of the IDCT2D.

6. Still Image Codec: LAR Case of Study

The LAR is a still-image coder [26] developed at the IETR/INSA of Rennes laboratory. It is based on the idea that the spatial coding can be locally dependent on the activity in the image. Thus, the higher the activity the lower the resolution is. This activity is dependent from the variation or the uniformity of the local luminance which can be detected using a morphological gradient. In the following, we detail

TABLE 5: IDCT2D area consumption.

Criterion	Xilinx design	Transformed design	Optimized design	VHDL design
Slice flip flops	1415/55296 (2%)	4002/55296 (7%)	2113/55296 (3%)	*
Occupied slices	1308/27648 (4%)	5238/27648 (18%)	2523/27648 (9%)	3571/27648 (12%)
4 input LUTs	2260/55296 (4%)	9861/55296 (17%)	4777/55296 (8%)	4640/55296 (8%)
Bonded IOBs	48/489 (9%)	49/489 (10%)	49/489 (10%)	*

*Not mentioned in the literature.

TABLE 6: IDCT2D area consumption with GAUT.

Criterion	GAUT design	Optimized design
Slice flip flops	2.080/135.168 (2%)	1.988/135.168 (2%)
Occupied slices	2.477/67.584 (3%)	2.353/67.584 (3%)
4 input LUTs	4.243/135.168 (3%)	4.458/135.168 (3%)
Bonded IOBs	627/768 (81%)	49/768 (6%)

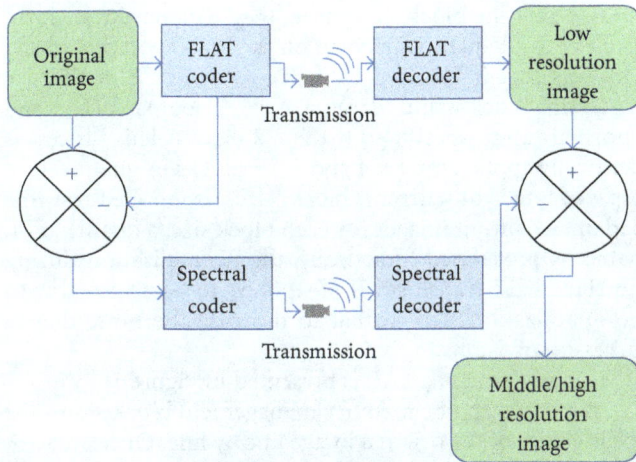

FIGURE 13: LAR concept.

coding principle of the LAR and we present the implementation techniques and results using the automatic transformation approach.

6.1. Concept. The LAR coding is based on considering that an image is a superposition of a global information image (mean blocks image), and the local texture image, which is given by the difference between the original image and the global one. This principle is modeled by

$$I = \bar{I} + \left(\underbrace{I - \bar{I}}_{E} \right), \qquad (5)$$

where I is the original image, $\bar{I}$ is the global information image and $I - \bar{I}$ is the error image, E. The dynamic range of the error image is consequently dependent on the local activity. In uniform regions, $\bar{I}$ values are close or equal to I consequently $I - \bar{I}$ values are around zero with a low dynamic range.

Considering these principles, the LAR coder concept (Figure 13) is composed of two parts: the FLAT LAR [27]

which is the part insuring the global information coding and the spectral part which is the error spectral coder.

Different profiles have been designed to fit with different types of application. In this paper, we focus on the baseline coder. Its mechanisms are detailed in the following.

The FLAT LAR. The Flat LAR is composed of 3 main parts: the partitioning, the block mean value computation and the DPCM (Differential Pulse Coding Modulation). In our work, only the DPCM is not yet developed with RVC-CAL.

(i) Partitioning: in this part, a Quad-tree partitioning is applied on the image pixels. The principle is to consider the lowest block size (2×2) then to compare the difference between the maximum (MAX) and the minimum (MIN) values of the block with a threshold (THD) defined as a generic variable for the design. If (MAX−MIN) > THD then the actual block size is considered. In the other case, the ($N \times 2$)$\times$($N \times 2$) size block is required. this process is recursively applied on the whole image blocks. The output of the overall is the block size image.

(ii) Block mean values computation process: this process is based on the Quad-tree output image. For each block of the variable size image, a mean value is put in the block as presented in the example of Figure 14.

(iii) The DPCM: the DPCM process is based on the prediction of neighbor values and the quantization of the block mean value image. The observation that a pixel value is mostly equal to a neighbor one led to the following estimation algorithm. If we consider the pixels in Figure 15, X value is estimated with the following algorithm:

If $|B - C| < |A - B|$ then $X = A$ else $X = C$.

The spectral coder, also called the texture coder, is composed of a variable block size Hadamard transform [28] and the Golomb-Rice [29, 30] entropy coder. The Golomb-Rice coder is still in development with the RVC-CAL specifications.

The Hadamard transform derives from a generalized class of the Fourier transform. It consists of a multiplication of a $2^m \times 2^m$ matrix by an Hadamard matrix (H_m) that has the same size. The transform is defined as follows.

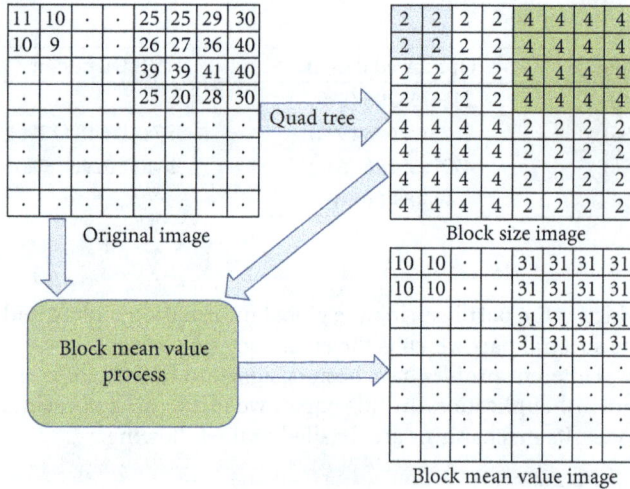

FIGURE 14: Block mean value process example.

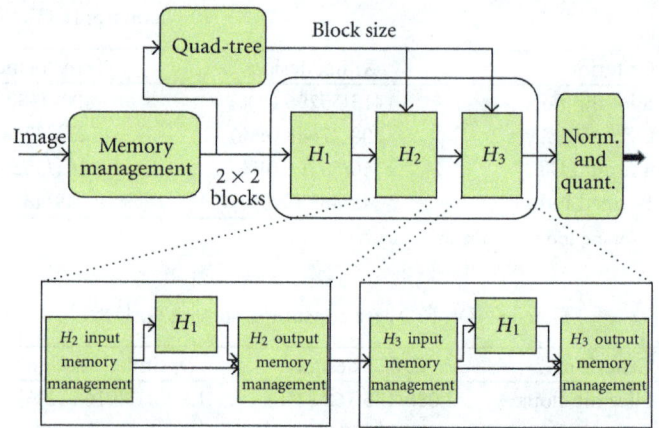

FIGURE 16: LAR baseline developed model.

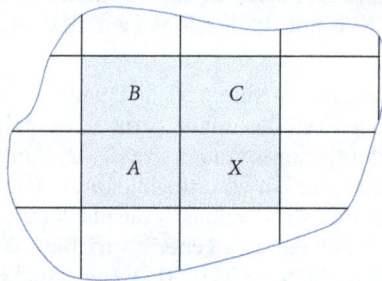

FIGURE 15: DPCM prediction of neighbor pixels.

H_0 is the identity matrix so $H_0 = 1$. For any $m > 0$, H_m is then deducted recursively by:

$$H_m = \frac{1}{\sqrt{2}} \begin{vmatrix} H_{m-1} & H_{m-1} \\ H_{m-1} & -H_{m-1} \end{vmatrix}. \tag{6}$$

Here are examples of Hadamard matrices:

$$H_0 = 1,$$

$$H_1 = \frac{1}{\sqrt{2}} \begin{vmatrix} 1 & 1 \\ 1 & -1 \end{vmatrix},$$

$$H_2 = \frac{1}{\sqrt{2}} \begin{vmatrix} 1 & 1 & 1 & 1 \\ 1 & -1 & 1 & -1 \\ 1 & 1 & -1 & -1 \\ 1 & -1 & -1 & 1 \end{vmatrix}, \text{ and so forth.} \tag{7}$$

6.2. Implementation and Results. This Section explains the mechanisms of the Hadamard transform and the Quad-tree used in the implementation.

6.2.1. Hardware Implementation. The LAR coding is dependent from the content of the image. It applies in the Quad-Tree a morphological gradient to extract information about the local activity on the image. The output is the block size image represented by variable size blocks: 2×2, 4×4 or

8×8. Using the block size image, the Hadamard transform applies the adequate transform on the corresponding block. It means that if we have a block size of 2×2 in the size image this block will undergo a 2×2 Hadamard (H_1) and a normalization specific to the 2×2 blocks. This process is identically applied for 4×4 and 8×8 blocks. A quantization step, adapted to current block size, is applied on the Hadamard output image. For each block size, a quantization matrix is predefined. Practically, the normalization during the Hadamard transform is postponed to be achieved with the quantization step so that to decrease the noise due to successive divisions.

The implemented LAR is presented in Figure 16.

As a first step, the memory management block stores the pixels values of the original image line by line. Once an 8×8 block is obtained, the actor divides it into sixteen 2×2 blocks and sends them in a specific order as presented in Figure 18.

This order is very important to improve the performance of remaining actors. In fact, considering the Figure 18, when the tokens are so ordered the first 4 tokens correspond to the first 2×2 block, the first 16 tokens to the first 4×4 block, and so forth Consequently, and as presented in Figure 16, the output of the H_1 is automatically the input of the H_2 and the output of the H_2 is automatically the input of the H_3.

In the Quad-tree, this order is also crucial. As presented in Figure 17, the superposition of the same actor (max for example) three times provides in the output of the first actor the maximum values of 2×2 blocks, in the output of the second actor the maximum values of 4×4 block and finally the maximum values of 8×8 blocks in the output of the third one. Using the maximum values and the minimums the morphological gradient in the Gradstep actors can process to extract the block size image. The same tip is used to calculate the block sums with three superposed sum actors. The block mean value actor considers the sums and the sizes to build the block mean value image.

We also notice that an (H_2) transform can be achieved using the (H_1) results of the four 2×2 blocks constituting the 4×4 block. The same observation can be made for the (H_3) one. This ascertainment is very important to decrease the complexity of the process. In fact, the Hadamard transform

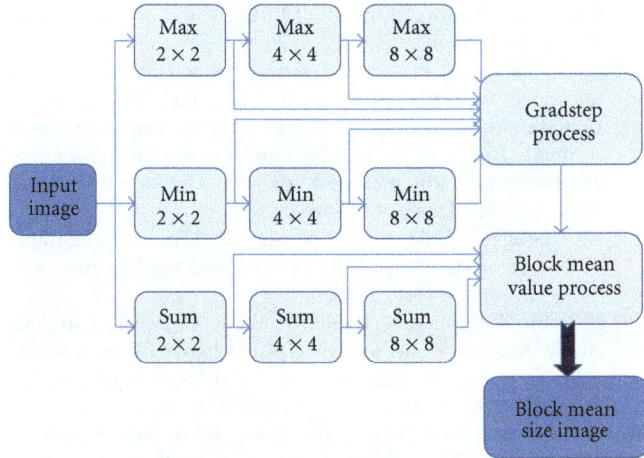

FIGURE 17: Quad-tree design.

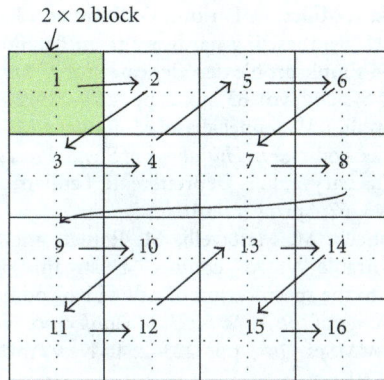

FIGURE 18: Memory management unit output order.

TABLE 7: LAR coder area consumption.

Transformation	Automatic	Manual
Slice flip flops	20.452/135.168 (15%)	12.157/135.168 (8%)
Occupied slices	47.576/67.584 (70%)	43.602/67.584 (67%)
4 input LUTs	59.868/135.168 (44%)	53.417/135.168 (39%)
Bonded IOBs	41/768 (5%)	41/768 (5%)

TABLE 8: LAR timing results.

Transformation	Automatic	Manual
Development time	30%	100%
Maximum frequency (MHz)	61.43	85.27
Latency (ms)	0.42	0.12
Throughput frequency (MHz)	3.5	5.6
Processing time (ms/image)	35	19
Global image processing (FPS)	34	53

of the LAR applies an (H_1) transform for the whole image then it applies the (H_2) transform only for the 4×4 and 8×8 blocks and the (H_3) transform only for the 8×8 blocks. The (H_2) and the (H_3) transforms are different from the full transforms as they are much less complex. Consequently, as shown in Figure 16, we designed the H_2 and the H_3 using H_1 actors associated with memory management units. They sort tokens in the adequate order and, considering the block size, whether the block is going to undergo the transform or not.

It is very important to mention that almost actors have been developed with generic variables for memory sizes or gradsteps which means that the design are flexible for easy transformation from an image size to another or for adding higher Hadamard process (H_4, H_5, etc.).

In [15], we added some optimizations on the processes using a Ping-Pong memory management algorithm [31] to pipeline the process.

6.2.2. Results and Comparison. As mentioned above, this work aims at comparing hardware implementation performances of the same LAR architecture generated with the optimized automatic transformation and with a manual transformation. The achieved automatic transformation was applied on the 23 actors of the LAR using Orcc. The HDL

generated code was implemented on a virtex4 (xc4vlx160-12ff1148). The area consumption results obtained are presented with those of manual transformations in Table 7.

After the synthesis of the design, we applied a simulation stream of compressed videos. Table 8 below presents the timing results of a CIF (352×288) image size video.

For area consumption, the difference is not considerable for LUTs and occupied slices and it can be explained by the fact that the transformation applies a general modification whatever the complexity of the actor. Also, the fact of creating an internal buffer for every input port involves more area consumption.

Concerning the timing results, the automatic and the manual transformed designs performances remain close and acceptable. The latency difference is explained by the fact that the untagged actions, as always given priority over the rest of actions, promote the data reading. It means that, as long as there is data in the FIFO, the untagged action fires even if there are enough data to fire the processing actions. This problem will also be resolved by further optimizations of the buffer size.

7. Conclusion

This paper presented an automatic transformation of RVC-CAL from high- to low-level description. The purpose of this work is to find a general solution to automate the whole hardware generation flow from system level. This transformation allows avoiding structures that are not understandable by RVC-CAL hardware compilers. We applied this automatic transformation on the 29 actors of MPEG4 part2 video intradecoder and successfully obtained the same behavior of the multitoken design and a synthesizable hardware implementation. To change the test context, we automatically transformed a high-level design of the LAR still image codec and obtained relatively acceptable results.

Several optimization processes were added to the transformation to reduce the area consumption about 50%. The transformation process is currently generalized for all actors.

The most important in this work is that we contributed in making RVC-CAL hardware generation very rapid with an average gain of 75% of conception, development, and validation time compared with manual approach. We insured that the generation is applicable at system level whatever the complexity of the actor.

Currently, improvements are also in progress to customize the transformation depending on the actor complexity analysis. A future work will be the study of the impact of the transformation on the power consumption of the generated implementation.

Acknowledgments

Special thanks to Matthieu Wipliez, Damien De Saint-Jorre, and Hervé Yviquel for their relevant contributions in the source code.

References

[1] B. Bailey, G. Martin, and A. Piziali, *ESL Design and Verification: A Prescription for Electronic System-Level Methodology*, The Morgan Kaufmann Series in Systems on Silicon, Morgan Kaufmann, 2007.

[2] "Synopsys: Synphony C compiler," In ESL design and verification: a prescription for electronic system-level methodology, http://www.synopsys.com/systems/blockDesign/HLS/pages/SynphonyC-Compiler.aspx

[3] "Mentor Graphics: Designing High-Performance DSP Hardware Using Catapult C Synthesis and the Altera Accelerated Libraries," In ESL design and verification: a prescription for electronic system-level methodology, http://www.altera.com/literature/wp/wp-01039.pdf.

[4] "Mentor Graphics: Catapult C," In ESL design and verification: a prescription for electronic system-level methodology, 2010, http://www.mentor.com/esl/catapult/overview.

[5] F. Plavec, Z. Vranesic, and S. Brown, "Towards compilation of streaming programs into FPGA hardware," in *Proceedings of the Forum on Specification, Verification and Design Languages (FDL '08)*, pp. 67–72, September 2008.

[6] D. Lau, O. Pritchard, and P. Molson, "Automated generation of hardware accelerators with direct memory access from ANSI/ISO standard C functions," in *Proceedings of the 14th Annual IEEE Symposium on Field-Programmable Custom Computing Machines (FCCM '06)*, pp. 45–56, April 2006.

[7] T. M. Bhatt and D. McCain, "Matlab as a development environment for FPGA design," in *Proceedings of the 42nd annual Design Automation Conference (DAC '05)*, ACM, New York, NY, USA, 2005.

[8] P. Coussy, D. D. Gajski, M. Meredith, and A. Takach, "An introduction to high-level synthesis," *IEEE Design and Test of Computers*, vol. 26, no. 4, pp. 8–17, 2009.

[9] P. Coussy, C. Chavet, P. Bomel, D. Heller, E. Senn, and E. Martin, "GAUT: a high-level synthesis tool for DSP applications," in *High-Level Synthesis: From Algorithm to Digital Circuits*, P. C. A. Morawiec, Ed., Springer, 2008.

[10] J. Eker and J. Janneck, "CAL language report," ERL Technical Memo UCB/ERL M03/48, University of California at Berkeley, 2003.

[11] C. Brooks, E. Lee, X. Liu, S. Neuendorer, and Y. Zhao, Eds., "HZ: PtolemyII—heterogeneous concurrent modeling and design in Java (Volume 1: introduction to ptolemyII),"

[12] E. A. Lee and T. M. Parks, "Dataow process networks," *Proceedings of the IEEE*, vol. 83, no. 5, Article ID 773801, 1995.

[13] G. Kahn, in *Proceedings of the IFIP Congress The Semantics of a Simple Language for Parallel Programming. In Information Processing*, J. L. Rosenfeld, Ed., pp. 471–475, North-Holland, New York, NY, USA, 1974.

[14] M. Mattavelli, I. Amer, and M. Raulet, "The reconfigurable video coding standard," *IEEE Signal Processing Magazine*, vol. 27, no. 3, pp. 159–167, 2010.

[15] K. Jerbi, M. Wipliez, M. Raulet, M. Babel, O. Deforges, and M. Abid, "Automatic method for efficient hardware implementation from rvc-cal dataflow: a lar coder baseline case study," *Journal Of Convergence*, vol. 1, Article ID 8592, 2010.

[16] K. Jerbi, M. Raulet, O. Deforges, and M. Abid, "Automatic generation of synthesizable hardware implementation from high level RVC-CAL design," in *Proceedings of the 37th International Conference on Acoustics Speech and Signal Processing (ICASSP '12)*, pp. 1597–1600, 2012.

[17] J. Janneck, I. Miller, D. Parlour, G. Roquier, M. Wipliez, and M. Raulet, "Synthesizing hardware from dataflow programs: an mpeg-4 simple profile decoder case study," *Journal of Signal Processing Systems*, vol. 63, no. 2, pp. 241–249, 2009.

[18] M. Mattavelli, J. W. Janneck, and M. Raulet, "MPEG reconfigurable video coding," in *Handbook of Signal Processing Systems*, S. S. Bhattacharyya, E. F. Deprettere, R. Leupers, and J. Takala, Eds., pp. 43–67, Springer, 2010.

[19] J. W. Janneck, M. Mattavelli, M. Raulet, and M. Wipliez, "Reconfigurable video coding: a stream programming approach to the specification of new video coding standards," in *Proceedings of the ACM SIGMM Conference on Multimedia Systems (MMSys '10)*, pp. 223–234, New York, NY, USA, February 2010.

[20] S. Bhattacharyya, G. Brebner, J. Eker et al., "OpenDF—a dataflow toolset for reconfigurable hardware and multicore systems," in *1st Swedish Workshop on MultiCore Computing (MCC '08)*, Ronneby, Sweden, November 2008.

[21] R. Gu, J. W. Janneck, S. S. Bhattacharyya, M. Raulet, M. Wipliez, and W. Plishker, "Exploring the concurrency of an MPEG RVC decoder based on dataflow program analysis," *IEEE Transactions on Circuits and Systems for Video Technology*, vol. 19, no. 11, pp. 1646–1657, 2009.

[22] G. Roquier, M. Wipliez, M. Raulet, J. W. Janneck, I. D. Miller, and D. B. Parlour, "Automatic software synthesis of dataflow program: an MPEG-4 simplpe profile decoder case study," in *Proceedings of IEEE Workshop on Signal Processing Systems (SiPS '08)*, pp. 281–286, Washington, DC, USA, October 2008.

[23] M. Wipliez, G. Roquier, and J. F. Nezan, "Software code generation for the RVC-CAL language," *Journal of Signal Processing Systems*, vol. 63, no. 2, pp. 203–213, 2011.

[24] J. Eker and J. W. Janneck, "A structured description of dataflow actors and its application," Technical Memorandum UCB/ERL M03/13, Electronics Research Laboratory, University of California at Berkeley, 2003.

[25] R. K. Megalingam, K. B. Venkat, S. V. Vineeth, M. Mithun, and R. Srikumar, "Hardware implementation of low power, high speed DCT/IDCT based digital image watermarking," in *Proceedings of the International Conference on Computer Technology and Development (ICCTD '09)*, pp. 535–539, November 2009.

[26] O. Déforges, M. Babel, L. Bédat, and J. Ronsin, "Color LAR codec: a color image representation and compression scheme based on local resolution adjustment and self-extracting

region representation," *IEEE Transactions on Circuits and Systems for Video Technology*, vol. 17, no. 8, pp. 974–987, 2007.

[27] O. Deforges and M. Babel, "LAR method: from algorithm to synthesis for an embedded low complexity image coder," in *Inproceedings of the 3rd International Design and Test Workshop (IDT '08)*, pp. 187–192, December 2008.

[28] J. Poncin, "Utilisation de la transformation de Hadamard pour le codage et la compression de signaux d'images," *Annales des Télécommunications*, vol. 26, no. 7-8, pp. 235–252, 1971.

[29] R. F. Rice, "Some practical universal noiseless coding techniques," Technical Report 79–22, 1979.

[30] S. W. Golomb, "Run length codings," *IEEE Transactions on Information Theory*, vol. 12, no. 7, Article ID 399401, 1966.

[31] C. H. Chang, M. H. Chang, and W. Hwang, "A flexible two-layer external memory management for H.264/AVC decoder," in *Inproceedings of the 20th Anniversary IEEE International SOC Conference*, pp. 219–222, September 2007.

11

A Graph-Based Approach to Optimal Scan Chain Stitching Using RTL Design Descriptions

Lilia Zaourar,[1] Yann Kieffer,[2] and Chouki Aktouf[3]

[1] SOC Department, LIP6 Laboratory, University Pierre and Marie Curie, 4 Place Jussieu, 75252 Paris Cedex 05, France
[2] LCIS, Grenoble Institute of Technology and University of Grenoble, 50 Rue Barthélémy de Laffémas, 26000 Valence Cedex, France
[3] DeFacTo Technologies, 167 Rue de Mayoussard, 38430 Moirans, France

Correspondence should be addressed to Lilia Zaourar, lilia.zaourar@lip6.fr

Academic Editor: Shantanu Dutt

The scan chain insertion problem is one of the mandatory logic insertion design tasks. The scanning of designs is a very efficient way of improving their testability. But it does impact size and performance, depending on the stitching ordering of the scan chain. In this paper, we propose a graph-based approach to a stitching algorithm for automatic and optimal scan chain insertion at the RTL. Our method is divided into two main steps. The first one builds graph models for inferring logical proximity information from the design, and then the second one uses classic approximation algorithms for the traveling salesman problem to determine the best scan-stitching ordering. We show how this algorithm allows the decrease of the cost of both scan analysis and implementation, by measuring total wirelength on placed and routed benchmark designs, both academic and industrial.

1. Introduction

The design flow of an integrated circuit (IC), meaning the software applications that allows the designer to move from its specification to its concrete realization, involves many stages of optimization problems, usually from system level to layout [1–3]. Indeed, the realization of an IC is a costly process both in time and money, and requires large engineering resources. In addition to technological advances, which continuously improve the efficiency of chip manufacturing techniques, as the fine prints on silicon for example, the continuous increase of the power of computers offers great opportunities for improving the process of designing and manufacturing complex ICs.

With both recent advances of the semiconductor industry and new market constraints, Time To Market (TTM) and product quality are becoming major issues. The circuit must meet flawlessly customer expectations in terms of functionality, speed, quality, reliability, and cost. In such a challenging economic environment, and given the significant level of complexity which is reached by the IC, manufacturing testing is more than ever an important factor in the design problem.

Today, chip testing should be short, efficient, and cost-effective. A significant amount of research work is ongoing with a focus on complex design-for-test problems at both universities and industry. Design For Test (DFT) techniques are becoming a key since they are considered during the chip design process and flow. Cost of manufacturing chips averages the cost of testing them. So the semiconductors community needs low cost and high quality test solutions. New and efficient DFT solutions are greeted with higher expectations than ever. Most current DFT solutions result in unpredictable design development time and development costs and directly impact (TTM).

In this context, DFT tools are receiving growing attention with the advent of core based System On Chip (SOC) design. In particular, when cores from different vendors are integrated together on the system on chip (as it is often done nowadays), the difficulty level of testing grows rapidly. Among the most apparent issues are core access, system diagnosis, test reuse, test compaction, tester qualification and Intellectual Property (IP) protection. The ITRS (International Technical Roadmap for Semiconductor) identifies key technological challenges and needs facing the semiconductor industry through the end of the next decade. Difficult near-term and long-term testing and test equipment challenges were reported in [4].

Several of the DFT solutions for ASIC and ASIP designs are based on the internal Scan DFT technique. Scan is potentially an efficient technique, if used properly. Currently, full Scan is the most widely used structured DFT approach where all the design sequential elements belong to the scan architecture or scan chains [5]. A Scan-based DFT architecture provides the ability to shift information by scanning the set of states of the circuit. All flip-flops are chained to each other to allow the introduction of test vectors as input and retrieval of test results as output. Thus, the resulting shift register is fully controllable and observable through the primary inputs and outputs of the circuit. This makes Scan testing widely adopted for manufacturing testing and other purposes such as silicon debug. However, we do care about hardware overhead (area, pin count, and so on), performance penalty, and extra design effort that may be associated with full scan insertion during the chip design process. Therefore, the primary objective in DFT is to automate the insertion of the test logic and to minimize the impact of test circuitry on chip performance and cost. Thus it is important and essential to optimize the scan chain insertion process.

However, traditional Scan solutions make such an extension very difficult since engineers need to handle gate-level netlist without taking benefit from what happens during synthesis optimization. Also, Scan implementation decisions are considered post synthesis. This is too late in comparison to RTL design decisions. Adopting Scan at the Register Transfer Level (RTL) will cover new design and manufacturing needs, strengthen verification, and consolidate reusable design methodologies by closing the gap between RTL and design for test. There are many advantages to inserting scan at the RTL level like: benefiting from the synthesis process (i.e., better optimization in terms of area and timing), the ability to debug testability issues early in the design flow, and leveraging the optimization done by the synthesis tool. The possibility to insert scan at the RTL dates back to the late nineties [6–8]. Although this idea did not get a widespread attention in the meantime, an EDA tool that lets one do it, namely HiDFT-SIGNOFF by DeFacTo Technologies, has appeared and is commercially available. Its main use today is to help debug testability issues early on in the design flow, thus helping avoid costly iterations around the synthesis step. To achieve this purpose, Scan chains are introduced at the RTL, bringing to light any problems in doing so—like noncontrollable Flip-Flops (FFs), for example. Then the scanned design can be used in other estimates, but will be discarded as far as the main flow is concerned: the real scan chains that will end up being implemented in the chip are still inserted at gate-level, using a traditional design flow. Before one can insert the actual scan chains at the RTL, one big hurdle has to be overcome. The impact of scan insertion on the design cost and performance can be important if the ordering of the FFs in the scan chain is not picked carefully. But the information allowing for such a choice, namely placement information of the FFs, is only available at the back-end of the design flow, far away from the point where

the RTL code for scan must be finalized. Our goal in this work is to develop a tool to automatically generate optimal scan chains at RTL in terms of area and additional test time for a given circuit, while respecting a number of electronics constraints.

To analyze the optimal design location where full scan chains need to be inserted, the following considerations are required. First, to implement a scan chain, one has to have knowledge of the Flip-Flops (FFs) of the design. Second, the best place to insert any new task in a flow is at the earliest. The motivations for an early scan insertion are threefold: the complexity of the objects grow as one goes forward in a flow, so data handling gets more costly; whatever is added to the design is better if integration happens earlier in the flow; and finally, should the treatment lead to iterations—either redesigning or iterations in the flow—the sooner the iterations, the lower the cost. Third, at the point where insertion is finalized, all the information required for the insertion has to be available. If this information depends at least partly on the insertion itself, this leads to iterations—see below. Fourth, insertion can start at some level in the flow, and end at a later point in same. Again, this can be undesirable, but also unavoidable once some design decisions on the insertion process have been taken. Since FFs are usually known after synthesis, once a gatel-level netlist is available, this sounds like a reasonable point in the flow to insert scan. But the replacement of normal FFs by scan FFs has to be done before synthesis ends, since timing closure is affected by it. We will see in the next section that FFs detection can happen before synthesis.

On the other hand, we will argue in Section 3 that the scan stitching ordering is an important part of making scan insertion seamless. To compute an optimized ordering, the best information to have is that of the placement of the FFs in the layout. If one uses placement information, scan insertion should be finalized during placement and routing (at the earliest). This is the classical scheme for scan insertion, and it is this one that is used in industrial EDA tools for scan insertion. This is why all scan insertion tools today either provide a costly ordering, or must decide the ordering in several phases. The next paragraph reviews the state of the art, while keeping an eye on these different criteria.

The traditional way to insert full scan in a design is to replace the FFs by scanned FFs during synthesis; possibly connect them into a chain; and then during place and route iterations, (re)connect them to try to minimize the total Manhattan length of the added connections. The exact place where one should reoptimize the ordering of the chain is a matter of debate, and heavily depends on the flow used. For an in-depth discussion of this topic in the case of the Synopsys flow, see [9].

Also, scan implementation decisions are considered post synthesis. This is too late in comparison to RTL design decisions. Adopting scan at the RTL level will cover new design and manufacturing needs, strengthen verification and consolidate reusable design methodologies by closing the gap between RTL and design for test. There are many advantages to inserting scan at the RTL level like benefiting from the

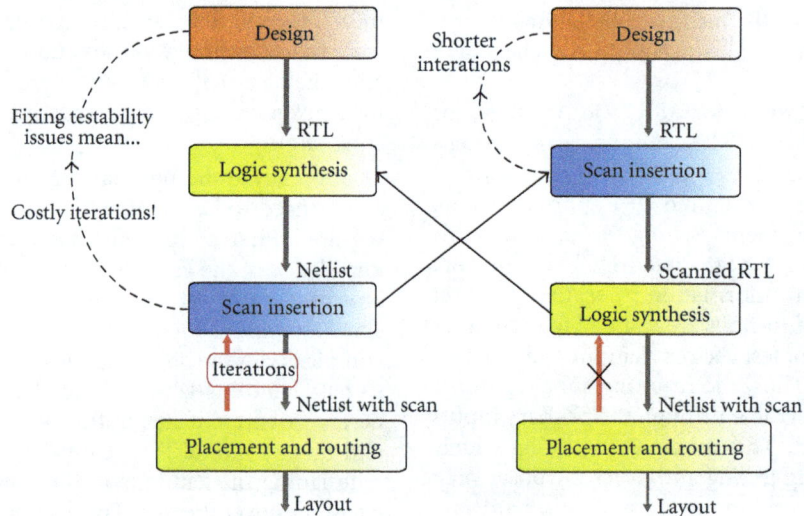

FIGURE 1: Classical scan insertion versus our method.

synthesis process (i.e., better optimization in terms of area and timing).

Most of the work done on scan chains insertion and stitching ordering optimization assumes that placement information is available [9–14], meaning that it is actually reordering optimization. In all the cases, scan (re)ordering is reduced to the problem of the Traveling Salesman Problem in a suitable graph. An additional assumption is added in [15]. The author proposes a procedure to find and break intersections. This is also done to reduce the overal wire length.

The knowledge of FFs is necessary to implement a scan chain. It has been suggested several times independently that one does not need a netlist to have knowledge of FFs of the design; this is the basis of higher-level scan insertion [6–8, 16, 17]. Although fault models are defined in terms of gates and nets, scan insertion itself can actually be done at the RTL. This is achieved by describing the behaviour of the scan chain directly at the RTL. It will then be translated by the synthesis tool into added nets between FFs, and multiplexers in front of FFs. No scanned FFs are used in this case; we give a more detailed account in the Section 3. It would seem that inserting scan at the RTL makes the problem of scan stitching reoptimization much worse. This may be one reason why in that case, the reoptimization possibility is commonly dropped, in favour of a single optimization at the time of insertion.

Most studies on higher-level scan either do not mention optimization [7], or mention local optimizations only, meaning trying to reuse bits of functional paths for the scan path, thus reducing the need for added nets and additional multiplexing logic [6, 16–18]. In [8] an attempt is made to achieve both local and global optimizations by carefully delineating a graphic model for the scan ordering problem. This graphic model rests on an analysis of the RTL source code. Unfortunately, the algorithm they propose was never implemented [19], and thus could be tested only on small designs.

In this paper, we try to close the gap between RTL Scan insertion and actual scan stitching optimization, while retaining the model of one-time insertion without any later reoptimization see Figure 1.

More specifically, we tackle the following optimization problem: given an RTL description of the design (Verilog or VHDL) we have to find a stitching ordering of the memory elements in the scan chain, which minimizes the impact of test circuitry on chip features (area, power) while keeping testing time at its minimum.

The remaining of this paper is organized as follows. Section 2 presents how our approach to RT-level scan is novel. Section 3 describes the problem of scan chain insertion at the RTL. It explains how RTL scan is implemented and gives arguments as to why wirelength is an efficient global measure of the scan insertion cost. In Section 4, we propose our RTL scan stitching algorithm together with the graph models on which it is based. Implementation, experimental results on both academic and industrial designs are given in Section 5, followed by discussions that RTL scan is a viable alternative to gate-level scan. Finally, Section 6 concludes the paper and points out some research directions for future work.

2. Original Algorithmic Content of This Work

To the best of our knowledge, this work is the first to offer a formal treatment of the scan stitching ordering problem as a discrete optimization problem in the case of RT-level scan insertion. We formalize the problem; give reductions from the several-chains problem to the one-chain problem; and solve the one-chain problem in two steps. The first step, which can be seen as a kind of preprocessing, allows us to build a graph representative of the actual optimization. Then in a second step, algorithms for the TSP are adapted to actually build the chain.

A part of this work has been presented before in [20] and recenlty in [21, 22]. The present paper is aimed at

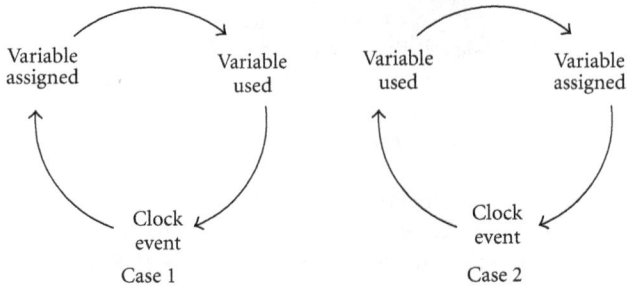

FIGURE 2: Two use cases for variables in VHDL.

```
process(clock)
begin
    if clock' event and clock='1' then
        Z <= X or Y;
        Y <= X and Y;
    end if;
end process;
```

ALGORITHM 1: Example VHDL process.

```
process(clock)
begin
    if clock' event and clock='1' then
        if scan_en='1' then
            Y <= X;
            Z <= Y;
        else
            Z <= X or Y;
            Y <= X and Y;
        end if;
    end if;
end process;
```

ALGORITHM 2: VHDL process enriched with scan code.

giving a more complete presentation for both theoretical and experimental contributions. In fact, we give here all of the theoretical basis and algorithmic elements of this work including detailed implementation that has not been presented before. Also some numerical results and comparisons resulting are added.

The main contributions of the paper are summarized below:

(i) we analyze what is a suitable objective for measuring the quality of scan stitching orderings,

(ii) we give a mathematical formulation of the problem of scan insertion at the register transfer level (RTL),

(iii) we give two reduction procedures to solve the several-chains variants based on a routine solving the one-chain case,

(iv) we solve the one-chain case in a two-steps approach,

(v) we evaluate our algorithm on both academic and industrial designs.

Our algorithm provides a basis for considering the implementation of scan chains as soon as the RTL of a block is available; the authors think that the lack of optimization has been a big obstacle to its adoption in design flows. The methodology has been validated by our industrial partner DeFacTo Technologies in the tool HiDFT-SIGNOFF, it is a first step towards considering the integration of scan at the RTL.

3. Scan Insertion at the RT-Level

We illustrate Scan insertion at RTL with the VHDL language. Scan insertion is a three steps process: first identify which signals and variables will give rise to flip-flops in the netlist; second, decide which ordering will be used to chain the FFs together; third, change the RTL code to offer a (new) testing mode for the design.

Since the object of this work is the second step of this whole process, we now restrict our attention to its first and third steps. This section is included mainly to make this article self-contained; for a more thorough presentation, the reader is referred to [6–8].

We first present how flip-flops are identified; then we illustrate RTL edition for introducing scan at RTL; finally, we examine what could be a good measure of the impact of scan insertion on a design.

3.1. Flip-Flops Identification. FFs identification is realized process by process. Each process is searched for variables and signals. All signals in a process will be translated into a FF in the netlist. To determine which variables will give rise to FFs, one has to first identify clocking signals. Then two cases can happen. To determine which applies in a particular case, one has to recall that processes execute cyclically—see Figure 2.

Either the variable is assigned between the clocking event in the process and the point where the variable is used (case 1). In that case, no FF is generated. Or the clocking event lies between the point where the variable is assigned and that where the variable is used (case 2). In that case, memorization has to occur, and a FF is generated.

3.2. Test-Mode Introduction by RTL Edition. To illustrate the introduction of a scan chain at RTL through RTL code edition, we consider the simple process in Algorithm 1.

In order to introduce a testing mode behavior to the design, we simply describe in VHDL what the process is going to do in test mode. The additional VHDL code is shown in Algorithm 2 in boldface; it consists of a conditional statement on the value of the scan_en signal, and assignments expressing the functional chaining of the registers of the process. Note that this (too simple) example is misleading in showing a doubling of the size of the RTL code; code growth for realistic designs is much less than that.

3.3. Wirelength as the Prime Parameter for Stitching Ordering. In most of the literature about scan ordering optimization, wirelength is the objective used to guide the optimization process. In these works, placement of the FFs is known; so wirelength is a quantity that is directly available during optimization. We will give our own argument for considering wirelength the most useful parameter to optimize stitching orderings.

When adding logic to a design for non-functional reasons, one tries to minimize the increase in size of the design. Additional area due to scan comes in two parts. First, FFs have to be instrumented either into scanned FFs, or with the help of a multiplexer at their input. In the former case, the area cost depends only on the number of FFs in the design; in the latter, the area cost can be less than the maximum if some optimization takes place during synthesis. In both cases, that cost is bounded independently from the stitching ordering. Second, wires have to be added between the output net of a FF and the next FF in the scan chain—either the scanned FF, or one input of the multiplexer in front of it. This does not represent an area cost in itself, since routing happens in the upper metallic layers. But it does add to the difficulty of placement and routing, with the possibility to have a very degraded situation if the stitching order is not chosen with care. In that case, the area can grow if routing is not possible anymore with the available space. We consider this a much more important factor than the additional cell area; therefore our measure of ordering quality will be based on it.

It is not possible to express the impact of added wires on placement and routing as a simple function of these wires; the impact will depend also on the sparsity of the design. But since we take a worst-case approach to the impact of wires, we will use the added wirelength due to scan insertion as our optimization function.

This choice will be validated by the variability that is observed on this parameter—see Section 5.

We now review quickly all the other parameters scan insertion could have an impact on.

In order to help ensuring timing closure, the maximum of the length of the added wires would be a reasonable measure. But since in our case synthesis happens after scan insertion, it will be up to the synthesis-and later on, placement and routing-tools, to ensure timing closure, using all the flexibility of gate sizing to achieve it. Note also that ensuring a low total added wirelength means that not too many added wires will be long, hence lowering the impact on timing, and the additional load for the synthesis tool.

Power while in functional mode grows with wirelength; so minimizing wirelength will also lower the impact on functional power. Power while in test mode is dominated by switching: the values shifted through the scan chain force the values in FFs to change more often than they would in functional mode, leading to power supply issues. A common remedy to this condition is to take benefit of don't-care values in test vectors to fill them in a way to minimize switching. This also works with RTL-scan. There have been many attempts in other works at minimizing power during test mode by changing the stitching ordering. We have not followed this trail; it would be worthwhile to try combining it with our approach based on minimizing wirelength.

Finally, observability and controllability are not impacted by the stitching ordering. We note here that although combinatorial parts of the circuit have the same observability and controllability in a full scan design implemented both at gate and at RT-level, in the case of RT-level scan, there will be more stuck-at faults, since the multiplexers are now no more combined with FFs. Hence fault coverage values tend to differ, although not much, between both methods for scan insertion.

Having established that wirelength should be minimized, we now turn to the precise description of the optimization problem we will consider in order to try to find low wirelength stitching orderings.

4. Stitching Algorithm for Optimal Scan Chain Insertion at the RT Level

To determine the scan chains that best meets the needs of different integrated circuit designers, while maintaining maximum constraints and restrictions related to the electronic problem, we propose a new scan stitching algorithm for the automatic insertion of optimal scan chains at the RTL. Our algorithm is structured into two phases, as shown in Figure 3. The first phase aims to build a graph-based model to translate the problem and its constraints from an RTL description to a generic mathematical model, in the form of an undirected graph. In a second phase, scan chains are determined through computations in that graph. The stages in each phase are outlined in the next paragraphs.

4.1. Phase 1: Graph Extraction. The challenge here is to be able to take into consideration still at high level what is going to happen later on during the placement and routing (P&R) steps. To build this model, we extract from the RTL description information on the expected proximity of the memory elements in the layout.

This phase is devised into three steps, as follows.

4.1.1. Design Elaboration (Lightweight Synthesis). First, we perform a preprocessing called Elaboration. Although scan logic can be described at RTL, the very elements of which fault models are talking—nets, gates and FFs—do not exist yet at this level. Hence we need first to translate the RTL code into these elements. This is what synthesis does. But it is not feasible to have a full synthesis at this step in the flow. Once RTL-scan is inserted, the synthesis step still has to be done; we do not want to duplicate that effort.

Our solution relies on a lightweight synthesis as the first step of the RTL scan insertion. This synthesis is done in terms of a virtual library of generic (non-physical) gates; it does not try to optimize logic, timing or gate sizes; it does not need the user to do any fine-tuning; in short, it is done transparently. The user of the scan insertion tool only provides his RTL code, and will get back a scanned RTL code: he will never see the netlist that comes out of the lightweight synthesis. Indeed, in our method, this netlist serves only one purpose, namely graph extraction.

4.1.2. Building the D-Graph. The second step after the design elaboration is to build the undirected D-graph (for Design graph), denoted by $G_D = (V_d, E_d)$. It represents the RTL description of the design. It is constructed as follows (cf.

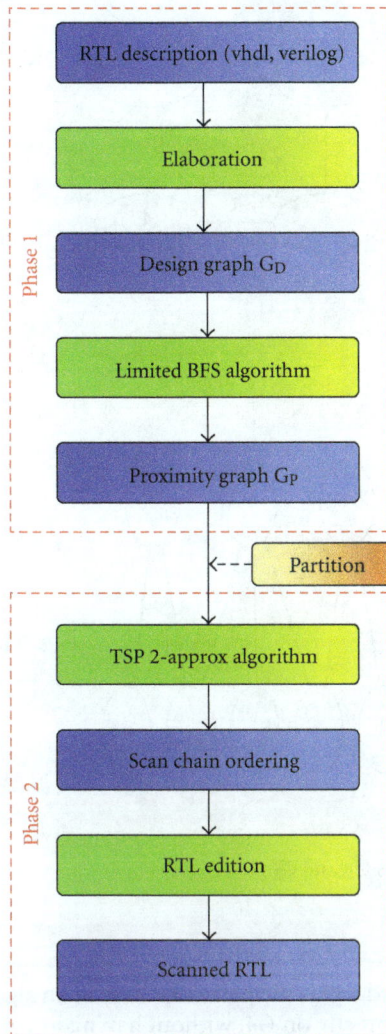

FIGURE 3: Stitching algorithm for optimal scan chain at RTL.

Figure 4). Each memory element, gate and net from the elaborated netlist corresponds to a vertex (V_D) of the graph G_D. There is an edge (E_D) between two vertices if and only if the corresponding elements in the RTL description or in the netlist are connected.

Note that the vertices are partitioned into two sets: nets on one hand, and gates and memory elements on the other; every edge in the graph has one end on each side. Therefore, G_D is a bipartite graph.

The graph G_D is undirected because it only aims at giving indications on how close memory elements might be located at the placement step; the direction of information flow along the nets is of no consequence for the decision in our algorithm to stitch together two FFs.

4.1.3. Building the P-Graph for Memory Elements Proximity.
The third step is to extract information from the design that is necessary for scan chain stitching optimization (second phase). This information will be given in the form of an edge-valued Proximity Graph (in short P-graph), to which we will now refer to as G_P.

The P-graph G_P will be built from the graph G_D as follows: the vertices (V_P) are the memory elements of the design; there is an edge (E_P) between two vertices if there is direct electronic connection in the design between the memory elements they represent; this connection is not allowed to go through other FFs. Additionally, the edges are labeled by the length of a shortest non-oriented path between the corresponding vertices in the graph G_D. The weight on edge ij will be denoted w_{ij}. It represents the Path lengths between the pair of memrory elements that it matches.

Path lengths are restricted to a threshold value t which is a fixed parameter of our algorithm: for a longer (shortest) path between two memory elements, no edge is put in G_P.

To compute G_P, we use a variant of the algorithm of Breadth First Search (BFS) [23] that limits the depth of the exploration to find the limited shortest path. This algorithm is called Limited-BFS in the following. The graph computed with the help of Limited-BFS is used to estimate a probable geometric proximity of memory elements in the Layout.

Figure 4 illustrate the construction of G_D and G_P on an example design.

4.2. Phase 2: Construction of Scan Chains.
Using this new formalism, the chaining of memory elements in the circuit corresponds to a partition of the vertices of G_P (representing the FFs of the design) into sequences of vertices, each sequence representing one scan chain. An important constraint in scan chaining asks for all chains to have the same number of FFs. Also, the sequences should preferably use edges from the graph G_P, although adding connections is a possibility. The cost of going from v_i to v_j in the sequence is counted as the length of a shortest path from v_i to v_j in the graph G_P; hence it will be w_{ij} if the edge ij is present in G_P.

In the case of a single scan chain, our problem reduces to the Traveling Salesman Problem [23]. Still, to apply TSP algorithms, the whole cost matrix for each pair of vertices of the graphs has to be precomputed; this is not feasible for designs with more than a couple tens of thousands FFs.

Also, single scan chains are not an option for big designs, where testing time would be prohibitive if scanning were not done using more than one chain.

Before explainning the second phase, we present two algorithmic devices to reduce the general chaining problem to the single-chain case. Then we show how our problem can be reduced to the Traveling Salesman Problem and some algorithms to solve it. Finally, we describe the two steps of the second phase.

Chain Splitting.
The simplest way of reducing the several-chains scanning problem to its single-chain subcase is by appropriate post-processing. Once one has computed a single scan chain for the whole design, this chain can be split into the desired number of segments (the actual scan chains). This device is really fast, and it is not expected that the quality of the output of the whole process will be much degraded as compared to that of the single-chaining algorithm. Also, it is very easy to implement, and is our recommendation for the cases where fast enough single-chaining algorithms are available.

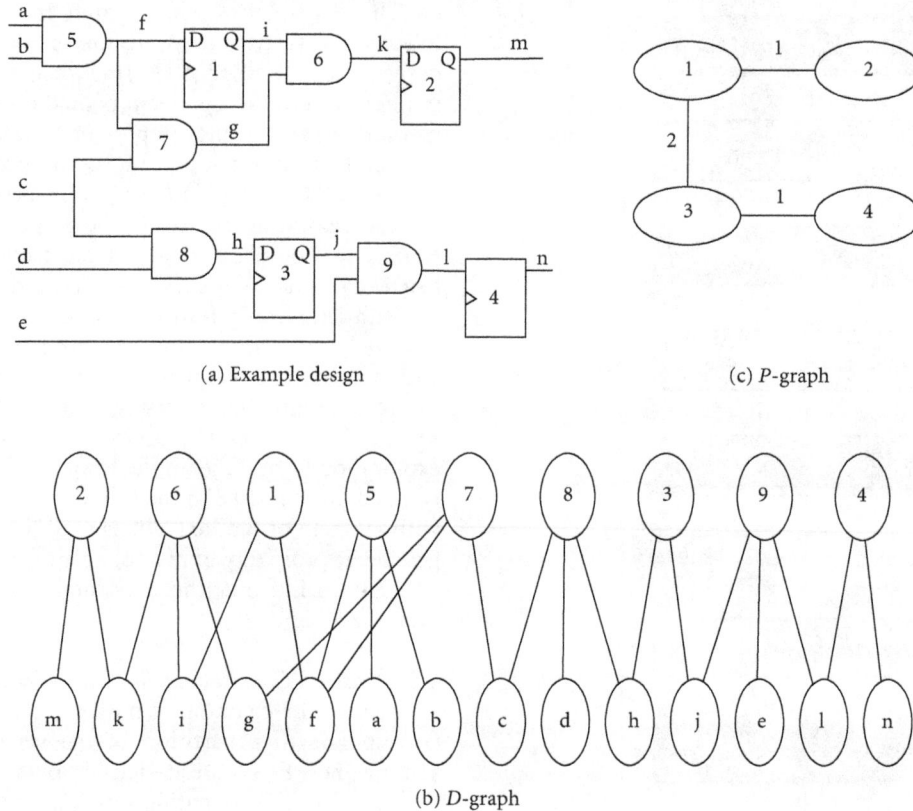

(a) Example design

(c) P-graph

(b) D-graph

FIGURE 4: Example design and associated graphs G_P and G_D.

Graph Partitioning. A more elaborate device is the preprocessing of the graph G_P with the help of a graph partitioning algorithm. If s scan chains are called for, the graph G_P should be partitioned into s parts having equal or similar numbers of vertices. Then finding scan chains for the whole design reduces to finding a single chain for each part of G_P.

Even if the partitioning is handled by an appropriate algorithmic package, implementing this solution is not as easy as chain splitting. But using it brings with it another benefit: the TSP problems to solve in this case are restricted to the length of the scan chains. For test application reasons, these are in actuality limited; although technically feasible, one seldom meets scan chains of 10000 FFs.

Hence this solution helps ease up the problem of computing the costs matrix for the input of the TSP, which is actually the longest step of our methodology.

Another benefit is that when using partitioning, the whole method scales with only a linear increase in running times if the maximum size of scan chains is kept constant.

TSP Algorithms. We now discuss how TSP algorithms can be applied for the single-chain stitching problem.

There are two distinct frameworks for applying TSP algorithms to our problem. The first one allows any algorithm to be used, but it puts constraints onto the size of input graphs that can be fed to it.

The second one is the particular case of an algorithm that can be used directly on G_P, without having to compute costs for edges that are not present in G_P.

Using a Generic TSP Algorithm. The input of a TSP routine is a symmetric matrix representing the costs for edges of a complete graph. In order to apply a TSP algorithm to solve the chaining problem, one needs first to compute the whole costs matrix (except diagonal elements). In our model, we attribute (as cost) to a pair ij that is not an edge of G_P the length of a shortest path between i and j in G_P. The postprocessing is very simple: just turn the hamiltonian cycle returned by the TSP routine into a hamiltonian path by removing any edge.

It is the preprocessing step that imposes a serious limitation on the possibility to use this scheme. Indeed, the G_P graph obtained even with small values of the threshold parameter for the Limited-BFS step tends to be quite dense; hence any all-pairs shortest path algorithm puts limit to the input size either through space or time complexity, or both. In practice, it may be feasible to handle 5000 FFs or 10000 FFs blocks, but it is difficult to go much higher and keep a reasonable computation time for design scanning.

If only designs smaller than this are to be treated, then this method is definitely worth trying, all the more that one has then the ability to test and compare different TSP algorithms.

4.2.1. Scan Chain Ordering Using the 2-Approximation for the TSP. A famous textbook algorithm for the TSP, and also a typical example of an approximation algorithm with guaranteed quality, the 2-approximation for the TSP also has nice properties when used on the scan chaining problem.

An algorithm for a minimization problem is called an α-approximation when the solution returned by the algorithm is guaranteed to be no more than α times higher than the optimal solution. Usually, such algorithms presented in the literature are polynomial-time algorithms for NP-hard problems: since the requirement to find an optimal solution is dropped, polynomiality can be recovered.

It may seem unintuitive that one can prove a quality bound without even knowing the value of the optimal solution. Actually, such a proof can be derived through the use of a lower bound on the value of the optimal solutions, through showing that the solution given is less than α times this lower bound.

In the case of the TSP, two approximation algorithms are found in every textbook on the subject: one has an approximation factor of 2, (we call it "the 2-approximation for the TSP"), and the other is Christofides' algorithm, with an approximation factor of 1.5. Both use in their proof, as a lower bound on the optimal tour length, the value of a minimum spanning tree of the graph.

Please note that these approximation ratios are only proven theoretical bounds, and are not enough to compare the empirical behavior of these algorithms. Still, Christofides is bound to give solutions that are not more than 50% away from the optimal one; this would entice one to use this algorithm.

Alas, Christofides uses Weighted Perfect Matching in bipartite graphs as a subroutine, which has to impacts on its use. First, weights have to be precomputed, and we stumble against the same blocks as mentioned in the previous section. Second, Weighted Perfect Matching needs $O(n^3)$ operations, seriously limitting the size of its input.

Without any regret, we turn now to the 2-approximation algorithm, which is seldom considered a useful practical choice because of its loose guaranty, and not-so-good performance on common benches.

The algorithm is in two steps. First, a minimum spanning tree is computed; then a root is chosen, and the tree is explored and vertices ouput with postfix ordering.

Only the first step actually looks at the input graph; and its only requirement is that the graph be connected. If G_P is disconnected, we reconnect it by adding arbitrarily edges between connected components until it is connected.

Thus, using the 2-approximation, we bypass the precomputation of the cost matrix. This means that bigger input graphs can be considered if this algorithm is used, as compared to other TSP algorithms.

Although this algorithm was meant for the TSP, that is for an input being a complete graph with values on all edges, one can prove that the 2-approximation guarantee still holds when the algorithm is applied to a connected graph. In this case, the tour cost has to be understood as the sum of the cost of the edges, where edge costs for inexistent edges are length of shortest paths in the input graphs—hence our choice to model the cost of inexistent edges in this way.

4.2.2. RTL Edition. The final step is the edition of the RTL code of the design. In this step, we insert in the original RTL code/description the additional RTL constructions required to implement the scan testing logic.

5. Implementation and Experimental Results

5.1. Implementation. The method we propose to optimize the stitching ordering of RTL scan chains has been implemented in an experimental version of HiDFT-SIGNOFF. This tool relies on a commercial software library for the parsing and elaboration of the design. This elaboration step is what we called "lightweight synthesis" in Section 4.1.1. The generic library of gates used for the elaboration is the one that comes with the software library.

HiDFT-SIGNOFF has the ability to stitch FFs into several scan chains. This possibility is important from the practical point of view: making several scan chains is the easiest way to reduce test application time. HiDFT-SIGNOFF implements the two strategies presented in Section 4 for handling several chains. One is to stitch a unique scan chain, and then split it into even parts. The other is to first partition the design, and then to stitch a scan chain in each part.

For graph partitioning, the METIS library [24] is used. METIS is one of the most widely used libraries for graph partitioning. It includes two strategies P_{MET} and K_{MET}. P_{MET} uses a multilevel recursive algorithm. The second one, K_{MET}, is implemented with an algorithm of K-partition such that the number of vertices per partition is equal. We use here the second one to partition the graph G_p into K parts.

5.2. Experimental Setup. In order to validate our method, we did experiments on a number of designs, both in VHDL and Verilog. We used the benchmarks 99 ITC [25] which are reference designs for testing integrated circuits. Indeed, these designs have been used extensively in the literature for integrated circuits testing. For each design, a complete description of the circuit with its value is presented on the website. We also used 3 opencore designs [26]: Simple_spi which is a SPI core, Biquad which is a DSP core, and Ac-97, a controller core. All 3 are written in Verilog. These data sets were also used many times in articles on Design and Test.

Table 1 describes these benchmarks. The first line gives the name of the circuit (b09,…, b19) for ITC 99 and three OpenCore (single-spi, biquad, Ac-97). The following lines give the number of memory elements number of FF and the RTL description language (VHDL or Verilog) for each design.

Since our optimization criterion is wirelength, and since we consider congestion during routing an important issue, experimentations were conducted till the place and route step. Two flows were considered, according to Figure 1. In both cases, synthesis, placement and routing were done by Magma's Talus integrated flow. In the case of the gate-level scan insertion flow, the scan chain was also inserted by Talus. No effort options were set for gate-level scan insertion, since

Table 1: Designs description.

Design	No. of FF	Language
b09	28	vhdl
b10	17	vhdl
b11	31	vhdl
b12	121	vhdl
b13	53	vhdl
b14	245	vhdl
b15	449	vhdl
b17	1415	vhdl
b18	3320	vhdl
b19	6642	vhdl
Simple-Spi (SS)	132	verilog
Biquad	204	verilog
Ac-97	2289	verilog

Talus does not offer any. Besides those flows, the designs scanned at the RT level were also fed into Tetramax, in order to check fault coverage rates.

5.3. Numerical Results. Table 2 presents results of the first phase (extraction graphs $G_D = (V_D, E_D)$ and $G_p = (V_p, E_p)$). The columns 1 and 2 give the name of circuits tested (*Design*) and the size in number of memory elements (no. of FF). The following columns give the size of the design graph G_D respectively the number of vertices $|V_D|$ (column 3) and the number of edges $|E_D|$ (column 4). Columns 5 and 6 present the parameters of the proximity graph G_P with the number of vertices $|V_P|$ and the number of edges $|E_P|$. We observe that the number of vertices $|V_P|$ is equal to the number of memory elements of the design (column 2). The last column gives the CPU time for the extraction of the graph G_P from the graph G_D (Times).

The computation of the graph G_P from G_D (cf. Section 4.1.3) involves the threshold parameter t. The practical value of t has been determined empirically. It has to be big enough in order not to put arbitrary limitations on the choices of the algorithm; and it has to be small enough for the design-graph not to fill the whole memory of the computer. For all designs that we tested, the value $t = 4$ was a suitable choice according to those two criteria. This value was used in all our experiments.

The figures for both wirelength after place and route, and computation times, have been gathered in Table 2. The column number of SC gives the number of scan chains for each design. In the wirelength column, GLS denotes the flow with gate-level scan; RTL-scan gives the values for the RTL-scan flow. The slack column gives the ratio (in percent) by which RTL-scan wirelength exceeds that for gate-level scan.

Computation times are given only for the two steps of the stitching optimization method, discarding the time for parsing and elaboration. The column G_P gives the time for constructing the proximity-graph; TSP is the time for the 2-approximation algorithm for the TSP.

We base our analysis of the comparison of wirelength between gate-level and RTL scan on the slack column

of Table 3, since the absolute values cannot be directly interpreted. The slack varies between −23% (meaning RTL-scan wirelength is shorter) and 6%. A closer analysis shows that gate-level scan is better mainly for small designs; our method is always better for the bigger designs, with only negative values of slack for all designs with more than 300 FFs.

Table 4 gives the fault coverage obtained by Tetramax for the designs scanned at the RT-level. Let us note that all values are above 98%, and only two lay below 99.8%. We conclude that RTL scan insertion cannot degrade fault coverage too severely as compared to traditional scan.

Computation times are mainly dominated by the setup of the graph G_P; at worst, the whole optimization process takes a little more than 5 minutes, in the case of the b19 design which has 6000 FFs.

6. Conclusion and Perspectives

After giving some motivations for inserting scan at the RT-level, we have exposed what we believe is an important challenge for RTL-scan, which is finding a good stitching in one single pass, working only at the RT-level. Therefore, the purpose of the present work is to solve the following problem: given an RT-level description of the design (Verilog or Vhdl) we have to find a stitching ordering of the memory elements in the scan chain, which minimizes the impact of test circuitry on chip performance (area, power) and testing time. Then, we have motivated our choice of selecting wire length as the prime parameter to optimize.

To solve this problem, we proposed a new scan stitching algorithm for optimal scan chain insertion at RTL. The techniques used are derived from the combinatorial optimization and operations research domains. Indeed, our algorithm is divided into two main steps. The first step proposes a mathematical model describing the electronic problem. The second one offers a resolution methodology.

The model we propose is based on graph theory. To build it, we extract from the RTL description information on the proximity of the memory elements (existing paths between flip-flops, clock domains, and various other relations extracted from hierarchical analysis) and translate them in two graphs G_D and G_P using the Limited-BFS algorithm. Then, we have shown that the scan stitching decision problem is equivalent to the Traveling Salesman Problem and therefore *NP-Complete*. However, the 2-approximation for the TSP can be used to find near optimal solutions in polynomial time. From this, we developed our stitching algorithm.

Finally, we integrated our tool in an industrial design flow and performed experiments over several academic and industrial design benchmarks. Numerical evidence showed that we are able to limit such a cost due to scan insertion in a reasonable computing time, without impacting DFT quality, especially fault coverage. The method seems better than the traditional one for middle-sized designs. The industrial interest for our algorithms and tools is confirmed by our industrial partners. Our RT-level scan optimization algorithm has been incorporated into the tool HiDFT-SIGNOFF by DeFacTo Technologies. In case new flip-flops

TABLE 2: Characteristics of the graphs G_D and G_p.

Design	No. of FF	$|V_D|$	$|E_D|$	$|V_P|$	$|E_P|$
b09	28	401	719	28	378
b10	17	525	947	17	136
b11	31	1094	1845	31	465
b12	121	3879	6993	121	7260
b13	53	633	1042	53	1378
b14	245	27269	46778	245	29890
b15	449	33179	56798	449	100576
b17	1415	100358	171759	1415	697990
b18	3320	271320	462326	3320	2330318
b19	6642	531161	906201	6642	7061713
Simple-Spi	132	1427	8646	132	8778
Biquad	204	357	20706	204	20910
Ac-97	2289	18727	2381495	2289	707911

TABLE 3: Wirelengths and insertion times.

Design	No. of SC	Wirelength		Slack (%)	Insertion time (ms)	
		GL-S	RTL-S		P-graph	TSP
ITC 99 Benchmarks (VHDL)						
b09	2	2.06	1.63	−20.59	<1	<1
b10	2	1.94	2.06	5.97	10	<1
b11	3	4.77	5.03	5.38	20	10
b12	10	12.6	12.92	2.59	200	20
b13	5	3.39	3.32	−2.25	20	<1
b14	24	63.64	64.2	0.96	3 s	110
b15	24	126.3	122	−3.39	6 s	320
b17	70	395.6	370.4	−6.37	30 s	3 s
b18	300	1187.2	922.6	−22.29	3 m	10 s
b19	600	2329.6	1858.2	−20.24	5 m	20 s
Opencore designs (Verilog)						
Simple-Spi	10	11.5	10.4	−8.95	100	20
Biquad	20	34.1	31.3	−8.14	350	80
Ac-97	200	247.7	238.5	−3.71	30 s	8 s

TABLE 4: Fault coverage for RTL scan.

Design name	Fault coverage
ITC 99 Benchmarks (VHDL)	
b09	99.86%
b10	99.85%
b11	99.93%
b12	99.97%
b13	99.92%
b14	99.99%
b15	99.97%
b17	99.47%
b18	99.81%
b19	99.81%
Opencore designs (Verilog)	
Simple Spi	98.35%
Biquad	99.96%
Ac-97	99.80%

are added or removed to existing scan chains, it is important to goldenize the RTL code and reflect such changes directly into the RTL code. The DeFacTo tool HiDFT-SIGNOFF allows that by introducing the new scan architecture and by including the flip-flops new ordering. In this way, our work represents a progress in the state of the art, as previous works are not automated and/or were evaluated only for small designs.

Last but not least, an important contribution of our work is to make a neat separation between the DFT problem and the mathematical model. This separation allows the same software to work for several successive technology nodes.

Our initial results give rise to a number of new directions for further research. These are summarized below. First, one could investigate whether cumulating local and global optimization in the manner of [8] would improve our results. Another aspect that could be added to our model is the question of power during testing: can we take it into account without degrading ou results? Finally, design analysis at a higher-level gives an opportunity to reduce both area overhead and test time application by adopting partial scan instead of full scan [16–18]. Deciding whether the combination of design analysis with our method could reduce even more the cost of scan would be worthwhile.

References

[1] B. Korte, L. Lovasz, H. J. Promel, and A. Schrijver, *Paths, Flows, and VLSI-Layout*, Springer, New York, NY, USA, 1990.

[2] S. H. Gerez, *Algorithms for VLSI Design Automation*, John Wiley & Sons, Chichester, UK, 1998.

[3] L. T. Wang, Y. W. Chang, and K. T. Cheng, *Electronic Design Automation: Synthesis, Verification, and Test*, Morgan Kaufmann, Boston, Mass, USA, 2009.

[4] Semiconductors Industry Association, *Test and Test Equipment. Update*, International Technical Roadmap for Semiconductor, 2010.

[5] L. T. Wang, C. E. Stroud, and N. A. Touba, *System-on-Chip Test Architectures: Nanometer Design For Testability*, Morgan Kaufmann, Boston, Mass, USA, 2008.

[6] S. Roy, "RTL based scan BIST," in *Proceedings of the VHDL International Users' Forum (VIUF '97)*, pp. 117–121, October 1997.

[7] C. Aktouf, H. Fleury, and C. Robach, "Inserting scan at the behavioral level," *IEEE Design and Test of Computers*, vol. 17, no. 3, pp. 34–42, 2000.

[8] Y. Huang, C. C. Tsai, N. Mukherjee, O. Samman, W. T. Cheng, and S. M. Reddy, "Synthesis of scan chains for netlist descriptions at RT-Level," *Journal of Electronic Testing*, vol. 18, no. 2, pp. 189–201, 2002.

[9] M. Hirech, J. Beausang, and X. Gu, "New approach to scan chain reordering using physical design information," in *Proceedings of the IEEE International Test Conference (ITC '98)*, pp. 348–355, October 1998.

[10] D. Berthelot, S. Chaudhuri, and H. Savoj, "An efficient linear time algorithm for scan chain optimization and repartitioning," in *Proceedings of the International Test Conference (ITC '02)*, pp. 781–787, Washington, DC, USA, October 2002.

[11] P. Gupta, A. B. Kahng, and S. Mantik, "Routing-aware scan chain ordering," *ACM Transactions on Design Automation of Electronic Systems*, vol. 10, no. 3, pp. 546–560, 2005.

[12] P. Gupta, A. B. Kahng, I. I. Măndoiu, and P. Sharma, "Layout-aware scan chain synthesis for improved path delay fault coverage," *IEEE Transactions on Computer-Aided Design of Integrated Circuits and Systems*, vol. 24, no. 7, pp. 1104–1114, 2005.

[13] B. B. Paul, R. Mukhopadhyay, and I. S. Gupta, "Genetic algorithm based scan chain optimization and test power reduction using physical information," in *Proceedings of the IEEE Region 10 Conference (TENCON '06)*, Hong Kong, November 2006.

[14] K. D. Boese, A. B. Kahng, and R. S. Tsay, "Scan chain optimization: heuristic and optimal solutions," Internal Report CS Department, University of California at Los Angeles, Los Angeles, Calif, USA, 1994.

[15] S. Makar, "Layout-based approach for ordering scan chain flip-flops," in *Proceedings of the IEEE International Test Conference (ITC '98)*, pp. 341–347, October 1998.

[16] S. Bhattacharya and S. Dey, "H-SCAN: a high level alternative to full-scan testing with reduced area and test application overheads," in *Proceedings of the 14th IEEE VLSI Test Symposium*, pp. 74–80, Princeton, NJ, May 1996.

[17] T. Asaka, S. Bhattacharya, S. Dey, and M. Yoshida, "H-SCAN+: a practical low-overhead RTL design-for-testability technique for industrial designs," in *Proceedings of the IEEE International Test Conference (ITC '97)*, pp. 265–274, Washington, DC, USA, November 1997.

[18] R. B. Norwood and E. J. McCluskey, "High-level synthesis for orthogonal scan," in *Proceedings of the 15th VLSI Test Symposium*, pp. 370–375, May 1997.

[19] Private communication.

[20] L. Zaourar and Y. Kieffer, "Scan chain optimization for integrated circuits testing : an operations research perspective," in *Proceedings of the 23rd European Conference on Operational Research (EURO '09)*, p. 289, Bonn, Germany, July 2009.

[21] L. Zaourar, Y. Kieffer, C. Aktouf, and V. Julliard, "global optimization for scan chain insertion at the RT-level," in *Proceedings of the IEEE Computer Society Annual Symposium VLSI (ISVLSI '11)*, pp. 321–322, University Pierre and Marie Curie, Paris, France, July 2011.

[22] L. Zaourar, Y. Kieffer, and C. Aktouf, "An innovative methodology for scan chain insertion and analysis at RTL," in *Proceedings of the Asian Test Symposium (ATS '11)*, pp. 66–71, IEEE Computer Society, Washington, DC, USA, November 2011.

[23] W. J. Cook, W. H. Cunningham, and W. R. Pulleyblank, *Combinatorial Optimization*, Wiley-Interscience, New York, NY, USA, 1997.

[24] http://glaros.dtc.umn.edu/gkhome/views/metis/.

[25] http://www.cerc.utexas.edu/itc99-benchmarks/bench.html.

[26] http://www.opencores.org/.

12

A 0.6-V to 1-V Audio ΔΣ Modulator in 65 nm CMOS with 90.2 dB SNDR at 0.6-V

Liyuan Liu,[1,2] Dongmei Li,[1] and Zhihua Wang[1]

[1] *Tsinghua University, Beijing 100084, China*
[2] *Institute of Semiconductors, Chinese Academy of Sciences, Beijing 100083, China*

Correspondence should be addressed to Liyuan Liu; liuly@semi.ac.cn

Academic Editor: Jose Silva-Martinez

This paper presents a discrete time, single loop, third order ΔΣ modulator. The input feed forward technique combined with 5-bit quantizer is adopted to suppress swings of integrators. Harmonic distortions as well as the noise mixture due to the nonlinear amplifier gain are prevented. The design of amplifiers is hence relaxed. To reduce the area and power cost of the 5-bit quantizer, the successive approximation quantizer with only a single comparator instead of traditional flash quantizer is employed. Fabricated in 65 nm CMOS, the modulator achieves 95 dB peak SNDR at 1-V supply with 24 kHz. Thanks to low swing circuit techniques and low threshold voltages of devices, the peak SNDR maintains 90.2 dB under 0.6-V low supply. The total power dissipation is 371 μW at 1-V and drops to only 133 μW at 0.6-V.

1. Introduction

CMOS technology has progressed into sub-100 nm era. Advanced technology offers speed power and area benefits for digital circuits design. However the sub 1-V low supply voltage, reduction of intrinsic gain, deteriorated device matching, and dramatic increase of flicker noise have brought difficulties to precision analog designs [1]. The ΔΣ modulator is a popular building block which is strongly demanded in high accurate analog signal conditioning such as digital audio. Several designs have been reported with sub-100 nm technology. Although large dynamic range has been achieved, the SNDR performance is lower than 80 dB due to large flicker noise and high harmonic distortion [2–5].

In [4, 5], continuous time ΔΣ modulator (CT-DSM) is adopted for low power implementation. Although the bandwidth requirement of the amplifier is relaxed in CT-DSM than that in discrete-time ΔΣ modulator (DT-DSM), CT-DSMs suffer from problems such as sensitivity to clock jitter, performance degradation due to excessive loop delay, and needing tuning circuits to deal with RC time constant variations.

In this work, we propose a high performance DT-DSM. The feedforward path together with multibit quantizer leads to very small swings at integrators' outputs. Both gain and bandwidth requirements of amplifiers are relaxed. To overcome large area and power dissipation problems associated with traditional flash multibit quantizer and conventional analog summing, in this work we propose a self-timing successive approximation (SAR) quantizer with embedded analog summing circuitry. The prototype modulator is fabricated in 65 nm CMOS. Measurement shows that peak signal to noise and distortion ratio (SNDR) of 95 dB is achieved across 24 kHz under 1-V with 371 μW. Under 0.6-V supply, the peak SNDR is still 90.2 dB. The power consumption reduces to only 133 μW.

In Section 2, the architecture of the modulator is described. Section 3 discusses detail circuits' implementations. Section 4 shows the measurement results and Section 5 concludes the paper.

2. Low Voltage Modulator Architecture

2.1. Design Considerations of the Modulator. A major challenge in low voltage design is that circuits must operate in limited voltage headroom conditions. Modulator topology with low swing is thus necessary. Shown in Figure 1 is

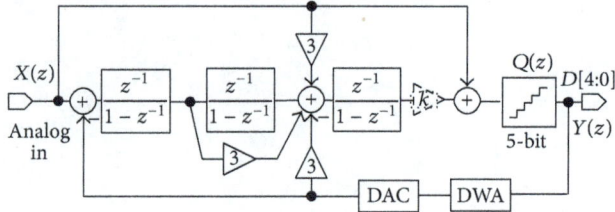

FIGURE 1: The 3rd order CIFB modulator with 5-bit quantizer.

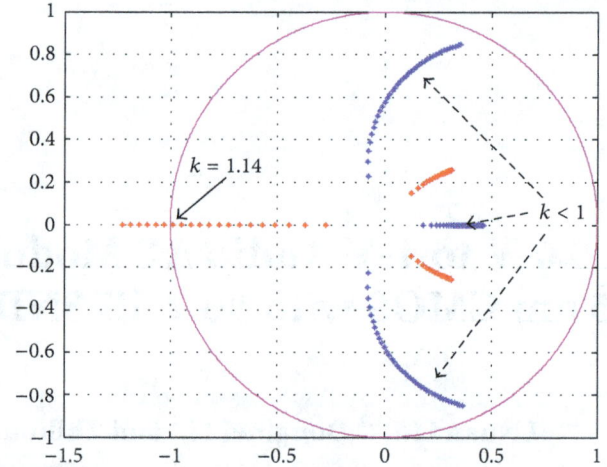

FIGURE 2: Poles location of the NTFs considering gain factor of k.

the modulator architecture. It is a 3rd order CIFB modulator reported in [6, 7]. By introducing a directly input feedforward path, signal components are removed from outputs of integrators. Swings of integrators are independent of input signal's amplitude and are only determined by quantization error. By employing a 5-bit quantizer, the quantization step is fine and signal swings are even smaller.

The modulator coefficients are chosen as shown in Figure 1. Ideally the noise transfer function (NTF) is $(1-z^{-1})^3$. Such kind of NTF is aggressive and modulator stability must be considered. Although a 5-bit quantizer can guarantee that the modulator is stable with an overload level of 0.9, the gain variation of the last integrator may do harmful to the modulator stability. Such variation is caused by nonideality of the analog summation which will be described later.

Suppose a gain factor of k is introduced as depicted in Figure 1. The transfer function of the modulator can be derived as (1) with the quantizer replaced by its linear model,

$$Y(z) = X(z) + \underbrace{\frac{(z-1)^3}{(z-1)^3 + 3k(z-1)^2 + 3k(z-1) + k}}_{\text{NTF}} Q(z).$$

$$(1)$$

Poles of the NTF can be found in

$$p_1 = -\sqrt[3]{k(k-1)^2} - \sqrt[3]{k^2(k-1)} - (k-1),$$

$$p_2 = \frac{\sqrt[3]{k(k-1)^2}}{2} + \frac{\sqrt[3]{k^2(k-1)}}{2} + (1-k)$$
$$+ \frac{\sqrt{3}j}{2}\left(\sqrt[3]{k^2(k-1)} - \sqrt[3]{k(k-1)^2}\right), \quad (2)$$

$$p_3 = \frac{\sqrt[3]{k(k-1)^2}}{2} + \frac{\sqrt[3]{k^2(k-1)}}{2} - (1-k)$$
$$+ \frac{\sqrt{3}j}{2}\left(\sqrt[3]{k^2(k-1)} - \sqrt[3]{k(k-1)^2}\right).$$

Figure 2 shows the location of poles with k varying from 0.7 to 1.2. If k is slightly larger than its critical value 1.14, one pole of the NTF will move out of the unit circle which means that the modulator is unstable regardless of the input signal amplitude. To prevent such a problem, in circuits' implementations the gain factor of last integrator is shrunk deliberately to compensate k.

The gain factor of the directly input feedforward path must be strictly set to unity so that signal components inside the loop filter are forced to be zero. Otherwise, there is signal leakage in the loop filter and the swings of integrators increase. Furthermore, because of nonlinear gain of OTAs, harmonic distortions will be observed in the output spectrum. This problem will be discussed in the design of the quantizer.

2.2. *Multibit Quantizer.* Employing multibit quantizer is popular in high performance modulator design. One benefit is that adoption of multibit can greatly enlarge the overload level (OL). Another advantage is the fine step in quantization which leads to low quantization noise power. This feature is helpful for alleviating the quantization noise mixture effect due to nonlinear OTA gain.

Although with advantages mentioned above, there are also drawbacks in traditional multibit quantizer design. Usually flash quantizer is preferred because of its low latency which is important for the stability of the modulator. Such implementation suffers from problems such as large area and power consumption cost, because there is an exponential dependence of the hardware and power budget on the bits of the quantizer. Several methods have been proposed to overcome these drawbacks. The first method is employing interpolated multilevel quantizer reported in [8]. The key idea is that comparison procedure is separated into coarse and fine steps. However such a method only achieves 50% reduction of the number of comparators. In [8] fifteen comparators are used to implement a 5-bit quantizer which traditionally requires 31 comparators. The second solution is called tracking ADC in [3] which only needs 3 comparators to realize a 4-bit quantizer. In [9] a technique is proposed to reduce the number of comparators to only one. Although reducing the amount of comparators to the minimum, the tracking quantizer suffers from problems such as needing properly defined initial operation point and sensitivity to high-frequency signal leakage. The third way is replacing flash quantizer with successive approximation (SAR) type

[10]. This quantizer also has only a single comparator, but it is free from the problem encountering in tracking quantizer. The problem with SAR quantizer is that serial operation needs fast clock. In this work, a SAR based 5-bit quantizer is adopted to reduce the area and power cost. To mitigate the requirement of fast clock generation, asynchronous control logic is used.

Ahead of the quantizer, an analog summer is needed. Classic capacitive charge sharing based summation is popular, but signal is attenuated by a factor equal to the number of capacitors. For a single bit modulator such attenuation is not important because the quantizer only needs to determine the sign of its input. For a multibit quantizer, this attenuation must be compensated by amplification or deliberated reference voltages scaling. Otherwise the transfer function of the modulator will be changed. In this work by introducing additional switches and clock phase, analog summation can be embedded before normal operation of the SAR quantizer without significant signal attenuation.

3. Circuits Design

Figure 3 illustrates the switched capacitor scheme of the ΔΣ modulator. Compared to Figure 1, the gain factor from output of the 1st integrator to the last integrator output is scaled down to 31/12, the factor from output of the 2nd integrator to the last integrator output is scaled down to 10/12, and gain of the last integrator is now set to 10/12. Such modification of coefficients is important to ensure the stability of the modulator as discussed in Section 2.1. The overall circuits contain three switched capacitor integrators and a 5-bit SAR quantizer. The input sampling capacitor is divided into 31 units to implement the 5-bit feedback DAC. Each unit has a value of 527 fF and the total sampling capacitance is 16.3 pF for low thermal noise. The unit capacitor for the 2nd and 3rd stages is 54 fF because of much less noise contribution. The first and last integrators share the switch network which reduces amount of switches. The directly feedforward input signal and the output of the last integrator are summed and quantized into 5-bit digital codes by SAR quantizer. The digital output is decoded to thermometer codes and feedback through a 5-bit DAC. To alleviate distortion due to mismatch among unit capacitor in the DAC, dynamic element matching (DEM) employing data weighted averaging (DWA) is adopted. Integrators are controlled by double phase nonoverlapping clock. The first stage OTA is chopped to remove the flicker noise out of the signal band. Chopping operation happens in the middle of sampling phase Φ_1 to prevent interference [11]. The control clocks of the input and output chopper circuits are also nonoverlapped. The SAR quantizer gives out binary codes which are firstly converted into thermometer code and are then used as switches control signals in the DAC.

3.1. OTA. Figure 4 depicts the current mirror OTA for integrators in this design. Looking like a cascade stages amplifier, the first stage contributes nondominant pole and dominant pole locates at the output. The OTA functions like a single stage amplifier which is stabilized by load compensation. Comparing to a miller amplifier, output stages in such an amplifier do not demand large current for stabilization and hence power is saved.

The drawback of the current OTA is low dc gain. Two cross coupled NMOS transistors $M_{ca,b}$ are placed in parallel with the diode connected NMOS load $M_{2a,b}$ in order to enhance the gain of the amplifier to approximately 40 dB [12]. Benefit from feedforward path and 5-bit quantizer, swings of integrators are small. Simulation indicates that maximum voltages of integrators are within ±0.124 V, ±0.064 V, and ±0.218 V which are 6.2%, 3.2%, and 10.9% of full swing ($2V_{pp}$), respectively. Although the OTA gain varies nonlinearly as shown in (3), small swings make such variations less significant,

$$A_v = 94 \left(-8V_o^4 - 1.3V_o^2 + 1 \right). \tag{3}$$

In (3) only even order nonlinear coefficients are considered owing to the symmetric characteristic of fully differential OTA. With (3), behavior simulation shows that modulator can achieve ideal SNDR of 119 dB. Further transistor level simulation shows better than 110 dB SNDR is maintained. The OTAs in the first and the third stages have the same transistor sizes. The bias current is 25 μA for every branch. The load capacitance of the 1st amplifier is approximately the same as the sampling capacitance which is 16.3 pF, the GBW of is then 9.5 MHz which is three times larger than the sampling frequency in order that insufficient settling does not degrade the modulator performance much. For the 2nd stage, since the capacitive load is quite small, the transistors' size and bias currents of the amplifier are half of those in the 1st and 3rd stages. The input common mode voltage (VCMI) is set by the biasing circuits in Figure 4. The benefit of such biasing method is that it can track the process, supply voltage, quiescent current variation so that $M_{1a,b}$ and tail current transistor M_t are always in the saturation range. In high resolution modulator design, circuit noise must be carefully treated. The first stage contributes the most part of noise and can be expressed as

$$N_T = \frac{4kT \left(1 + 2\beta\gamma/3 \right)}{OSRC_S}. \tag{4}$$

In the above expression, k is the Boltzmann constant (1.38×10^{-23} J/K) and T is the absolute room temperature (300 K). The factor γ is the noise excessive factor of OTA and is found as (5) by noise analysis,

$$\gamma = 1 + \frac{g_{m2a} + g_{mca}}{g_{m1a}} + \frac{(g_{m3a} + g_{m4a}) g_{m2a}^2}{g_{m1a} g_{m3a}^2}. \tag{5}$$

All the transistors in Figure 4 are sized to have identical overdriver voltage of 80 mV and thus γ is calculated to be 2.4. Besides thermal noise, flicker noise should also be considered because in audio bandwidth from 20 Hz to 24 kHz the power of such noise is significant. Figure 5 shows the noise figure of the first OTA. Below 10 kHz, flicker noise dominates. Between 10 kHz and 100 kHz, contributions of thermal noise

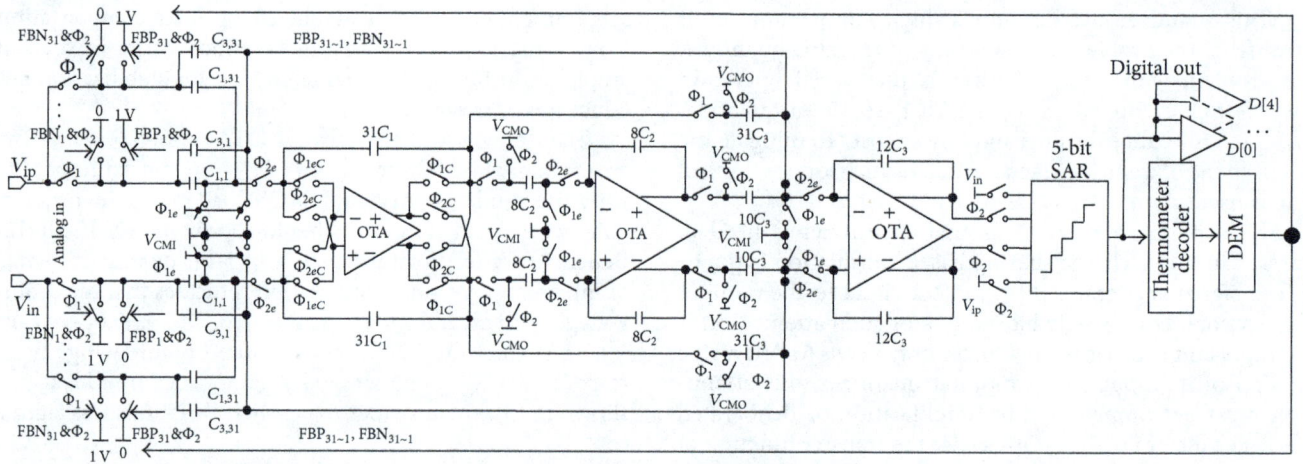

FIGURE 3: Switched capacitor scheme of the modulator.

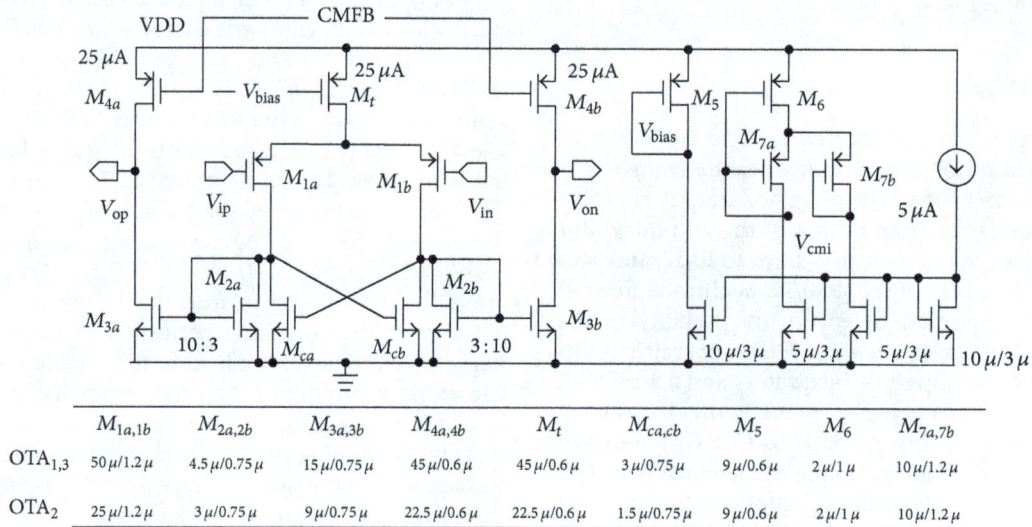

	$M_{1a,1b}$	$M_{2a,2b}$	$M_{3a,3b}$	$M_{4a,4b}$	M_t	$M_{ca,cb}$	M_5	M_6	$M_{7a,7b}$
OTA$_{1,3}$	$50\,\mu/1.2\,\mu$	$4.5\,\mu/0.75\,\mu$	$15\,\mu/0.75\,\mu$	$45\,\mu/0.6\,\mu$	$45\,\mu/0.6\,\mu$	$3\,\mu/0.75\,\mu$	$9\,\mu/0.6\,\mu$	$2\,\mu/1\,\mu$	$10\,\mu/1.2\,\mu$
OTA$_2$	$25\,\mu/1.2\,\mu$	$3\,\mu/0.75\,\mu$	$9\,\mu/0.75\,\mu$	$22.5\,\mu/0.6\,\mu$	$22.5\,\mu/0.6\,\mu$	$1.5\,\mu/0.75\,\mu$	$9\,\mu/0.6\,\mu$	$2\,\mu/1\,\mu$	$10\,\mu/1.2\,\mu$

FIGURE 4: OTAs for integrators.

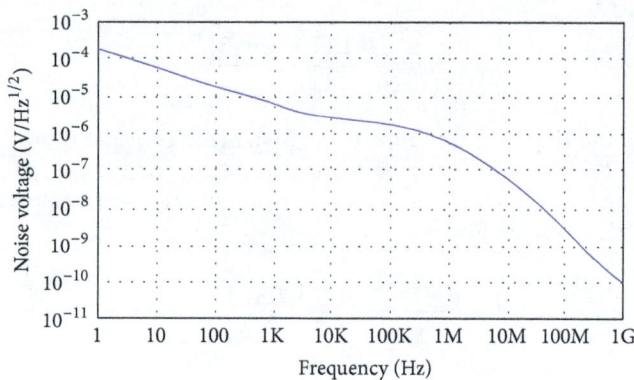

FIGURE 5: Noise figure of the first OTA.

to Figure 5 a pessimistic estimation of the OTA corner frequency is around 100 kHz. To remove the flicker noise in low frequency range, the first OTA is chopped at 48 kHz. Although large portion of such noise can be moved to base and odd order harmonics of chopping frequency, there is still residue noise. This effect is modeled by factor β in (4). According to analysis in [13], the residue noise and β can be expressed as

$$\beta = \left(1 + 0.8525 f_k T_C\right), \tag{6}$$

in which f_k is the corner frequency and T_C is the chopping frequency. It is found to be 2.8 in this design. The total noise N_T can now be evaluated. With 16.3 pF sampling capacitance, theoretical maximum signal-to-noise ratio is better than 100 dB which is enough for high fidelity audio applications.

3.2. SAR Quantizer. Figure 6 depicts the 5-bit SAR quantizer. It is composed of a charge sharing DAC, a comparator, and

gradually get larger. Above 100 kHz the noise power density drops because of OTA's frequency characteristic. According

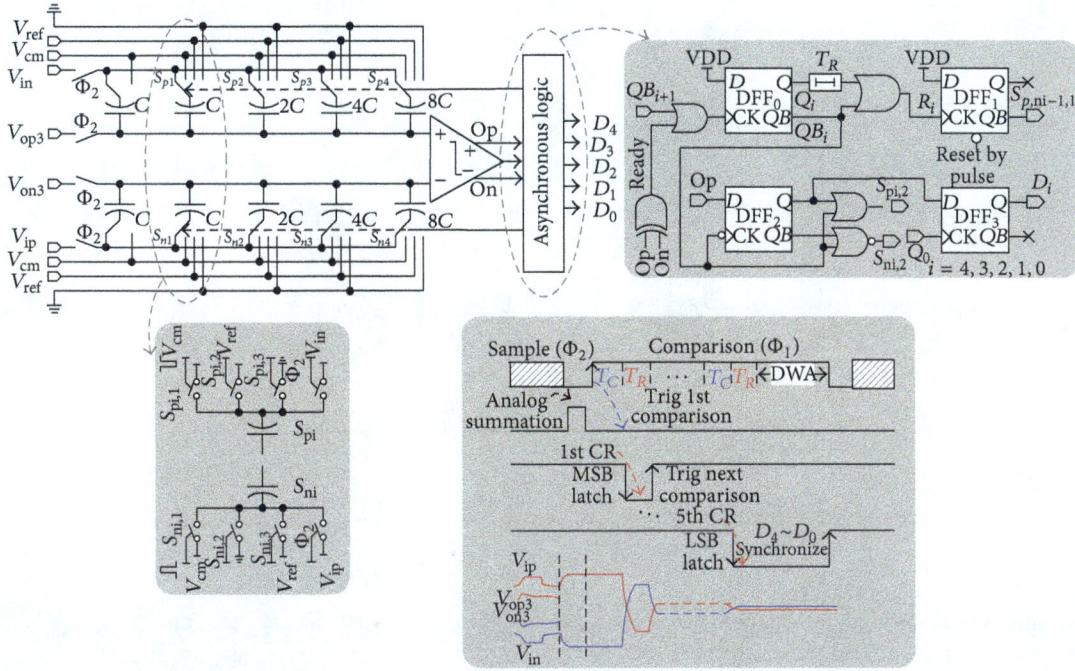

FIGURE 6: The self-timing 5-bit successive approximation quantizer.

asynchronous control logic. During Φ_2, the top plates of capacitors are tracking the last integrator outputs. At the same time, the bottom plates are connected to input signal V_{in} and V_{ip}, respectively. At the end of Φ_2, V_{op3}, V_{on3} and V_{in}, V_{ip} are sampled. Charges stored on the capacitor array are $(V_{op3} - V_{in}) C_{total}$ and $(V_{on3} - V_{ip}) C_{total}$, where $C_{total} = 16\,C$. In the nonoverlapped time between the falling edge of Φ_2 and rising edge of Φ_1 a short pulse is generated. The bottom plates are connected to a common mode voltage (V_{CMO}) which is in the middle of reference voltage (V_{ref}) and ground. Regardless of the parasites, voltages at both terminals of the comparator become $V_{op3} - V_{in} + V_{CMO}$ and $V_{on3} - V_{ip} + V_{CMO}$, respectively. The voltage difference is hence

$$V_{diff} = \left(V_{op3} - V_{on3}\right) + \left(V_{ip} - V_{in}\right). \qquad (7)$$

Parasitic capacitances at the input terminals of the comparators are composed of gate-source capacitance of input transistors, parasitic capacitance from top plates of capacitor to the substrate, and parasitic contributed by switches. Denoted as C_P herein, voltage difference after analog summation is

$$V_{diff} = \left(V_{op3} - V_{on3}\right) + \frac{C_{total}}{C_{total} + C_P}\left(V_{ip} - V_{in}\right). \qquad (8)$$

Clearly, the input signal is attenuated after the summation which causes signal leakage in the modulator without any compensation. However, V_{ref} is also connected to the bottom plates of capacitors. In the quantization phase, the parasitic also brings the same attenuation to V_{ref}. Scaling down of V_{ref} implies that an equivalent gain of $(C_{total} + C_P)/C_{total}$ exists

in the quantizer. As a result, direct input signal path still has a unity gain which prevents leakage problem. However an extra gain is introduced to the last integrator's output. According to the analysis in Section 2, the modulator may become unstable. Ideally, the integration capacitor of the last stage should compose 10 unit capacitors in order that the noise shaping function is $(1 - z^{-1})^3$. To mitigating instability, in the circuits design, we choose 12 unit capacitors to provide an attenuation which can compensate the equivalent gain by the quantizer.

After the analog summation, the SAR quantizer generates 5-bit digital codes serially. Asynchronous timing control logic is used. The operation is triggered by the rising edge of Φ_1. The comparator is activated and makes decision of the sign of the voltage difference between its two inputs after its intrinsic delay of T_c.

Because both outputs of the comparator are forced to ground before activation, the XOR gate outputs "0." When the comparison is completed, XOR gate outputs "1" and this rising edge will trigger $DFF_{0,i}$, flip. All the flip-flops are reset when Φ_1 is low. Hence the flipping of $DFF_{0,i}$ causes R_i go low immediately which simultaneously deactivate the comparator. Because of the delay, R_i will keep at low for T_R. During T_R, the charge redistribution happens. $S_{pi,2}$ and $S_{ni,2}$ are generated according to comparator's decision which makes correspondence bottom plates of capacitors connect to V_{ref} or ground. The purpose of $DFF_{1,i}$ is ensuring the bottom plates of correspondence capacitors keep connecting to V_{CMO} when the preceding charge redistribution is ongoing. After all operation cycles, comparison results D_i ($i = 4, 3, 2, 1, 0$) are synchronized and latched on $DFF_{3,i}$. Using the

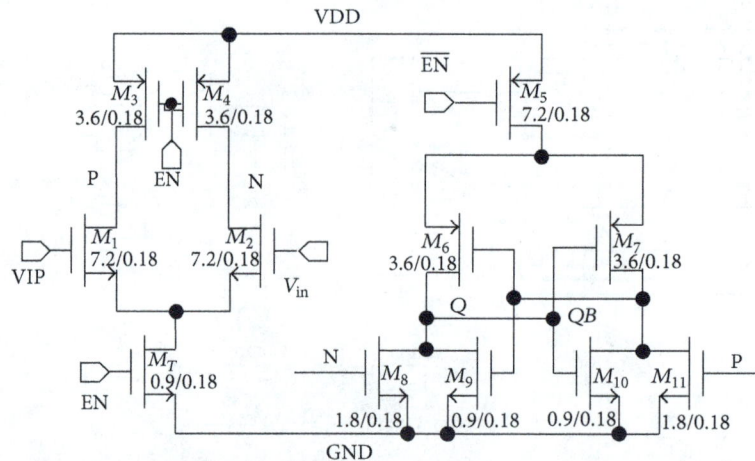

FIGURE 7: Comparator.

asynchronous logic, Φ_2 is divided into comparison phases and charge redistribution phases automatically without fast clock generation.

The advantages of a SAR quantizer over a flash quantizer are as follows. Firstly, the number of comparator in the quantizer is reduced to only one. Hence area cost is greatly decreased. Although the comparator must work N times faster than that in an N-bit flash quantizer, the number of comparator is reduced by 2^N and thus the overall power dissipation is saved. Secondly, compared to a flash quantizer, the offset of comparator in a SAR quantizer only brings an offset voltage to the output spectrum without disturbing the noise floor. For a flash quantizer, however, distribution of different offset voltages due to MOS mismatch among comparators alters the quantization step. Some steps become larger and others get smaller. Enlarging of steps causes the quantization error power to increase. With nonlinear gain as mentioned above, noise floor will increase. To prevent this, the offset of each comparator must be well controlled and preamplifier with offset cancellation is required which further increases power consumption. Although in a SAR quantizer, the mismatch among capacitors will also disturb the quantization step, such variation of step is much smaller because of excellent capacitor matching which is exactly the case under advanced technology owing to improved etching precision.

The comparator in the design is shown in Figure 7 [14]. When the comparator is reset, nodes N and P are charged to supply. The enable signal makes M_T conduct. Discharging current of M_1 and M_2 will pull low the common mode voltage of N and P which makes the next stage latch functional. In a very short time zone, $M_1 \sim M_4$ are all in saturate range and the voltage difference between V_{ip} and V_{in} can be magnified which finally causes the latch flip and generates the decision result.

The noise of such comparator is analyzed in detail in [15] which shows that the noise power is inversely proportional to the parasitic capacitance at nodes N and P. Thanks to the 3-stage integration before the quantizer, the equivalent noise

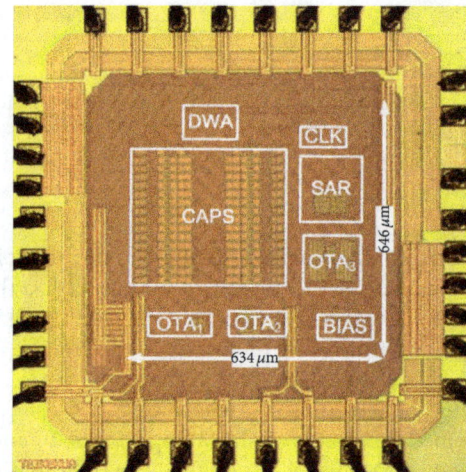

FIGURE 8: Chip microphotograph.

contribution of the comparator at the input of the modulator is greatly suppressed. Hence the sizes of transistors in the comparator can be quite small which saves silicon area.

4. Experiment Results

The prototype modulator is fabricated in 65 nm general purpose 1P9M CMOS. Only regular threshold voltage nMOS and pMOS transistors are used. The chip micrograph is shown in Figure 8. The total die area is 1.28×1.33 mm^2 with bonding pads and the active core only occupies 0.41 mm^2.

4.1. Measurement Setup. We designed a four-layer printed circuit board (PCB) to evaluate the performance of the chip. An integral ground plane is included. Decoupling capacitors with value of 3.3 μF are connected between supplies and ground for voltage regularity. At all the digital pads, resistors of 50 Ω are connected so as to reduce signal ringing. Input signal is provided by DS360 which is an ultra low noise and

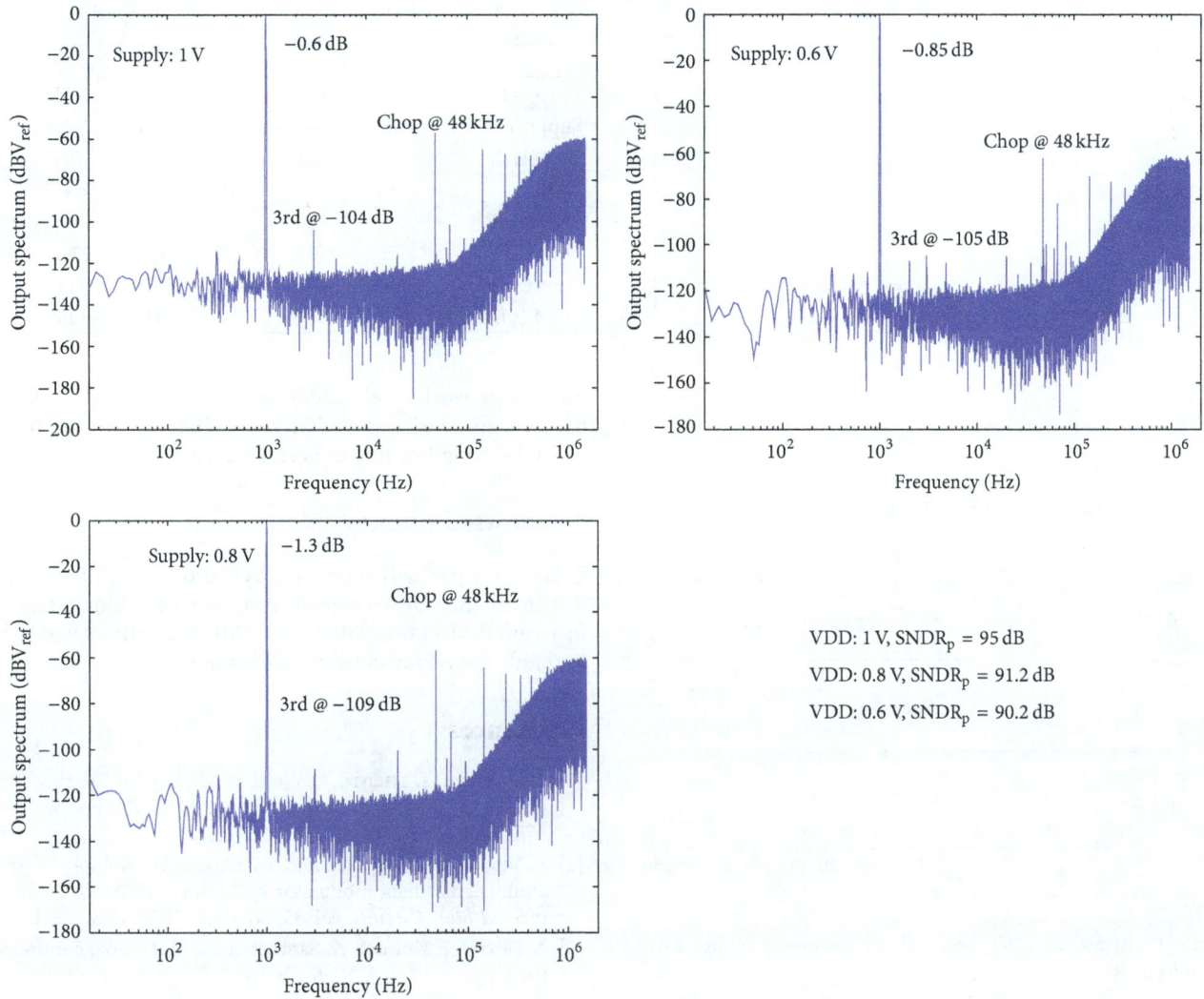

VDD: 1 V, SNDR$_p$ = 95 dB

VDD: 0.8 V, SNDR$_p$ = 91.2 dB

VDD: 0.6 V, SNDR$_p$ = 90.2 dB

FIGURE 9: Measured output spectrum.

low distortion signal generator. If the output sine wave has amplitude less than 1.26 V, the noise floor of the equipment is less than $15\,\mathrm{nV}/\sqrt{\mathrm{Hz}}$ which leads to integrated noise of only $2.3\,\mu V_{\mathrm{rms}}$ with 24 kHz bandwidth. The total harmonic distortion (THD) is less than −110 dBc. The five-bit digital codes are captured by logic analyzer and MATLAB program is used to do the spectrum analysis.

4.2. Performance Measurement. Figure 9 shows the output spectrum. The frequency of the input is chosen as 997 Hz which is a standard in audio measurement. Measurements are done under different supply voltages. The measured peak SNDR is 95 dB, 91.2 dB, and 90.2 dB corresponding to 1 V, 0.8 V, and 0.6 V supply. Our design operates well with only 0.6 V supply. The reasons are firstly input feedforward combined with 5-bit quantizer which makes the signal swing at each integrator output quite small. Although the supply is lowered down, there is still enough voltage headroom for internal amplifiers. Secondly, low device threshold voltage

under 65 nm helps amplifiers maintain their proper operational range.

Tones are observed at odd times of Nyquist frequency which shows the effectiveness of chopper. Thermal noise dominates the performance. Figure 10 gives the SNDR performance versus varying input signal amplitude. The measured dynamic range of the modulator is 96.4 dB, 92.4 dB, and 91.9 dB, respectively.

The total power consumption of the modulator is measured with the help of pico-ampere instrument because the voltage drop introduced by the instrument is less than $200\,\mu V$ which does not cause any disturbance to the normal operation of the circuits. Under 1 V supply, measured power dissipation of the modulator is $371\,\mu W$ and it drops $133\,\mu W$ under 0.6 V.

The SAR quantizer inside the loop can be measured separately. Figure 11 shows the differential nonlinearity (DNL) and integral nonlinearity (INL). According to the measurement DNL of the quantizer is less than 0.18 LSB and INL is within 0.23 LSB.

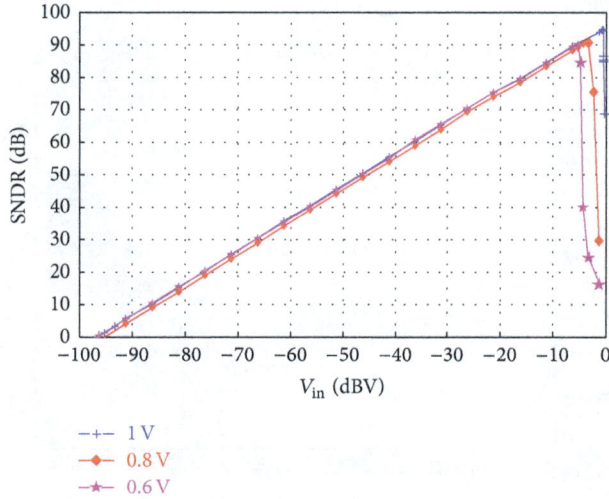

FIGURE 10: Measured SNDR versus input amplitude.

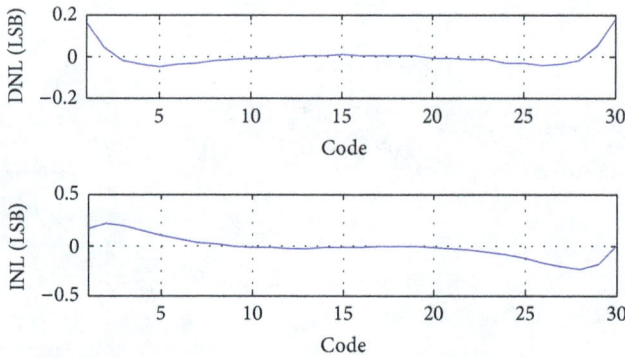

FIGURE 11: Measured DNL and INL performance of the asynchronous SAR.

Figure-of-merit (FOM) is popular in comparison between designs with different specifications. It is defined by

$$\text{FOM} = \frac{P}{2^{(\text{SNDR}_{\text{peak}}-1.76)/6.02} f_N},\qquad(9)$$

where P denotes the total power dissipation and f_N is the Nyquist frequency which is twice the signal bandwidth. Our design achieves 0.17 pJ/converstion step under 1 V. This value improves to 0.10 pJ/converstion step under 0.6 V. The performance is summarized and compared with published works in Table 1.

5. Conclusion

In this paper, design, implementation, and measurement of a high performance audio ΔΣ modulator under 65 nm standard CMOS technology are presented. Low voltage and low power modulator architecture is employed which greatly relaxes the design requirement of the amplifier and comparator. The prototype modulator achieves 95 dB peak SNDR with 24 kHz under 1 V supply while only consumes 371 μW. With the help of low threshold voltage of devices under 65 nm, the

TABLE 1: Performance summary and comparison.

	[3]	[4]	[9]	This work		
Process (nm)	65LP	45LP	180	65GP		
Supply (V)	1.2	1.1	0.7	1.0	0.8	0.6
f_s (MHz)	12	12	5	3.072		
OSR	300	300	100	64		
SNDR (dB)	74	76.5	95	95	91.2	90.2
DR (dB)	95	91.7	100	96.4	92.4	91.9
Power (μW)	2200	1200	870	371	224	133
FOM (pJ/step)	11.2	4.57	0.37	0.17	0.16	0.10

modulator works well under voltage of only 0.6 V. FOM is used in comparisons which show that our design has an overwhelming low power performance.

Acknowledgments

This work is partially supported by National High Technology Research and Development Program of China (The 863 Program Project no. 2008AA010700). The authors would like to thank TSMC for the chip fabrication.

References

[1] L. L. Lewyn, T. Ytterdal, C. Wulff, and K. Martin, "Analog circuit design in nanoscale CMOS technologies," *Proceedings of the IEEE*, vol. 97, no. 10, pp. 1687–1714, 2009.

[2] L. Yao, M. S. J. Steyaert, and W. Sansen, "A 1-V 140-μW 88-dB audio sigma-delta modulator in 90-nm CMOS," *IEEE Journal of Solid-State Circuits*, vol. 39, no. 11, pp. 1809–1818, 2004.

[3] L. Dörrer, F. Kuttner, A. Santner et al., "A 2.2 mW, continuous-time sigma-delta ADC for voice coding with 95 dB dynamic range in a 65 nm CMOS process," in *Proceedings of the 32nd European Solid-State Circuits Conference (ESSCIRC '06)*, pp. 195–198, September 2006.

[4] L. Dorrer, F. Kuttner, A. Santner et al., "A continuous time DS ADC for voice coding with 92 dB DR in 45 nm CMOS," in *Proceedings of the Solid-State Circuits Conference IEEE ISSCC Digest of Technical Papers*, pp. 502–503, 2008.

[5] J. Zhang, Y. Lian, L. Yao, and B. Shi, "A 0.6-V 82-dB 28.6-mW continuous-time audio delta-sigma modulator," *IEEE Journal of Solid-State Circuits*, vol. 46, no. 10, pp. 2326–2335, 2011.

[6] P. Benabes, A. Gauthier, and D. Billet, "New wideband sigma-delta convertor," *IET Electronics Letters*, vol. 29, no. 17, pp. 1575–1577, 1993.

[7] K. Lee and G. C. Temes, "Improved low-distortion ΔΣ DS ADC topology," in *Proceedings of the IEEE International Symposium on Circuits and Systems (ISCAS '09)*, pp. 1341–1344, May 2009.

[8] Y. Yang, T. Sculley, and J. Abraham, "A single-die 124 dB stereo audio delta-sigma ADC with 111 dB THD," *IEEE Journal of Solid-State Circuits*, vol. 43, no. 7, pp. 1657–1665, 2008.

[9] H. Park, K. Nam, D. K. Su, K. Vleugels, and B. A. Wooley, "A 0.7-V 870-μW digital-audio CMOS sigma-delta modulator," *IEEE Journal of Solid-State Circuits*, vol. 44, no. 4, pp. 1078–1088, 2009.

[10] L. Samid and Y. Manoli, "A multibit continuous time sigma delta modulator with successive-approximation quantizer," in *Proceedings of the IEEE International Symposium on Circuits and Systems (ISCAS '06)*, pp. 2965–2968, May 2006.

[11] Y. Yang, A. Chokhawala, M. Alexander, J. Melanson, and D. Hester, "A 114-dB 68-mW chopper-stabilized stereo multibit audio ADC in 5.62 mm^2," *IEEE Journal of Solid-State Circuits*, vol. 38, no. 12, pp. 2061–2068, 2003.

[12] J. Roh, S. Byun, Y. Choi, H. Roh, Y. G. Kim, and J. K. Kwon, "A 0.9-V 60-μW 1-bit fourth-order delta-sigma modulator with 83-dB dynamic range," *IEEE Journal of Solid-State Circuits*, vol. 43, no. 2, pp. 361–370, 2008.

[13] C. C. Enz and G. C. Temes, "Circuit techniques for reducing the effects of Op-Amp imperfections: autozeroing, correlated double sampling, and chopper stabilization," *Proceedings of the IEEE*, vol. 84, no. 11, pp. 1584–1614, 1996.

[14] D. Schinkel, E. Mensink, E. Klumperink, E. Van Tuijl, and B. Nauta, "A double-tail latch-type voltage sense amplifier with 18ps setup+hold time," in *Proceedings of the 54th IEEE International Solid-State Circuits Conference (ISSCC '07)*, pp. 314–315, February 2007.

[15] M. Van Elzakker, E. Van Tuijl, P. Geraedts, D. Schinkel, E. A. M. Klumperink, and B. Nauta, "A 10-bit charge-redistribution ADC consuming 1.9 μW at 1 MS/s," *IEEE Journal of Solid-State Circuits*, vol. 45, no. 5, pp. 1007–1015, 2010.

13

A New Length-Based Algebraic Multigrid Clustering Algorithm

L. Rakai, A. Farshidi, L. Behjat, and D. Westwick

Department of Electrical and Computer Engineering, University of Calgary, 2500 University Drive NW Calgary, AB, Canada T2N 1N4

Correspondence should be addressed to L. Behjat, laleh@ucalgary.ca

Academic Editor: Rached Tourki

Clustering algorithms have been used to improve the speed and quality of placement. Traditionally, clustering focuses on the local connections between cells. In this paper, a new clustering algorithm that is based on the estimated lengths of circuit interconnects and the connectivity is proposed. In the proposed algorithm, first an a priori length estimation technique is used to estimate the lengths of nets. Then, the estimated lengths are used in a clustering framework to modify a clustering technique based on algebraic multigrid (AMG), that finds the cells with the highest connectivity. Finally, based on the results from the AMG-based process, clusters are made. In addition, a new physical unclustering technique is proposed. The results show a significant improvement, reductions of up to 40%, in wire length can be achieved when using the proposed technique with three academic placers on industry-based circuits. Moreover, the runtime is not significantly degraded and can even be improved.

1. Introduction

Clustering is usually employed during the large-scale placement problems encountered in today's circuits, to speed up the placement process and improve the solution quality. The clustering algorithms used today are based on finding small groups of cells with high connectivity and putting each of them in a cluster. This scheme has proven effective, to a large extent, as there is a high correlation between the cells' connectivity and the lengths of the nets that connect them [1, 2]. Hence, by clustering cells that are close to one another, the lengths of the nets between them, and hence the total wire length will be reduced.

In this paper, a new perspective on how to cluster cells is proposed, where clustering decisions are made based on estimates of the lengths of nets connecting the cells. If a net is estimated to be short, then its cells are expected to be physically closer to each other, and the cells can be put in a cluster. The opposite is true for a net which has a high estimated length.

The proposed algorithm consists of three phases. In the first phase, a state-of-the-art length estimation technique is used to estimate the lengths of individual nets before placement.

Accurate length estimates are then used to direct the clustering decisions. These length estimates are used in the second phase to build a length-based proximity matrix. Then, it is proposed to use the proximity matrix in an AMG-based clustering framework to find clusters of the circuit. The AMG-based clustering finds seed cells and assigns scores based on the whole proximity matrix, not only the local connections. Hence, it can be more effective in finding a more globally optimal clustering configuration.

In the third phase, clusters are formed, and an initial placement is performed on the new clustered circuit. The initial placement solution is then refined in the unclustering phase, where a new technique for unclustering is proposed.

The major contributions of this paper are as follows:

(i) using estimated lengths when deciding which cells should be clustered;

(ii) designing and implementing a length-based AMG clustering algorithm;

(iii) proposing and implementing an unclustering refinement algorithm;

(iv) improving the runtime of the placement process;

(v) illustrating the effectiveness of multilevel length-based AMG clustering.

The rest of this paper is organized as follows. In Section 2, a literature review of existing net length estimation methods, an introduction to AMG, and a background of clustering techniques are presented. In Section 3, the proposed length-based clustering algorithm is described in detail. The proposed model efficacy is examined and validated by several experiments in Section 4. Conclusions are given in Section 5.

2. Background

In this section, a literature review of the three main components of the algorithm are given. As these components are distinct subjects not all normally used in the context of VLSI placement, the literature review is divided into three distinct sections: length estimation (2.1), AMG (2.2), and finally, clustering algorithms (2.3).

2.1. Net Length Estimation Techniques. Several a priori length estimation techniques, for example [3–6], have been proposed that try to estimate the length of individual nets. Older techniques, such as [7–12], can only estimate the average length of a set of nets or the distributions of net lengths.

In [6], a variable referred to as the intrinsic shortest path length (ISPL) is developed and used to estimate the individual net lengths. Although the estimation results obtained using this technique are exceptional, this approach is not very useful for modern mixed-size circuits since it only considers cells with unit area and nets with degree two.

Different properties of nets and cells are modeled by several variables in [3]. These variables are then used to make a third-order polynomial model for length estimation. The estimation results are well correlated for relatively small circuits that only include standard cells. However, none of these variables considers the effects of macro blocks on net lengths, and so the estimation results for mixed-size circuits are unreliable. In [5], this technique is further studied using a quadratic polynomial. Three new variables are proposed to account for the effects of macro blocks. The estimation results for modern mixed-size circuits show around 10% improvement over those obtained using [3].

Some applications of a priori net length estimation techniques are introduced in [4, 13, 14]. A variable called mutual contraction, is proposed and used for net length estimation, in [13, 14]. This variable is then used in placement. The estimated net lengths are utilized to quantify the quality of potential clusters. In [4], another application of preplacement length estimation is proposed where the negative effects of clustering are corrected based on the estimation results.

2.2. Algebraic Multigrid. A challenging and frequently encountered problem in various domains is to solve a large system of linear equations as in

$$\mathbf{Ax} = \mathbf{b}, \quad (1)$$

where $\mathbf{A}$ is a large, sparse square matrix, $\mathbf{x}$ is a vector containing the unknown variables, and $\mathbf{b}$ is the known right-hand side vector. When $\mathbf{A}$ is very large, solving this system of equations directly becomes very expensive, in general. In the circuit placement domain, this equation arises from minimizing a quadratic length objective where the dimensions of $\mathbf{A}$ are on the order of millions.

The algebraic multigrid (AMG) technique uses a multilevel framework to approximately solve (1), hence reducing the computational cost. Several AMG algorithms have been developed [15–21]. Each one of these methods is suited for a specific type of problem where the matrix $\mathbf{A}$ has certain properties. General methods such as [15, 17, 20] have sufficient convergence conditions and can easily be adapted to more problems.

A typical AMG technique consists of three steps: coarsening, direct solution, and interpolation. Coarsening is used to decrease the size of the matrix $\mathbf{A}$ by keeping the essential elements of it and disregarding the nonessential elements. Once the matrix size has been adequately reduced, the second step starts. In this step, the reduced problem is solved exactly using a direct method. In the final step, interpolation, the solution obtained after the direct method step is projected back onto the sequence of larger systems of equations obtained during coarsening to achieve a final solution.

Another difference between the AMG framework and other direct methods is that in AMG, instead of solving for $\mathbf{x}$ directly, the difference or the error, $\mathbf{e}^0$, between the initial guess for $\mathbf{x}^0$ and the actual value of $\mathbf{x}$ is driven to zero. This is accomplished by solving the residual equation $\mathbf{A}^l \mathbf{e}^l = \mathbf{r}^l$ for level l, where $\mathbf{r}$ is the residual that is calculated as $\mathbf{r}^l = \mathbf{b}^l - \mathbf{A}^l \mathbf{x}^l$. A multilevel AMG scheme is illustrated in Figure 1. In this figure, in each level of coarsening, a restriction matrix, $\mathbf{W}_l^{l+1}$, is calculated. This matrix contains fewer rows than columns. To reduce the size of the equation matrix at level l, $\mathbf{A}^l$, is premultiplied by $\mathbf{W}_l^{l+1}$ and postmultiplied by $(\mathbf{W}_l^{l+1})^T$, the interpolation matrix, and the equation matrix at level $l + 1$, $\mathbf{A}^{l+1} = \mathbf{W}_l^{l+1} \mathbf{A}^l (\mathbf{W}_l^{l+1})^T$ is obtained [16]. Thus, $\mathbf{A}^{l+1}$ has smaller dimensions than $\mathbf{A}^l$. At the lowest level, L, an exact solution for $\mathbf{e}^L$ is found, that is, $\mathbf{e}^L = (\mathbf{A}^L)^{-1} \mathbf{r}^L$ as shown in the figure. In the interpolation stage, the error is propagated to the higher levels using the appropriate interpolation matrix. Finally, using the calculated error in the highest level, the solution approximation $\mathbf{x}^0$ is updated: $\mathbf{x}^0 \leftarrow \mathbf{x}^0 + \mathbf{e}^0$. The relaxation steps shown in the figure refer to using a fast, iterative method to improve the solution at the current level.

2.3. Clustering Algorithms. In different stages of VLSI physical design, clustering algorithms are used to reduce the sizes of circuits. Several clustering algorithms have been proposed to handle the modern large-size circuits, for example, [1, 2, 22–26]. Most placement techniques, such as the ones that competed in the ISPD 2006 [27] placement contest, use clustering algorithms.

Clustering algorithms such as heavy-edge matching [22], Edge Coarsening [28], FirstChoice [26], and PinEC [23] are similar in how they choose and form clusters. In these algorithms, a cell is chosen randomly to become the *seed* for a cluster. A seed refers to the first cell of a cluster. Once a seed has been chosen, a cell that measures the highest in connectivity with the seed is added to the cluster. These

> Algorithm AMG-LE: AMG-based clustering with length estimation
> Input: Circuit net list, AMG parameters
> Output: Clustered circuit
> 1. Pre-placement individual net length estimation
> 2. Proximity matrix construction using the estimated lengths
> 3. AMG-based coarsening on the proximity matrix
> 4. Cluster seed cell and score assignment
> 5. Final cluster formation

ALGORITHM 1: The high-level flow of the AMG-LE clustering algorithm.

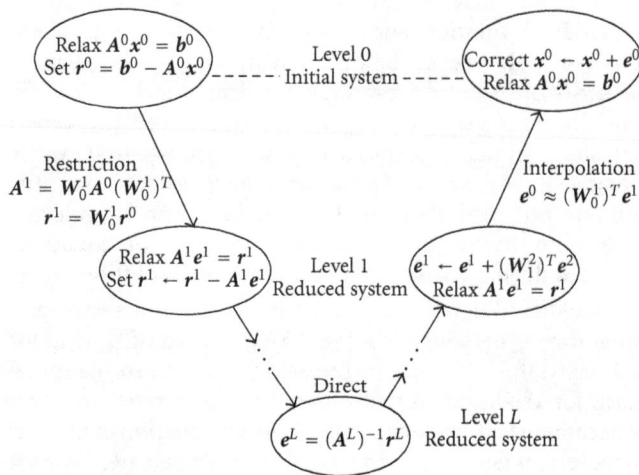

FIGURE 1: A multilevel AMG scheme.

algorithms are easy to implement, and the runtimes for them are typically low. However, as no comparisons between potential clusters are performed, the quality of the results can be low.

In an effort to improve results, algorithms, such as edge-separability [24], fine-granularity clustering (FGC) [25], best-choice clustering [1], and Net Cluster [2], first prepare a set of potential clusters and then either refine the potential clusters or only finalize potential clusters that have high scores.

In [29, 30], a clustering algorithm for wire length-driven placement, called SafeChoice, is proposed where a condition is used to check if potential clusters will degrade the placement solution. If a cluster passes the check, it is referred to as a safe cluster, and the cluster is finalized.

In [31], a linear-time AMG-based clustering technique is proposed. This AMG-based algorithm uses only the connectivity matrix of a circuit to form clusters.

3. The Proposed Length-Driven Multilevel Placement Framework

3.1. The Proposed Length Estimation-Based AMG Clustering Algorithm. The main objective of the placement phase in the physical design of circuits is to reduce the total wire length of the circuit. Clustering algorithms are normally

used during placement to improve the total wire length and runtime. However, clustering is performed based on circuit connectivity and not the wire length. In this paper, an algorithm that performs clustering based on estimated wire lengths is proposed. The novelty of the algorithm is in using the estimated lengths instead of connectivities in determining the best cells to be clustered.

The algorithm has five main stages: preplacement length estimation, proximity matrix construction, AMG-based coarsening, cluster seed cell and score assignment, and final cluster formation, as summarized in Algorithm 1.

3.1.1. Preplacement Length Estimation. In an estimation technique, first a set of model parameters needs to be calculated. In this work, the parameters used or developed in [5] are employed as these have been shown to be a comprehensive set of parameters that are well suited to the mixed-size circuits used today. These parameters include local net characteristics, such as the half perimeter of the cells of each net and global characteristics, such as the number of degree-two nets in the design.

Once the model parameters have been selected, an estimation technique should be used to fit the parameters into a model. The model parameters in [5] are used to fit a quadratic model. Any terms of the resulting model which have small coefficients, ineffective terms, are pruned from the model. The remaining terms, effective terms, are used to make the estimation model and calculate the length estimates.

It is worth noting that any other individual net length estimation technique could be used in this step. The model in [5] is selected as it includes several factors such as macro cells but is independent of the placer to be used allowing for broader application of this work.

To better illustrate the algorithm flow, a small example circuit is presented in Figure 2. In this figure, the nets are annotated with the estimated lengths. The estimated lengths are found using a simplified version of the technique in [5]. Due to the small size of the example circuit, only the variables associated with the cell dimensions and net degree are used. These lengths will be used in the following sections to perform clustering using the proposed AMG-LE algorithm.

3.1.2. Proximity Matrix Construction. To perform circuit clustering, the circuit's net list should be represented using

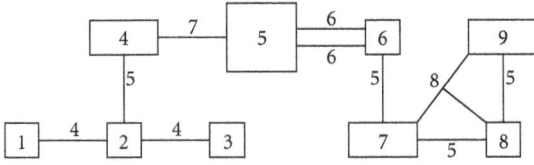

FIGURE 2: An example circuit with nets annotated with estimated lengths. The dimensions of all cells are 2×2 except cells 4, 7, and 9 that are 4×2, and cell 5 which is 4×4.

a matrix. Normally, this matrix is the connectivity matrix of the circuit which shows the total number of weighted connections between two cells. In this paper, a new matrix, called the proximity matrix, is designed which shows how close two cells are predicted to be. The proximity matrix is later used for the AMG-based clustering.

It is shown in [16] that for the best AMG performance the matrix $\mathbf{A}$ of (1), which in this case represents the net list, should be an *M-matrix*, that is, a symmetric, positive-definite matrix with positive diagonal and nonpositive offdiagonal elements. In the proposed AMG-based clustering technique, a proximity matrix $\mathbf{P}^0$, which is an M-matrix, is constructed. The rows and columns of $\mathbf{P}^0$ represent the cells of the circuit. The element $p_{i,j}^0$ shows the connectivity between cell i and cell j. Each $p_{i,j}^0$ is determined as follows: consider two cells, i and j, which are treated as points in AMG, and let $N_{i,j}$ be the set of nets, n_k, that contain both cells i and j. Then, element $p_{i,j}^0$ is calculated as

$$p_{i,j}^0 = \begin{cases} \sum_{n_k \in N_{i,j}} -\operatorname{il}(n_k), & i \neq j, N_{i,j} \neq \varnothing, \\ \sum_{n_k \in N_{i,i}} \operatorname{il}(n_k), & i = j, \\ 0, & \text{otherwise,} \end{cases} \quad (2)$$

where il($\cdot$) represents the inverse of the estimated length of a net and is defined as

$$\operatorname{il}(n_k) = \frac{1}{l_{\text{est}}(n_k)}, \quad (3)$$

where $l_{\text{est}}(n_k)$ denotes the estimated length of a net obtained from the model described in Section 3.1.1. The inverse of the estimate is used because nets whose estimated lengths are small have a high proximity and vice versa. Using the formulation in (2), any nonzero offdiagonal element of the proximity matrix is equal to the negative sum of the inverses of the estimated lengths of nets between i and j, and any diagonal element is equal to the sum of all the inverses of nets connected to i. This matrix is a positive-semidefinite M-matrix which is highly desirable for AMG. In addition, multiterminal nets and multiple connections between cells are handled by (2).

As an example, the proximity matrix for the example circuit in Figure 2 is given in (4). For any two cells i and j that are not connected, the $p_{i,j}^0$ entry is equal to zero. For any two cells that are connected, the $p_{i,j}^0$ entry is equal to the negative sum of the inverse of estimated lengths of

the nets connecting them. As an example, cells 5 and 6 are connected by two nets each with estimated length of six. Therefore, $p_{5,6}^0 = -(1/6 + 1/6) = -1/3$. Finally, the diagonal elements are the negative sum of the off-diagonal elements in the corresponding row of the matrix. For example, $p_{5,5}^0 = -(p_{5,4}^0 + p_{5,6}^0) = -(-1/7 - 1/3) = 10/21$.

$$\mathbf{P}^0 = \begin{pmatrix} \frac{1}{4} & -\frac{1}{4} & 0 & 0 & 0 & 0 & 0 & 0 & 0 \\ -\frac{1}{4} & \frac{7}{10} & -\frac{1}{4} & -\frac{1}{5} & 0 & 0 & 0 & 0 & 0 \\ 0 & -\frac{1}{4} & \frac{1}{4} & 0 & 0 & 0 & 0 & 0 & 0 \\ 0 & -\frac{1}{5} & 0 & \frac{12}{35} & -\frac{1}{7} & 0 & 0 & 0 & 0 \\ 0 & 0 & 0 & -\frac{1}{7} & \frac{10}{21} & -\frac{1}{3} & 0 & 0 & 0 \\ 0 & 0 & 0 & 0 & -\frac{1}{3} & \frac{8}{15} & -\frac{1}{5} & 0 & 0 \\ 0 & 0 & 0 & 0 & 0 & -\frac{1}{5} & \frac{13}{20} & -\frac{13}{40} & -\frac{1}{8} \\ 0 & 0 & 0 & 0 & 0 & 0 & -\frac{13}{40} & \frac{13}{20} & -\frac{13}{40} \\ 0 & 0 & 0 & 0 & 0 & 0 & -\frac{1}{8} & -\frac{13}{40} & \frac{9}{20} \end{pmatrix} \quad (4)$$

3.1.3. AMG-Based Coarsening on the Proximity Matrix.

Once the proximity matrix $\mathbf{P}^0$ has been constructed, the coarsening algorithm in [31] is used to reduce $\mathbf{P}^0$ and construct $\mathbf{P}^1$ and the associated restriction matrix $\mathbf{W}_0^1$. The coarsening can be loosely thought of as a heuristic for selecting a maximally independent subset of cells which are the most significant in $\mathbf{P}^0$. The significance of each cell is measured by the number of cells that *strongly connect* to it. A parameter, $\theta \in (0, 1]$, is used in the definition of the strength of a connection. Cell i is strongly connected to cell j if

$$\left| p_{i,j}^l \right| \geq \theta \times \max_{k \neq i} \left\{ \left| p_{i,k}^l \right| \right\}, \quad k = 1, \ldots, C^l, \quad (5)$$

where, $p_{i,j}^l$ is the $(i, j)^{th}$ element of $\mathbf{P}^l$, and C^l is the number of cells in level l. The relationship between the parameter θ and the amount coarsening performed is complex. If θ is close to one, more aggressive coarsening will be performed. However, more aggressive coarsening can overly simplify the reduced matrix and the associated reduced circuit. An interpretation of the strength of connection criteria in the context of the proposed proximity matrix is that if a cell is in several nets with small estimated length it is likely to be a part of the reduced circuit.

3.1.4. Seed Cell and Cluster Score Assignment.

After coarsening the proximity matrix, the cells that are selected to form the reduced circuit become seed cells used for clustering. The cells which are not selected to form the reduced circuit are considered to be clustered with the selected seed cells. This is the same technique used in [31]. In this technique, the entries of the AMG interpolation matrix $(\mathbf{W}_l^{l+1})^T$ are used to rank each seed cell that a non-selected cell connects to. A large entry in the interpolation matrix means the seed cell is representative of the non-selected cell in the reduced circuit.

$$(\mathbf{W}_0^1)^T = \begin{pmatrix} 1 & 0 & 0 \\ 1 & 0 & 0 \\ 1 & 0 & 0 \\ \frac{7}{12} & \frac{5}{12} & 0 \\ 0 & 1 & 0 \\ 0 & \frac{5}{8} & \frac{3}{8} \\ 0 & 0 & 1 \\ 0 & 0 & 1 \\ 0 & 0 & 1 \end{pmatrix}$$

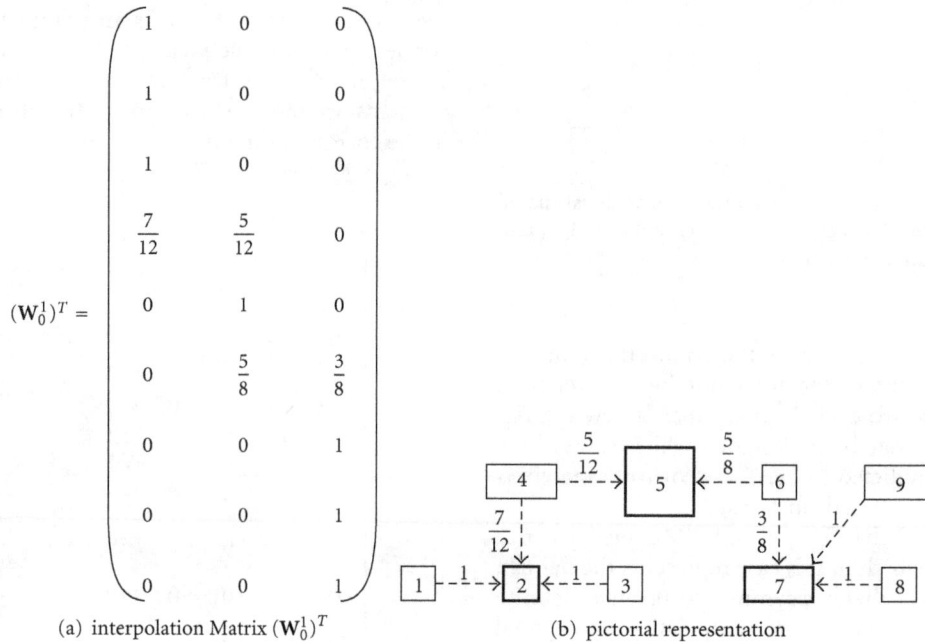

(a) interpolation Matrix $(\mathbf{W}_0^1)^T$ (b) pictorial representation

FIGURE 3: The interpolation matrix shown in matrix form and pictorial form for the example circuit in Figure 2. In the pictorial representation, the selected seed cells are shown in bold, and arrows are annotated with the interpolation weights of non-selected cells.

The selected seed cells for the example in Figure 2 are cells 2, 5, and 7. Therefore, the interpolation matrix has three columns, one for each seed cell with the first column corresponding to cell 2, and so forth. The number of rows is nine, which is the total number of cells and row 1 corresponds to cell 1, and so forth. The interpolation matrix is shown in Figure 3(a). Each seed cell interpolates from itself directly, so entries $w_{2,1}$, $w_{5,2}$ and $w_{7,3}$ are all one, where $w_{i,j}$ is the (i, j)th element of $(\mathbf{W}_0^1)^T$. The other non-zero entries represent the interpolation weight of non-selected cells to seed cells. As an example, cell 4 is connected to seed cells 2 and 5. The interpolation weights of cell 4 with seed cells 2 and 5 are given by entries $w_{4,1} = 7/12$ and $w_{4,2} = 5/12$, respectively. To better visualize the seed cell and cluster score assignment using the interpolation matrix, a pictorial representation of $(\mathbf{W}_0^1)^T$ is presented in Figure 3(b). In this figure, the seed cells are shown with a bold border. Each non-selected cell has an interpolation weight to the seed cells that they connect to. These interpolation weights annotate arrows pointing from non-selected cells to seed cells in Figure 3(b).

3.1.5. Final Cluster Formation.
The final step of the clustering algorithm is to cluster non-selected cells to the seed cell which they interpolate from most strongly. A cluster is finalized as long as the area of the cluster is not greater than five times the average standard cell area and its interpolation weight is more than 0.9. The cluster area is restricted to prevent the formation of macro-sized clusters which require special treatment in placement algorithms. The interpolation weight threshold is set high to ensure that only the best quality clusters are formed. Each cluster meeting the constraints is given a physical shape in the

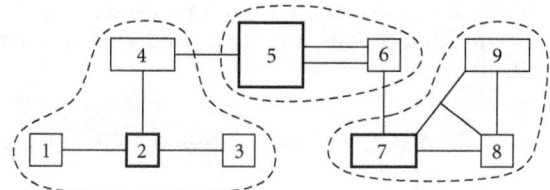

FIGURE 4: The final clusters for the example circuit in Figure 2 shown by dashed lines.

clustered circuit with a height equal to the maximum height of all of its cells and a width equal to the sum of the widths of its cells.

The final clusters for the example in Figure 2 are shown with dashed lines in Figure 4. The clusters are formed by clustering each non-selected cell with the seed cell that it has the maximum interpolation weight with. Assuming each cluster has an area of less than or equal to five times the average standard cell area, the clusters are finalized.

3.2. The Proposed Length-Driven Unclustering Technique.
The description of a clustering algorithm alone is not enough to fully specify the implementation of a clustering algorithm used for placement. Just as important is the algorithm for unclustering, or determining the locations of cells within a cluster after the cluster's location has been determined. The description of this step is rarely mentioned in the literature of clustering algorithms.

In this work, a length-driven unclustering algorithm is proposed. The chief benefit of using the proposed technique is that a legal solution is preserved after unclustering.

(a) clustered placement

(b) locations of the cluster's cells determined by minimizing (6)

(c) final ordering of the cluster's cells

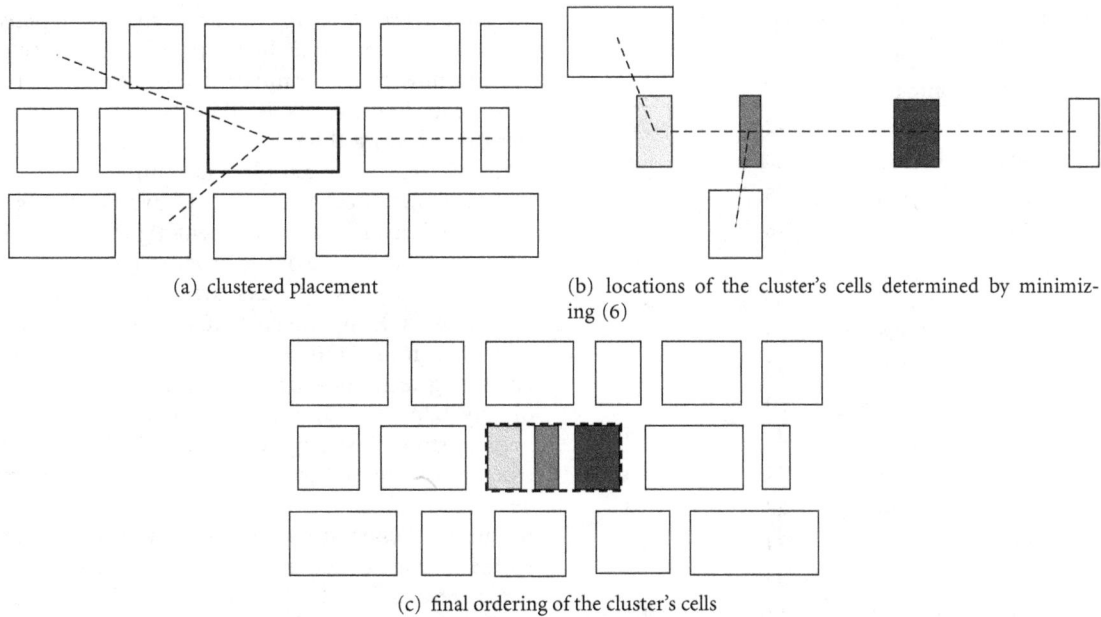

FIGURE 5: An illustration of the length-driven unclustering technique.

Global placement algorithms often produce legal solutions before performing detailed placement. A new unclustering technique which preserves the legality of the globally-placed solution improves the runtime in multilevel placement. With the legality restriction, the problem of unclustering reduces to ordering the cluster's cells. The proposed technique finds the ordering by solving a relaxed length-driven minimization and using the result to determine the ordering of the cells. In addition, the combination of preserving legality and restricting cluster areas means that the row in which the unclustered cells will be placed is also preserved. Consequently, the vertical components of the lengths can be ignored. This helps to reduce the disruption in the wire length before and after unclustering which is a significant problem for multilevel placement as discussed in [32].

The locations of cells, $\mathbf{x}_{C_i}$, in cluster C_i are determined by minimizing the quadratic matrix length objective

$$\mathbf{x}_{C_i}^T \mathbf{U} \mathbf{x}_{C_i} + \mathbf{t}^T \mathbf{x}_{C_i} + v, \qquad (6)$$

where $\mathbf{U}$ is a weighted connectivity matrix between the cells of the cluster, that is, $i, j \in C_i$, and is defined as

$$u_{i,j} = \begin{cases} \sum_{n_k \in N_{i,j}} -\dfrac{1}{|n_k|}, & i \neq j, N_{i,j} \neq \varnothing, \\ \sum_{n_k \in N_{i,i}} \dfrac{1}{|n_k|}, & i = j, \\ 0, & \text{otherwise}, \end{cases} \qquad (7)$$

where $|n_k|$ is the degree of net n_k, that is, the number of cells in n_k. The vector $\mathbf{t}$ contains the weighted horizontal cell

locations of cells outside of the cluster that each clustered cell connects to, that is, $i \in C_i$ and $j \notin C_i$ defined as

$$t_i = \begin{cases} \sum_{n_k \in N_{i,j}} -\dfrac{x_j}{|n_k|}, & i \neq j, N_{i,j} \neq \varnothing, \\ 0, & \text{otherwise}, \end{cases} \qquad (8)$$

where x_j is the location of the cell or the cluster that a cell belongs to. The scalar v represents the length of all the nets not involving the cells in C_i and can be ignored. After minimizing (6), the cells are inserted into the area occupied by the cluster in the order of their locations in $\mathbf{x}_{C_i}$. An illustration of the length-driven unclustering is given in Figure 5. In Figure 5(a), a clustered placement is given. The highlighted cluster is the focus of the example and contains three cells (not shown in Figure 5(a)). The cluster is connected to three other cells via degree-two nets, shown by dashed lines in the illustration. The cells contained in the cluster are shown in shades of grey in Figure 5(b). The three external connections as well as the two internal connections are used to determine the non-zero values in the matrix $\mathbf{U}$. In this example all of the values for $|n_k|$ are equal to 2, the degree of each net. The locations of the cells not in the cluster, shown in white, are used in forming the vector $\mathbf{t}$. The locations of the cluster's cells in Figure 5(b) are determined by minimizing (6). These locations are used to determine the order of the cells inside of the cluster's area to complete the unclustering, illustrated in Figure 5(c). Because the unclustered cells remain inside of the cluster's area, the placement remains legal.

This technique does not take advantage of the locations of already unclustered cells as it iterates through the clusters. However, this also means that the technique is not sensitive to the order in which the clusters are visited. Furthermore, because the clusters are restricted to have an area of less

TABLE 1: Statistics of ICCAD04 benchmarks.

Circuit	Nets number	Cells number	$|n_{max}|$	$|c_{max}|$
ibm01	14,111	12,752	42	39
ibm02	19,584	19,601	134	69
ibm03	27,401	23,136	55	100
ibm04	31,970	27,507	46	425
ibm05	28,446	29,347	17	9
ibm06	34,826	32,498	35	91
ibm07	48,117	45,926	25	98
ibm08	50,513	51,309	75	1165
ibm09	60,962	53,395	39	173
ibm10	75,196	69,429	41	137
ibm11	81,454	70,558	24	174
ibm12	77,240	71,076	28	473
ibm13	99,666	84,199	24	180
ibm14	152,772	147,605	33	270
ibm15	186,608	161,570	33	306
ibm16	190,048	183,484	40	177
ibm17	189,581	185,495	36	81
ibm18	201,920	210,613	66	97

TABLE 2: Statistics of ISPD05 benchmarks.

Circuit	Nets number	Cells number	$|n_{max}|$	$|c_{max}|$
adaptec1	221,142	211,447	2,271	448
adaptec2	266,009	255,023	1,935	620
adaptec3	466,758	451,650	3,713	1224
adaptec4	515,951	496,045	3,974	416
bigblue1	284,479	278,164	2,621	388
bigblue2	577,235	557,866	11,869	119
bigblue3	1,123,170	1,096,812	7,623	1692

than five times the average standard cell area, the additional benefit of using the locations of already unclustered cells and a good ordering of the clusters is expected to be small.

4. Experimental Results

Several experiments have been carried out to verify the efficacy of the proposed clustering algorithm. These experiments are performed using the ICCAD04 benchmark circuits [33] released by IBM. The detailed statistics of these circuits are given in Table 1. In columns 2 and 3 of this table the number of nets and cells in each circuit are given, respectively. In columns 4 and 5, the maximum net degree, $|n_{max}|$, and maximum cell degree, $|c_{max}|$, are given.

To better show the scalability of the proposed clustering algorithm, the same experiments are also performed with the ISPD05 benchmark suite [34]. The statistics of these benchmark circuits are presented in Table 2. The benchmark circuit bigblue4 is not included in the experiments due to the memory limitations of the testing platform.

4.1. Multilevel Clustering Results Using the Proposed AMG-LE Technique. To evaluate the proposed clustering algorithm, its effectiveness in the context of multilevel circuit placement using multiple placers is illustrated. The experiments are performed using three high-quality academic placers: Capo 10.5 [35], Fastplace 3.0 [36], and mPL6 [37]. Fastplace and mPL are analytical placers and have clustering algorithms embedded inside their placement algorithms. To better evaluate the performance of the proposed technique, the internal clustering of these placers are disabled in the clustering experiments. If the internal clustering is enabled, the results will have noise and it will not be clear if the proposed clustering technique or the internal clustering is the cause of any benefits. Comparisons of the proposed technique with existing clustering techniques, including the technique used internally by Fastplace 3.0 and mPL6, are presented in Section 4.3. Capo is a partitioning-based placer and does not use internal clustering. Capo is, however, non-deterministic so the average of ten runs of each experiment are reported when results for Capo are given. The experiments are performed on a 64-bit dual core AMD opteron system running Linux with 8 GB of RAM. In addition, the parameter θ in (5) is set to 0.1 in the experiments.

Wire length and runtime results of the experiment are presented in Table 3(a) and 3(b), respectively. In both tables, Column 1 identifies the circuit that is placed. Columns 2–4, 5–7, and 8–10 show the results for Capo, Fastplace, and mPL, respectively. Each placer is used in three modes: baseline without clustering, one-level AMG-LE clustering with length-driven unclustering, and two-level AMG-LE clustering with length-driven unclustering. The columns showing one- and two-level AMG-LE are given in terms of the percentage improvement over the baseline. In this representation, positive percentages represent improvements.

The results in Table 3(a) show that AMG-LE is effective at improving wire length on average for all placers using one or two levels of clustering. A maximum improvement of 38.1% is achieved using one-level AMG-LE and Fastplace. Using one-level AMG-LE, every circuit for Capo, 16 circuits for Fastplace, and 10 circuits for mPL have improved wire length when compared to the baseline. When two-level AMG-LE is used, every circuit for Capo and Fastplace, and 12 circuits for mPL are improved.

The runtime results show that using the one-level scheme for mPL and the two-level scheme for all placers double the total runtime. For a placer, such as Fastplace, the increase in runtime may be more acceptable because it requires much less time than other placers. However, the increase for Capo and particularly for mPL are less acceptable. It should be noted that mPL does not offer an option to perform only detailed placement requiring full global and detailed placement to be performed between each level, which accounts for the significant increases compared to the other two placers. The experiment performed in the following section will illustrate how to improve the runtime results while maintaining, and even improving, the wire length results.

An interesting result arising from the experiment is that the placements resulting from using the AMG-LE framework

TABLE 3: Placement wire length and runtime results comparing one- and two-level AMG-LE clustering with length-driven unclustering to baseline placements without using clustering with three placers on the ICCAD04 benchmark suite.

(a) HPWL

Circuit	Capo			Fastplace			mPL		
		AMG-LE			AMG-LE			AMG-LE	
	Baseline ($\times 10^5$)	1Lvl (%)	2Lvl (%)	Baseline ($\times 10^5$)	1Lvl (%)	2Lvl (%)	Baseline ($\times 10^5$)	1Lvl (%)	2Lvl (%)
ibm01	25	**3.1**	**6.4**	24	−1.8	**1.3**	24	**4.6**	**4.2**
ibm02	51	**4.4**	**5.4**	54	**3.1**	**4.6**	52	**1.3**	**1.1**
ibm03	77	**1.8**	**9.8**	80	**5.1**	**6.8**	82	−7.0	−3.6
ibm04	93	**14.8**	**15.7**	86	**7.9**	**7.7**	109	−5.7	−4.4
ibm05	103	**1.1**	**2.1**	101	**0.7**	**0.5**	93	−5.2	−5.1
ibm06	67	**0.6**	**3.8**	99	**38.1**	**33.9**	88	**5.8**	**2.7**
ibm07	126	**9.3**	**11.9**	123	**7.4**	**12.2**	124	−3.1	−4.0
ibm08	137	**7.9**	**8.1**	147	−4.1	**10.1**	211	**4.7**	**4.2**
ibm09	145	**4.4**	**6.6**	155	**8.6**	**8.0**	189	**2.9**	**1.7**
ibm10	318	**4.0**	**4.9**	362	**10.8**	**12.2**	363	**2.0**	**1.8**
ibm11	210	**4.1**	**7.0**	225	**11.0**	**10.1**	243	−0.1	−2.6
ibm12	413	**10.9**	**14.5**	410	**11.4**	**14.4**	461	−1.1	**1.1**
ibm13	266	**3.4**	**7.6**	273	**12.1**	**11.8**	324	−0.4	**1.1**
ibm14	392	**2.6**	**4.0**	478	**14.5**	**21.7**	824	**6.5**	**6.6**
ibm15	544	**6.1**	**5.6**	577	**8.6**	**12.2**	1001	−2.9	−5.0
ibm16	629	**5.6**	**6.4**	680	**11.5**	**11.5**	931	**6.7**	**5.7**
ibm17	737	**2.7**	**3.8**	810	**9.3**	**12.2**	1144	**7.5**	**7.2**
ibm18	458	**2.1**	**3.3**	574	**19.0**	**19.8**	885	**11.0**	**10.7**
Average	—	4.9	7.1	—	9.6	11.7	—	1.5	1.3

(b) Runtime

Circuit	Capo			Fastplace			mPL		
		AMG-LE			AMG-LE			AMG-LE	
	Baseline (s)	1Lvl (%)	2Lvl (%)	Baseline (s)	1Lvl (%)	2Lvl (%)	Baseline (s)	1Lvl (%)	2Lvl (%)
ibm01	193	−3	−75	26	−5	−62	109	−69	−152
ibm02	329	−24	−112	46	−14	−72	236	−91	−171
ibm03	487	−21	−105	46	−20	−82	221	−108	−210
ibm04	512	−23	−110	51	−17	−90	250	−87	−195
ibm05	411	−25	−124	36	−35	−81	174	−118	−183
ibm06	580	−19	−115	96	29	−6	390	−83	−197
ibm07	920	−21	−105	96	−9	−82	387	−100	−198
ibm08	941	−20	−121	99	−31	−158	1109	−105	−197
ibm09	1198	−29	−119	122	−6	−66	898	−136	−221
ibm10	1868	−12	−120	286	4	−62	1450	−126	−210
ibm11	1834	−30	−121	135	−66	−139	1084	−84	−193
ibm12	1996	−22	−124	282	−12	−52	1517	−115	−222
ibm13	2387	−16	−120	239	−43	−118	1235	−110	−218
ibm14	3327	−29	−129	530	−42	−121	2189	−93	−213
ibm15	5781	−30	−142	624	−57	−197	3969	−101	−247
ibm16	4917	−35	−132	722	−75	−217	5457	−57	−118
ibm17	5286	−31	−135	1133	−33	−101	2788	−131	−228
ibm18	4215	−39	−142	1589	3	−57	3106	−125	−257
Average	—	−24	−119	—	−24	−98	—	−102	−202

TABLE 4: Placement wire length and runtime results comparing one- and two-level AMG-LE clustering with length-driven unclustering to baseline placements without using clustering with three placers on the ISPD05 benchmark suite.

(a) HPWL

Circuit	Capo			Fastplace			mPL		
	Baseline ($\times 10^6$)	AMG-LE		Baseline ($\times 10^6$)	AMG-LE		Baseline ($\times 10^6$)	AMG-LE	
		1Lvl (%)	2Lvl (%)		1Lvl (%)	2Lvl (%)		1Lvl (%)	2Lvl (%)
adaptec1	91	**1.5**	**2.3**	87	**8.6**	**8.7**	81	−1.1	−1.1
adaptec2	120	**14.2**	**14.5**	108	**4.0**	**9.6**	97	−0.1	−0.1
adaptec3	254	**6.4**	**5.7**	287	**15.4**	**17.2**	224	**0.6**	**0.6**
adaptec4	—	—	—	230	**8.6**	**12.7**	195	−0.5	−0.5
bigblue1	114	**4.0**	**3.3**	107	**5.5**	**7.8**	101	**0.0**	**0.0**
bigblue2	167	**2.7**	**2.3**	181	**8.5**	**12.3**	152	−0.3	−0.3
bigblue3	439	**3.6**	**6.4**	663	**37.0**	**40.1**	492	**1.7**	**1.7**
Average	—	**5.4**	**5.7**	—	**12.5**	**15.5**	—	**0.0**	**0.0**

(b) Runtime

Circuit	Capo			Fastplace			mPL		
	Baseline ($\times 10^2$ s)	AMG-LE		Baseline ($\times 10^2$ s)	AMG-LE		Baseline ($\times 10^2$ s)	AMG-LE	
		1Lvl (%)	2Lvl (%)		1Lvl (%)	2Lvl (%)		1Lvl (%)	2Lvl (%)
adaptec1	30	−24	−73	5	−79	−109	18	−80	−109
adaptec2	44	2	−49	13	−15	−4	17	−134	−211
adaptec3	81	−32	−107	28	−136	−77	65	−89	−135
adaptec4	—	—	—	21	−208	−184	45	−258	−199
bigblue1	54	11	−36	6	−158	−141	18	−128	−173
bigblue2	92	−64	−100	23	−299	−278	49	−235	−286
bigblue3	254	−24	−49	139	−45	−78	123	−323	−365
Average	—	−22	−69	—	−134	−125	—	−178	−211

are more correlated with the length estimates than the baseline placements. The correlation improvement is small, with a maximum improvement of 6% for Capo using two-level AMG-LE, but does suggest that the estimates are directing the placement.

In order to assess the scalability of the proposed clustering and unclustering algorithms, the same experiments are performed on the ISPD05 circuits. These circuits are bigger than the ICCAD04 benchmarks and have significantly larger maximum net degree. The results are tabulated in Table 4. The version of Capo used in the experiment could not place the adaptec4 benchmark on the test platform so its entries in Table 4 are not included. In general, the results resemble those of Table 3 with Capo and Fastplace obtaining significant wire length improvement for both one- and two-level AMG-LE. Meanwhile, only a negligible average improvement is achieved by mPL. The runtime increases for all placers with mPL having the largest increase in runtime, which is not surprising given that mPL must perform global placement in addition to detailed placement at each level. Capo has a more modest increase in runtime compared to the other two placers, on average. In this case, the placements for mPL are nearly identical whether one or two levels of AMG-LE clustering are performed. This means that the effects of the second level of clustering are almost entirely

removed by performing global placement on the unclustered circuit.

4.2. Length-Driven Unclustering as a Detailed Placer. In this section, the same multilevel experiments are performed with the difference of not performing any placement apart from placing the bottommost clustered circuit. The circuit is placed at the bottom level using the global and detailed placement functions of each placer (if the two are distinguished). As a result, the placement at the bottom level is legal, that is, free from overlaps. To improve upon the runtime results given in the previous section, it is proposed to only use the length-driven unclustering technique as the only refinement between levels. The placement will preserve its legality using the technique mentioned in Section 3.2. The results of this experiment are given in Table 5.

The format of the Tables 5(a) and 5(b) is the same as Table 3. It can be seen that all placers' wire lengths are improved on average using either one- or two-level AMG-LE. The best wire length improvement is again for Fastplace with one-level AMG-LE on ibm06 with a 35.8% improvement. Using one-level AMG-LE 16, 15, and 14 circuits improve for Capo, Fastplace, and mPL, respectively. Improvement is observed in 14, 16, and 15 circuits for Capo, Fastplace,

TABLE 5: Placement wire length and runtime results comparing one- and two-level AMG-LE clustering with length-driven unclustering without detailed placement to baseline placements without using clustering with three placers on the ICCAD04 benchmark suite.

(a) HPWL

Circuit	Capo			Fastplace			mPL		
		AMG-LE			AMG-LE			AMG-LE	
	Baseline ($\times 10^5$)	1Lvl (%)	2Lvl (%)	Baseline ($\times 10^5$)	1Lvl (%)	2Lvl (%)	Baseline ($\times 10^5$)	1Lvl (%)	2Lvl (%)
ibm01	25	−1.0	−1.7	24	−7.7	−4.9	24	−3.2	−3.4
ibm02	51	**1.6**	**1.2**	54	**1.3**	**2.5**	52	−3.8	−3.8
ibm03	77	**4.2**	**4.6**	80	**1.3**	**3.4**	82	**5.0**	**5.1**
ibm04	93	**10.6**	**10.4**	86	**4.0**	**4.7**	109	**11.3**	**15.9**
ibm05	103	**0.3**	−1.1	101	−1.9	−2.7	93	−10.1	−13.4
ibm06	67	**1.3**	**0.3**	99	**35.8**	**31.1**	88	**15.4**	**17.1**
ibm07	126	**10.2**	**9.9**	123	**3.3**	**7.0**	124	−4.4	**1.2**
ibm08	137	**2.6**	**1.9**	147	−3.0	**3.3**	211	**13.0**	**20.0**
ibm09	145	**0.6**	**1.5**	155	**4.3**	**3.9**	189	**11.9**	**15.9**
ibm10	318	−1.0	−0.5	362	**7.4**	**9.0**	363	**3.2**	**5.7**
ibm11	210	**1.9**	**1.9**	225	**7.6**	**6.3**	243	**4.3**	**4.1**
ibm12	413	**9.6**	**9.6**	410	**7.4**	**10.4**	461	**6.4**	**6.9**
ibm13	266	**3.3**	**3.4**	273	**7.7**	**7.3**	324	**9.7**	**14.3**
ibm14	392	**2.4**	**1.0**	478	**11.5**	**18.4**	824	**30.0**	**35.2**
ibm15	544	**2.6**	−0.1	577	**5.0**	**8.6**	1001	**23.3**	**23.4**
ibm16	629	**5.1**	**2.5**	680	**8.5**	**7.9**	931	**16.8**	**21.8**
ibm17	737	**1.9**	**1.2**	810	**7.0**	**9.0**	1144	**13.0**	**21.5**
ibm18	458	**0.9**	**0.0**	574	**15.3**	**15.8**	885	**18.8**	**28.9**
Average	—	**3.2**	**2.6**	—	**6.4**	**7.8**	—	**8.9**	**12.0**

(b) Runtime

Circuit	Capo			Fastplace			mPL		
		AMG-LE			AMG-LE			AMG-LE	
	Baseline (s)	1Lvl (%)	2Lvl (%)	Baseline (s)	1Lvl (%)	2Lvl (%)	Baseline (s)	1Lvl (%)	2Lvl (%)
ibm01	193	16	12	26	−12	−22	109	47	54
ibm02	329	−8	2	46	1	−28	236	27	38
ibm03	487	−5	5	46	−16	−36	221	10	32
ibm04	512	−1	−4	51	−13	−42	250	39	40
ibm05	411	−1	−1	36	−29	−63	174	−2	32
ibm06	580	12	5	96	34	15	390	39	51
ibm07	920	7	8	96	−8	−57	387	28	35
ibm08	941	6	−11	99	−22	−110	1109	62	67
ibm09	1198	−8	1	122	−3	−42	898	55	62
ibm10	1868	3	−3	286	8	−11	1450	8	38
ibm11	1834	−1	−5	135	−64	−94	1084	56	56
ibm12	1996	−1	−7	282	−3	−17	1517	29	39
ibm13	2387	−7	−6	239	−32	−81	1235	54	56
ibm14	3327	−2	−14	530	−51	−125	2189	27	49
ibm15	5781	0	−22	624	−38	−188	3969	32	43
ibm16	4917	−4	−11	722	−58	−167	5457	50	53
ibm17	5286	−4	−26	1133	−40	−84	2788	3	12
ibm18	4215	−8	−36	1589	2	−43	3106	2	14
Average	—	0	−6	—	−19	−66	—	32	43

TABLE 6: Placement wire length and runtime results comparing one- and two-level AMG-LE clustering with length-driven unclustering without detailed placement to baseline placements without using clustering with three placers on the ISPD05 benchmark suite.

(a) HPWL

Circuit	Capo			Fastplace			mPL		
		AMG-LE			AMG-LE			AMG-LE	
	Baseline ($\times 10^6$)	1Lvl (%)	2Lvl (%)	Baseline ($\times 10^6$)	1Lvl (%)	2Lvl (%)	Baseline ($\times 10^6$)	1Lvl (%)	2Lvl (%)
adaptec1	91	−1.4	−2.7	87	**1.8**	−1.6	81	**5.3**	**3.1**
adaptec2	120	**12.6**	**10.8**	108	−0.2	**3.6**	97	**13.7**	**12.9**
adaptec3	254	**4.0**	**2.0**	287	**12.0**	**10.8**	224	**16.4**	**15.3**
adaptec4	—	—	—	230	**4.9**	**7.4**	195	**8.0**	**7.6**
bigblue1	114	**1.1**	**0.1**	107	−1.7	−2.6	101	−0.4	−2.3
bigblue2	167	**1.0**	−0.4	181	**0.8**	**2.0**	152	**3.2**	**1.6**
bigblue3	439	**4.3**	**4.6**	663	**35.7**	**37.6**	492	**30.4**	**29.1**
Average	—	**3.6**	**2.4**	—	**7.6**	**8.2**	—	**10.9**	**9.6**

(b) Runtime

Circuit	Capo			Fastplace			mPL		
		AMG-LE			AMG-LE			AMG-LE	
	Baseline ($\times 10^2$ s)	1Lvl (%)	2Lvl (%)	Baseline ($\times 10^2$ s)	1Lvl (%)	2Lvl (%)	Baseline ($\times 10^2$ s)	1Lvl (%)	2Lvl (%)
adaptec1	30	21	17	5	−61	−65	18	32	3
adaptec2	44	34	41	13	−14	10	17	−34	−59
adaptec3	81	19	29	28	−47	−73	65	−22	16
adaptec4	—	—	—	21	−203	−195	45	−59	−52
bigblue1	54	39	42	6	−95	−74	18	−2	−36
bigblue2	92	−2	−16	23	−128	−114	49	−71	−91
bigblue3	254	11	−4	139	−29	−68	123	−90	−75
Average	—	20	18	—	−82	−83	—	−35	−42

and mPL, respectively, when using two-level AMG-LE. When compared to the results in Section 4.1, the average wire length for Capo and Fastplace degrades while mPL significantly improves. This improvement is because mPL has no option to perform just detailed placement. It performs global and detailed placement using the clustered placement as the starting point. During this process, the clusters become dispersed. Therefore, when the proposed length-driven unclustering is used as a detailed placer, the benefits of clustering and length-driven unclustering are preserved.

The runtime results show significant improvements over the results in Section 4.1. When compared to the baseline, Capo is not significantly changed while Fastplace is degraded by 19% and 66% for one- and two-level AMG-LE. The increases in runtime for Fastplace may be tolerated because of its comparatively low runtimes. However, mPL runtimes are improved significantly with more improvement seen with two-level AMG-LE, nearly cutting the average runtime by half.

In order to assess the scalability of the proposed length-driven unclustering as a detailed placer, the same experiments are performed on the ISPD05 circuits. The results are tabulated in Table 6. The results follow similar trends as those

of Table 5. All placers achieve wire length improvements and significantly improved running times compared to the results of Table 4 which are obtained by using each placers' detailed placement algorithm in between clustering levels. In this case, Capo achieves an overall runtime improvement compared to the baseline with 3.6% and 2.4% average improvements in wire length when performing one and two levels of AMG-LE clustering, respectively. mPL achieves substantial improvements in wire length with modest increases in average running time. Finally, for Fastplace, significant average wire length improvements are observed for both one and two levels of AMG-LE clustering, while the runtime has degraded which is acceptable considering the relatively low runtimes of Fastplace.

In the end, the circuit designers can decide which of the placement flows presented in Sections 4.1 and 4.2 is preferred for their application. Both improve wire length but to varying degrees depending on the choice of placer. When using a placer like Capo or Fastplace, the best wire length is obtained by an increase in runtime, while mPL achieves the best results with the fast detailed placement flow. If runtime is the most urgent concern, the proposed scheme of using length-driven unclustering as a detailed placer is the best option.

TABLE 7: Placement wire length and runtime results comparing one-level of existing clustering algorithms and AMG-LE with and without length-driven unclustering to baseline placements without using clustering with mPL6 on the ICCAD04 benchmark suite.

(a) HPWL

Circuit	Baseline ($\times 10^6$)	HEM (%)	BC (%)	AMGC (%)	AMG-LE 1Lvl	
					Regular (%)	Fast detailed (%)
ibm01	2.4	5.9	7.1	5.7	4.6	−3.2
ibm02	5.2	−6.3	−1.6	−9.9	1.3	−3.8
ibm03	8.2	2.2	6.1	5.8	−7	5
ibm04	10.9	0.4	−1.3	−1.9	−5.7	11.3
ibm05	9.3	−5.4	−5.3	−5.4	−5.2	−10.1
ibm06	8.8	10	12.5	11.9	5.8	15.4
ibm07	12.4	1.4	−0.1	2.2	−3.1	−4.4
ibm08	21.1	−2.7	−2.9	−2.9	4.7	13
ibm09	18.9	5.1	−2	3.7	2.9	11.9
ibm10	36.3	1.9	1.7	2.1	2	3.2
ibm11	24.3	−2.6	−4.6	−1.8	−0.1	4.3
ibm12	46.1	−5	−5.9	−4.2	−1.1	6.4
ibm13	32.4	−1.2	0.3	1.7	−0.4	9.7
ibm14	82.4	6.9	7.4	7.4	6.5	30
ibm15	100.1	−7.5	−3.3	1.7	−2.9	23.3
ibm16	93.1	2.5	2.4	2.6	6.7	16.8
ibm17	114.4	5.7	5.9	4.9	7.5	13
ibm18	88.5	11.6	12	12	11	18.8
Average	—	1.2	1.5	1.9	1.5	8.9

(b) Runtime

Circuit	Baseline (s)	HEM (%)	BC (%)	AMGC (%)	AMG-LE 1Lvl	
					Regular (%)	Fast detailed (%)
ibm01	150	−20	−20	−14	−69	47
ibm02	324	−39	−41	−38	−91	27
ibm03	281	−67	−52	−85	−108	10
ibm04	344	−40	−28	−31	−87	39
ibm05	232	−45	−29	−22	−118	−2
ibm06	475	−62	−32	−49	−83	39
ibm07	488	−51	−43	−49	−100	28
ibm08	1468	−29	−44	−26	−105	62
ibm09	1134	−26	−26	−24	−136	55
ibm10	1762	−45	−49	−45	−126	8
ibm11	1501	−15	−1	−4	−84	56
ibm12	2009	−41	−57	−36	−115	29
ibm13	1456	−28	−26	−23	−110	54
ibm14	2547	−39	−33	−29	−93	27
ibm15	4208	−48	−26	−23	−101	32
ibm16	5821	−33	−35	−26	−57	50
ibm17	3025	−52	−53	−59	−131	3
ibm18	3672	−47	−49	−49	−125	2
Average	—	−40	−36	−35	−102	31

4.3. Comparison to Existing Clustering Algorithms. The proposed AMG-LE clustering algorithm with length-driven unclustering is compared to other existing clustering algorithms in this section. Each technique is evaluated in terms of after placement wire length and total runtime using the mPL6 placer [37]. The clustering algorithms compared to are heavy-edge matching (HEM) [22], best-choice (BC) [1], and the AMG clustering (AMGC) algorithm in [31]. It is worth mentioning that best-choice is the clustering algorithm used internally by Fastplace 3.0 and mPL6. Placement is first performed without any clustering to establish a baseline. Then, each of the algorithms is used to produce a clustered circuit which is placed by mPL. The clustered placement is then unclustered and detailed placement is performed by mPL or by using length-driven unclustering as a detailed placer. Results of the experiment are given in Table 7.

The percentage improvement for AMG-LE in a regular clustered placement flow, as described in Section 4.1, is given under the subheading Regular. The column labeled Fast Detailed refers to the results using only length-driven unclustering as a detailed placer, described in Section 4.2. In terms of wire length, the regular AMG-LE scheme is competitive with the existing algorithms performing better in some circuits and worse in others. On the other hand, the runtime for regular AMG-LE degrades more than the other clustering techniques. This is because the runtime includes the time to perform pre-placement length estimation and the algorithm is implemented in the MATLAB environment Therefore, the runtime can be way more competitive if the procedure is implemented in a more efficient way. However, the fast detailed variant of AMG-LE is a clear winner in terms of average wire length and runtime improvement.

5. Conclusions

This paper presents a new clustering technique by utilizing length estimates in an algebraic multigrid-based clustering technique. In addition, a physical unclustering technique is proposed with the benefits of reducing wire length and preserving legality. Because of its legality-preserving nature, the technique eliminates the need to use detailed placement between levels in the framework. These techniques are shown to be effective in reducing wire length and can also be used to benefit runtime.

Acknowledgments

This research is supported by the Natural Sciences and Engineering Research Council of Canada and Alberta Innovates Technology Futures. This research has been enabled by the use of computing resources provided by WestGrid and Compute/Calcul Canada.

References

[1] C. Alpert, A. Kahng, G. J. Nam, S. Reda, and P. Villarrubia, "A semi-persistent clustering technique for vlsi circuit placement," in *International Symposium on Physical Design (ISPD '05)*, pp. 200–207, April 2005.

[2] J. Li, L. Behjat, and A. Kennings, "Net cluster: a net-reduction-based clustering preprocessing algorithm for partitioning and placement," *IEEE Transactions on Computer-aided Design of Integrated Circuits and Systems*, vol. 26, no. 4, pp. 669–679, 2007.

[3] S. Bodapati and F. N. Najm, "Prelayout estimation of individual wire lengths," *IEEE Transactions on VLSI Systems*, vol. 9, no. 6, pp. 943–958, 2001.

[4] A. Farshidi, L. Behjat, L. Rakai, and B. Fathi, "A pre-placement individual net length estimation model and an application for modern circuits," *Integration*, vol. 44, no. 2, pp. 111–122, 2011.

[5] B. Fathi, L. Behjat, and L. M. Rakai, "A pre-placement net length estimation technique for mixed-size circuits," in *ACM/IEEE Workshop on System Level Interconnect Prediction (SLIP '09)*, pp. 45–52, July 2009.

[6] A. B. Kahng and S. Reda, "Intrinsic shortest path length: a new, accurate a priori wirelength estimator," in *IEEE/ACM International Conference on Computer-Aided Design (ICCAD '05)*, pp. 173–180, November 2005.

[7] W. E. Donath, "Placement and average interconnection lengths of computer logic," *IEEE Trans Circuits Syst*, vol. 26, no. 4, pp. 272–277, 1979.

[8] W. E. Donath, "Wire length distribution for placements of computer logic," *Ibm Journal of Research and Development*, vol. 25, no. 2-3, pp. 152–155, 1981.

[9] M. Feuer, "Connectivity of random logic," *IEEE Transactions on Computers*, vol. C-31, no. 1, pp. 29–33, 1982.

[10] T. Hamada, C. K. Cheng, and P. M. Chau, "A wire length estimation technique utilizing neighborhood density equations," *IEEE Transactions on Computer-aided Design of Integrated Circuits and Systems*, vol. 15, no. 8, pp. 912–922, 1996.

[11] H. T. Heineken and W. Maly, "Standard cell interconnect length prediction from structural circuit attributes," in *IEEE Custom Integrated Circuits Conference*, pp. 167–170, May 1996.

[12] M. Pedram and B. Preas, "Interconnection length estimation for optimized standard cell layouts," in *IEEE International Conference on Computer-Aided Design (ICCAD '89)*, pp. 390–393, November 1989.

[13] B. Hu and M. Marek-Sadowska, "Wire length prediction based clustering and its application in placement," in *40th Design Automation Conference*, pp. 800–805, June 2003.

[14] Q. Liu, B. Hu, and M. Marek-Sadowska, "Wire length prediction in constraint driven placement," in *International Workshop on System Level Interconnect Prediction*, pp. 99–105, April 2003.

[15] A. Brandt, "Algebraic multigrid theory: the symmetric case," *Applied Mathematics and Computation*, vol. 19, no. 1–4, pp. 23–56, 1986.

[16] W. L. Briggs, V. E. Henson, and S. F. McCormick, *A Multigrid Tutorial*, SIAM, Philadelphia, Pa, USA, 2nd edition, 2000.

[17] S. McCormick, Ed., *Multigrid Methods*, vol. 5 of *Frontiers in Applied Mathematics*, SIAM, Philadelphia, Pa, USA, 1987.

[18] J. Ruge and K. Stüben, *Multigrid Methods*, chapter 4, SIAM, Philadelphia, Pa, USA, 1987.

[19] K. Stüben, "Algebraic multigrid (AMG): experiences and comparisons," *Applied Mathematics and Computation*, vol. 13, no. 3-4, pp. 419–451, 1983.

[20] U. Trottenberg, C. Oosterlee, and A. Schüller, *Multigrid*, Academic Press, 2001.

[21] P. Wesseling, *An Introduction to Multigrid Methods*, Pure and Applied Mathematics Series, John Wiley and Sons, New York, NY, USA, 1992.

[22] C. J. Alpert, J. H. Huang, and A. B. Kahng, "Multilevel circuit partitioning," *IEEE Transactions on Computer-aided Design of*

Integrated Circuits and Systems, vol. 17, no. 8, pp. 655–667, 1998.

[23] A. Caldwell, A. Kahng, and I. Markov, "Improved algorithms for hypergraph bi-partitioning," in *Asia and South Pacific Design Automation Conference (ASP-DAC '00)*, pp. 661–666, 2000.

[24] J. Cong and S. K. Lim, "Edge separability-based circuit clustering with application to multilevel circuit partitioning," *IEEE Transactions on Computer-aided Design of Integrated Circuits and Systems*, vol. 23, no. 3, pp. 346–357, 2004.

[25] B. Hu and M. Marek-Sadowska, "Fine granularity clustering for large scale placement problems," in *International Symposium on Physical Design*, pp. 67–74, April 2003.

[26] G. Karypis, R. Aggarwal, V. Kumar, and S. Shekhar, "Multilevel hypergraph partitioning: applications in vlsi domain," *IEEE Transactions on VLSI Systems*, vol. 7, no. 1, pp. 69–79, 1999.

[27] G. J. Nam, "ISPD 2006 placement contest: benchmark suite and results," in *International Symposium on Physical Design (ISPD '06)*, p. 167, April 2006.

[28] G. Karypis, R. Aggarwal, V. Kumar, and S. Shekhar, "Multilevel hypergraph partitioning: application in vlsi domain," in *34th Design Automation Conference*, pp. 526–529, June 1997.

[29] J. Z. Yan, C. Chu, and W. K. Mak, "Safechoice: a novel clustering algorithm for wirelength-driven placement," in *ACM International Symposium on Physical Design (ISPD '10)*, pp. 185–192, March 2010.

[30] J. Z. Yan, C. Chu, and W. K. Mak, "Safechoice: a novel approach to hypergraph clustering for wirelength-driven placement," *IEEE Transactions on Computer-aided Design of Integrated Circuits and Systems*, vol. 30, no. 7, pp. 1020–1033, 2011.

[31] L. Rakai, L. Behjat, S. Martin, and J. Aguado, "An algebraic multigrid-based algorithm for circuit clustering," *Applied Mathematics and Computation*, vol. 218, no. 9, pp. 5202–5216, 2012.

[32] A. B. Kahng and Q. Wang, "Implementation and extensibility of an analytic placer," *IEEE Transactions on Computer-aided Design of Integrated Circuits and Systems*, vol. 24, no. 5, pp. 734–747, 2005.

[33] S. N. Adya, S. Chaturvedi, J. A. Roy, D. A. Papa, and I. L. Markov, "Unification of partitioning, placement and floorplanning," in *IEEE/ACM International Conference on Computer-Aided Design (ICCAD '04)*, pp. 550–557, November 2004.

[34] G. J. Nam, C. J. Alpert, P. Villarrubia, B. Winter, and M. Yildiz, "The ispd2005 placement contest and benchmark suite," in *International Symposium on Physical Design (ISPD '05)*, pp. 216–220, April 2005.

[35] J. Roy and I. Markov, *Partitioning-driven Techniques for VLSI Placement. Handbook of Algorithms for VLSI Physical Design Automation*, CRC Press, 2008.

[36] N. Viswanathan, M. Pan, and C. Chu, "Fastplace 3.0: a fast multilevel quadratic placement algorithm with placement congestion control," in *Asia and South Pacific Design Automation Conference (ASP-DAC '07)*, pp. 135–140, January 2007.

[37] T. F. Chan, J. Cong, J. R. Shinnerl, K. Sze, and M. Xie, "Mpl6: enhanced multilevel mixed-size placement," in *International Symposium on Physical Design (ISPD '06)*, pp. 212–214, April 2006.

14

An Efficient Multi-Core SIMD Implementation for H.264/AVC Encoder

M. Bariani, P. Lambruschini, and M. Raggio

Department of Biophysical and Electronic Engineering, University of Genova, Via Opera Pia 11 A, 16145 Genova, Italy

Correspondence should be addressed to P. Lambruschini, lambruschini@dibe.unige.it

Academic Editor: Muhammad Shafique

The optimization process of a H.264/AVC encoder on three different architectures is presented. The architectures are multi- and singlecore and SIMD instruction sets have different vector registers size. The need of code optimization is fundamental when addressing HD resolutions with real-time constraints. The encoder is subdivided in functional modules in order to better understand where the optimization is a key factor and to evaluate in details the performance improvement. Common issues in both partitioning a video encoder into parallel architectures and SIMD optimization are described, and author solutions are presented for all the architectures. Besides showing efficient video encoder implementations, one of the main purposes of this paper is to discuss how the characteristics of different architectures and different set of SIMD instructions can impact on the target application performance. Results about the achieved speedup are provided in order to compare the different implementations and evaluate the more suitable solutions for present and next generation video-coding algorithms.

1. Introduction

In the last years the video compression algorithms have played an important role in the enjoying of multimedia contents. The passage from analog to digital world in multimedia environment cannot be performed without compression algorithms. DVDs, Blu-Ray, and Digital TV are typical examples. The compression algorithm used in DVDs is MPEG-2, and Blu-Ray supports VC-1 standardized with the name SMPTE 421M [1], in addition to MPEG-2 and H.264. In the digital television, the compression algorithms are used to reduce the transmission throughput. In DVB-T, the picture format for DVD and Standard Definition TV (SDTV) is 720 × 576, and this resolution is the most used in digital multimedia contents. The most recent standards for digital television as DVB-T2 and DVB-H support H.264/MPEG-4 AVC for coding video.

The H.264/AVC [2] video compression standard can cope with a large range of applications, reaching compression rate and video quality levels never accomplished by previous algorithms. Even if the initial H.264/AVC standard (completed in May 2003) was primarily focused on "entertainment-quality" video, not dealing with the highest video resolutions, the introduction of a new set of extensions

in July 2004 covered this lack. These extensions are known as "fidelity range extensions" (FRExt) and produced a set of new profiles, collectively called High Profiles. As described in [3], these profiles support all the Main Profile features and introduce additional characteristics such as adaptive transform block-size and perceptual quantization scaling matrices. Experimental results show that, when restricted to intra-only coding, H.264/AVC High Profile outperforms the state-of-the-art in still-image coding represented by JPEG2000 on a set of monochrome test images by 0.5 dB average PSNR [4].

It results that a H.264 encoder addressing high definition (HD) resolutions needs to support High Profiles in order to be part of an effective video application. On the other hand, the already great complexity of the H.264 algorithm is further increased by supporting FRExt. In particular, this leads to implement two new modules: the 8 × 8 intraprediction and the 8 × 8 transform.

In case of mobile devices, the H.264 complexity issues together with the constraints of limited power consumption and the typical need of real-time operations in video-based applications draw a difficult scenario for video application developers.

The HD resolution involves a large amount of data, and the compression algorithms are high computational demand applications, often used as benchmark to measure the processor performance. In order to support real-time video encoding and decoding, specific architectures are developed. Multicore architectures have the potential to meet the performance levels required by the real-time coding of HD video resolutions. But in order to exploit multicore architectures, several problems have to be faced. The first issue is the subdivision of an encoder application in modules that can be executed in parallel. In this case, the main difficult is the strong data dependency in video encoder algorithms. Parallel architectures can be more easily exploited using other kind of algorithm like computer graphics, rendering technology or cryptography, where the data dependency is not as strong as in video compression. Once a good partitioning is achieved, the optimization of a video encoder should take advantage of the data level parallelism to increase the performance of each encoder module running on the architecture's processing element. A common approach is to use the SIMD instructions to exploit the data level parallelism during the execution; otherwise, ASIC design can be adopted for critical kernel. SIMD architectures are widely used for their flexibility. SIMD ISAs are added at most market spread processor: Intel's MMX, SSE1, SSE2, SSE3, SSE4; Amd 3DNow!; ARM's NEON; Motorola's AltiVec (also known as Apple's Velocity Engine or IBM's VMX).

In this paper, we will show how the data level parallelism is exploited by SIMD and which instructions are more useful in video processing. Different instruction set architectures (ISAs) will be compared in order to show how the optimization can be driven and how different ISA features can lead to different performance. This paper is intended to be a great help to both software programmers that have to choose for the most suitable SIMD ISA for developing a video-based application and for ISA designers that want to create a generic instruction set being able to give good performance on video applications. In that regard, the authors will select a set of generic SIMD instructions that can speed up video codec applications, detailing the modules that will profit from the introduction of each instruction. Besides describing the optimization methods, the paper indicates a few guidelines that should be followed to partition the encoder in separate modules.

Even though the work focuses on H.264/AVC, most of the proposed solutions will also apply to the earlier mentioned standards as well as to more recent video compression algorithms as scalable video coding (SVC) [5]. Moreover, H.264/AVC tools will have a fundamental role in the emerging high efficiency video coding (HEVC) standardization project [6].

This paper is organized as follows. Section 2 gives an overview of the state-of-the-art SIMD-based architectures, giving particular attention to those targeting video-coding applications. A brief description of the three architectures used for the presented project is given in Section 3. Section 4 describes the H.264 optimized encoder, focusing on module partitioning and SIMD-based implementation. The performance results of both the C-pure implementation and the SIMD version are given in Section 5 together with an explanation about what are the key instructions for optimizing a video codec. Finally, the conclusion is drawn in Section 6.

2. Related Works

The basic concept of SIMD instructions is the possibility of fill vector registers with multiple data in order to execute the same operation on several elements. One of the major bottlenecks in the SIMD approach is the overhead due to the data handling needed to feed the vector registers. Typical required operations are extra memory accesses, packing data, element permutation inside vectors, and conversion from vector to scalar results. All these preliminary operations limit the vector dimension and the performance enhancement achievable with SIMD optimization.

In literature, several studies regarding the SIMD optimization of video-coding applications are available [7–11]. The scope driving these studies is the achievement of the maximum performance, adopting measures in order to reduce the known bottlenecks. Since its standardization, SIMD optimizations targeting the H.264 algorithm have been proposed as can be seen in [12, 13]. However, both the works only address the H.264 decoder and present a MMX optimization starting from the H.264 reference code. Besides addressing a more complex application, our aims were also to discuss how the characteristics of different architectures and different set of SIMD instructions can impact on the encoder performance.

In SIMD processors the memory access has an important impact on performance. The unaligned access is not usually possible in SIMD ISA, and when possible it is discouraged due to additional instruction latency. The programmers usually take care of handling the unaligned load adding further overhead to vector data organization. Moreover, the need of unaligned load is always present in video-coding algorithm especially in motion estimation (ME) and motion compensation (MC), where the pixel blocks selected by motion vectors are frequently at misaligned positions even if the start of a frame is memory aligned. Often, the position of a block we need to access cannot be known in advance, and this leads to unpredictable misalignment in data loaded from memory. In Intel's architectures, starting from SSE2 the support to unaligned load has been added, but the performance is strongly reduced either if the load operation crosses the cache boundary or, with SSE3, if the load instruction needs store-to-load forwarding. In AltiVec, it is necessary to load two adjacent positions and shift data in order to achieve one unaligned load, a usually adopted approach to overcome the misalignment access issue. This problem is common in digital signal processor (DSP) as well. Usually, DSP do not support unaligned loads, but due to the large use of DSP in video application several producers have added the support to this kind of operation. For example, Texas Instruments family TMS320C64x supports unaligned load and store operations of 32 and 64 bit element, but with only one of the two memory ports [14].

The MediaBreeze SIMD processor was proposed to reduce the bottlenecks in SIMD implementations [15]. The Breeze SIMD ISA uses a multidimensional vector able to speed up nested loops but at the cost of a very complicated instruction structure requiring a dedicated instruction memory. In [16], a specific SIMD ISA named VS-ISA was proposed in order to improve performance in video coding. The authors adopted specific solutions for sum of absolute difference (SAD), not aligned load applied to ME, interpolation, DCT-IDCT, and quantization dequantization.

Another typical approach to reduce the SIMD overhead is the usage of multibank vector memory where data is stored interleaved. The drawback is the increase of hardware cost for supporting the addresses generation.

An alternative to SIMD implementation on programmable processor architectures is the hardwired processor. Usually, it is only used when performance and low power consumption are essential requirements [7, 14, 17]. In fact, the lack of flexibility typical of hardwired processors reduces their applicability to a narrow segment of the market, where the programmability is either not required or considerably reduced.

3. SIMD ISA Description

In order to optimize the H.264 encoder, we chose three different ISAs. The adopted architectures are ST240, xStream, and P2012, all developed by STMicroelectronics. The former is a single-processor architecture, and the others are multicore platforms. In the following, the three architectures will be briefly described, giving special attention to the SIMD instruction set.

We chose these architectures for their novelty and for the possibility to have a complete toolchain (code generation, simulation, profiling, etc.) for developing an application in an optimal way. Each toolchain allowed a complete observability of the system. In this way, it was possible to evaluate the effectiveness of every author's solution. Observability is a very important characteristic when developing/optimizing an application. Using a real system it is not always possible to reach the degree of observability you have using a simulator and a suitable toolchain. Moreover, in an architecture under development as P2012 we had the possibility to contribute to the SIMD instruction set and, more important, to evaluate the contribution of each particular SIMD to the performance of the target video codec application. The three instruction sets present suitable characteristics for our research; they are generic instruction set, but ST240 includes a few video-specific instructions; we can analyse the impact of different vector register sizes; even if xStream and P2012 share many characteristics, only xStream supports horizontal SIMD (this is a special feature; e.g., other SIMD extensions as Intel SSE and ARM NEON do not have the same support); in P2012 platform, we were able to define and insert new SIMD instructions.

Besides the type of instructions, the SIMD extensions differ in both size and precision. These differences allow analyzing the impact of different architecture solutions on the global performance.

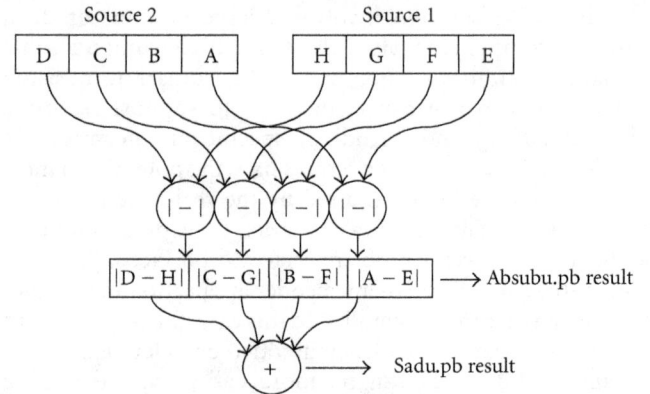

FIGURE 1: SAD operation.

3.1. ST240. The ST240 is a processor of STMicroelectronics ST200 family based on LX technology jointly developed with Hewlett Packard [18, 19]. The main ST240's features are the following:

(i) 4-issue Very Long Instruction Word (VLIW)

(ii) 64-32-bit general purpose registers

(iii) 32KB D-Cache and 32KB I-Cache

(iv) 450 MHz clock frequency

(v) 8-bit/16-bit arithmetic SIMD.

In the H.264 encoder SIMD optimization, the most significant instructions of the ST240 ISA are the following: the SIMD add.ph and sub.ph which perform, respectively, the packed 16-bit addition or subtraction; the perm.pb instruction which performs byte permutations and the muladdus.pb which multiplies an unsigned byte by a signed byte in each of the byte lanes and then sums across the four lanes to produce a single result. Furthermore, several data manipulation instructions are defined: pack.pb packs 16-bit values to byte elements ignoring the upper half; shuffeve.pb and shuffodd.pb, respectively, perform 8-bit shuffle of even and odd lanes. Two averaging operations (avg4u.pb and avgu.pb) are also defined in the instruction set.

One important operation in video-coding algorithms, the absolute value of the difference, abs (a-b), can be performed with the absubu.pb instruction (Figure 1) which works on each byte lane (treating each byte lane as an unsigned value) and returns the result in the corresponding byte lane of the destination register. The sadu.pb (Figure 1) performs the same operation and then sums the byte lanes value and returns the result.

3.2. xStream. xStream is a multiprocessor dataflow architecture for high-performance embedded multimedia streaming applications designed at STMicroelectronics [20, 21].

xStream is constituted by a parallel distributed and shared memory architecture. It is an array of processing elements connected by a Network on Chip (NoC) with specific hardware for management of communication [22], as depicted in Figure 2.

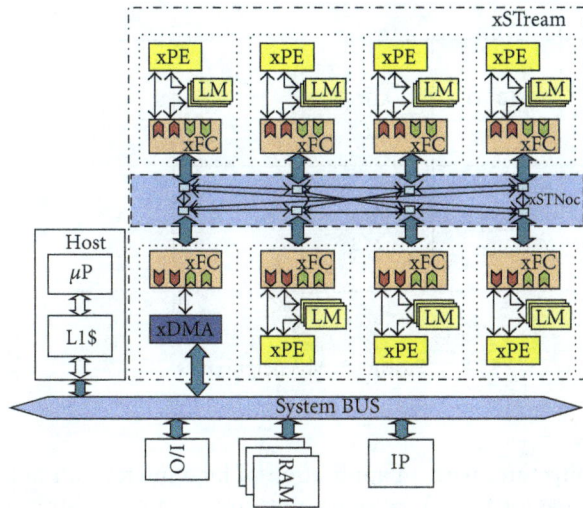

FIGURE 2: xSTream architecture.

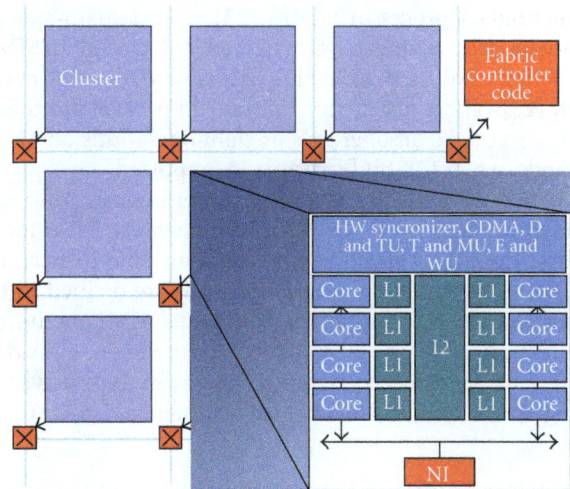

FIGURE 4: P2012 scheme.

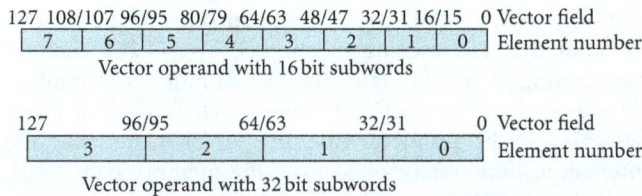

FIGURE 3: Vector operand.

The main elements in Figure 2 are the general purpose engine, the xSTreaming Processing Engines (XPEs) and the NoC interconnecting all components.

The XPEs are based on ST231 VLIW processors [22] of ST200 STMicroelectronics family [18, 19]. The main features can be resumed as

(i) 2-issue VLIW,

(ii) 128-bit vector registers,

(iii) up to 512 KB local memory cache,

(iv) up to 1 GHz clock frequency, and

(v) 16-bit/32-bit arithmetic SIMD.

In order to achieve excellent performance, the XPE core tries to exploit available parallelism at various levels. It supports a plethora of SIMD instructions to exploit available data-level parallelism. These instructions concurrently execute up to four operations on 32-bit operands or eight operations on 16-bit operands. The core supports wide 128-bit load/store.

The xSTream architecture handles scalar and vector operands.

Vector operands are 128-bit wide and consist of either eight 16-bit half-words or four 32-bit words, as shown in Figure 3.

In the xSTream ISA each SIMD instruction has an additional operand allowing permuting the result's element positions or replicating any element in the other positions.

This feature considerably increases the SIMD flexibility because the results have often to be reordered for further elaboration. This is especially true for video-coding algorithm with operations performed on several steps where the input of next step is usually the output of previous one. The permutation operand allows this with the cost of only one additional cycle. This leads to reduced costs to perform all the operation needed for data reordering.

The XPE supports horizontal SIMD as well. This kind of SIMD allows operations among elements in the same vector, and it is a key feature for speeding up execution in several H.264 functional units, as we will see in next sections.

3.3. *Platform 2012 (P2012).* Platform 2012 is a high-performance programmable architecture for high computational demanding embedded multimedia applications, currently under joint development by STMicroelectronics and Commissariat à l'énergie atomique et aux énergies alternatives (CEA) [23]. The goal of P2012 platform is to be reference architecture for next generation of multimedia product.

The P2012 architecture (Figure 4) is constituted by a large number of decoupled clusters of STxP70 processors interconnected by a Network on Chip (NoC). Each cluster can contain a number of computational elements ranging from 1 to 16. The main features of the STxP70 processor element are as follows:

(i) 32-bit RISC processor (up to 2 instructions per cycle),

(ii) 128-bit vector registers,

(iii) 256 KB of memory shared by all the processors (per cluster),

(iv) 600 MHz clock frequency, and

(v) 16-bit/32-bit arithmetic SIMD.

The P2012 basic modules can be easily replicated to provide scalability [24]. Each module is constituted by a computing cluster with cache memory hierarchy and a communication engine. The STxP70 is dual issue application-specific instruction-set processor (ASIP) [25] with domain-specific parameterized vector extension named VECx. STxP70 SIMD instructions are used to exploit available data level parallelism [26]. These instructions execute in parallel up to four operations on 32-bit operands or eight operations on 16-bit operands, while 128-bit load/store is supported.

Vector operands are 128-bit wide and consist of either eight 16-bit half-words or four 32-bit words. In order to increase the SIMD flexibility, instructions able to permute data positions inside the vector operands are defined in the instruction set. The support to horizontal SIMD is limited at operation involving only two adjacent elements inside a vector, but its presence is fundamental for typical video-coding operation like sum of absolute difference (SAD).

3.4. SIMD Instruction-Set Evaluation. Whatever platform we choose, we will have a limited number of SIMD instructions because of hardware constraints. For this reason, besides precision and size, one of the key issues while choosing a SIMD extension is generality versus application-specific instructions. The former can show good speedups for a large variety of applications. The latter can reach greater performance, but limited to a particular family of applications. Of course, there are a lot of solutions that lay in the middle.

The vector register size impacts performance, hardware reliability, and costs. The choice of the optimal size and precision of SIMD instructions is a key factor for reaching the desired performance for the target application. The axiom larger SIMD equal to better performance may be valid for applications having no constraints and data dependencies in either spatial or temporal field. It is not the H.264 encoder condition. In general, algorithms with a heavy control flow are very difficult to vectorize, and the SIMD optimization does not always lead to the desired performance enhancement.

The application developers should choose the dimension that best fit their needs, as well as ISA designers should take into account the requirements of the application families they are targeting. As stated in [7], in a processor designed to handle video-coding standards for which the theoretical worst-case video sequence will consist of a large number of 4×4 blocks, four-way SIMD parallelism makes full use of data paths. In this case, increasing the size will lead to little performance improvement. In contrast, if we focus on the H.264's fidelity range extensions, with their 8×8 transform and 8×8 intraprediction, an ISA with eight-way SIMD parallelism will yield to better performance. Next generation video-coding standards like HEVC will use wider ranges of block sizes for both prediction and transformation processes, making the choice of the optimal vector register size even more complicated.

4. H.264 Encoder Implementation

4.1. Software Partitioning. In order to support real-time video encoding addressing HD resolutions, multiprocessor architectures seem to be an optimal solution, as earlier explained. Moreover, we would like to test the multicore

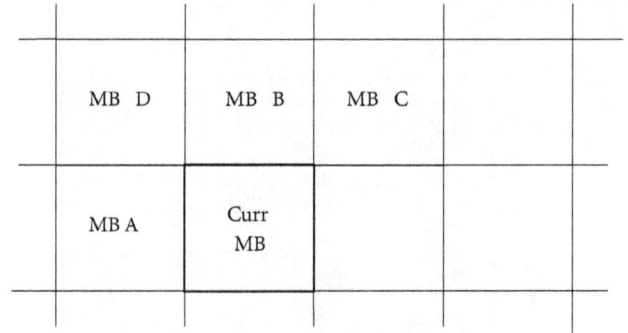

FIGURE 5: MB neighbours.

architectures with an application of high interest but not so suitable for these kind of architectures in order to stress the architecture design and to evaluate possible issues and finding solutions that could be also useful for other applications.

The first programmer's task dealing with this type of platforms is the subdivision of the encoder application in modules that can execute in parallel. The H.264 encoder partitioning plays a fundamental role in multicore architectures as xStream and P2012, where each functional block has to meet the resources of processor elements, and the interconnection system must fulfil the memory bandwidth needed to feed the modules. The designer choice becomes more complex when some modules can run in parallel avoiding stalls in pipeline [26].

Even if a detailed description of the encoder partitioning is beyond the scope of this paper, we can here depict some issues we faced approaching this process and the solutions we adopted.

First of all, it is worth to take into account the data dependency inside the H.264 encoder. Temporal data dependency is implicit in the Motion Estimation mechanism; the coding of the current frame always depends on the previously encoded frame(s) that are used as reference. Thus, there is always a temporal data dependency, except if the current frame is an I picture. Anyway, the encoding process also shows a spatial data dependency between macroblocks, that is, the basic encoding block comprising 16×16 pixel elements. While coding the current macroblock (MB), we need data from the previously encoded MBs belonging to the same frame, or, to be more precise, to the same slice (a sequence of MBs in which the frame can be segmented). Figure 5 shows the current MB together with the already reconstructed neighbours that are needed for its prediction. Specifically, MB A, B, C, and D are required for intraprediction, motion vector prediction, and spatial direct prediction (in the SVC-compatible version). Furthermore, MB A and B are used to check the skip mode in P frames.

Spatial data dependency can even occur inside a MB. The prediction of a 4×4 block may depend on the results of already-predicted neighbouring blocks. e.g., this occurs in Intra 4×4 or in the deblocking filter.

In this scenario, we cannot encode two frames in parallel, because of the temporal data dependency, and we cannot concurrently process different MBs, because of the spatial

TABLE 1: Encoder data flow.

Module	Input	Output	Input from local buffers
MV prediction	MV A	MV pred	MV B, C, D
Motion estimation	Search window, original MB, MV predictor	MV, cost, best intermode, MB predictor	
Intraprediction	Original MB, reconstructed MB A	Cost, best intramode, MB predictor	Reconstructed MB B, C, D
Residual coding	Original MB, MB predictor	Residual signal, coded MB parameters	
IDCT-DeQuant and reconstruction	Residual signal	Reconstructed MB	
Deblocking filter	Reconstructed MB A	Decoded MB	Reconstructed MB B
Entropy coding	Coded MB parameters	Output stream	

data dependency, unless the MBs belong to different slices. Thus, one opportunity is to concurrently process every slice, but this solution has two drawbacks; it strongly depends on the particular encoder configuration, and it requests to implement the whole encoder on every processor element. Therefore, the only chance to partition the encoder is during the MB processing. This does not mean to separately process 8×8 or 4×4 blocks, but to separately execute the encoder functional units at MB level.

The encoder partitioning should now derive from an evaluation of the functional units that can be concurrently computed, taking into account the amount of data that needs to be exchanged between the different cores.

If we suppose that each module will run on a different core, we must consider both the chunk of data each core needs to exchange with the interconnected cores and the frequency of such communications. Therefore, for an optimal module partitioning, it is important to analyse the encoder data flow. Basically, this analysis should result in a list of selected modules with a set of input and output data for every list's entry, as shown in Table 1. The Figure 5's notation is used to indicate the neighbouring MBs. This table allows identifying the dependencies between modules as well as the data flow, from which we can obtain the requested bandwidth for the communication mechanism between processor elements. This preliminary analysis also produces the partition diagram, shown in Figure 6.

Each module will keep local memory buffers containing the data required to process the current MB. For example, the Intraprediction module needs to store a row of reconstructed MBs plus one MB (the left MB) in order to be able to predict the current MB. The deblocking filter will need to store the same number of reconstructed MBs as well. These local storages are filled by producer modules as soon as they complete the respective tasks. In the previous example, "IDCT-DeQuant & Reconstruction" is the producer for intraprediction; when the MB reconstruction has completed for MB_n, the intraprediction of MB_{n+1} can start. It is worth noting that the intraprediction of MB_{n+1} can be concurrently executed with the motion estimation of MB_{n+1} and the deblocking filter of MB_n.

For the sake of simplicity we did not put into Figure 6 scheme all the project components. The buffer mechanism for passing reference-frame data to the ME and the decoded

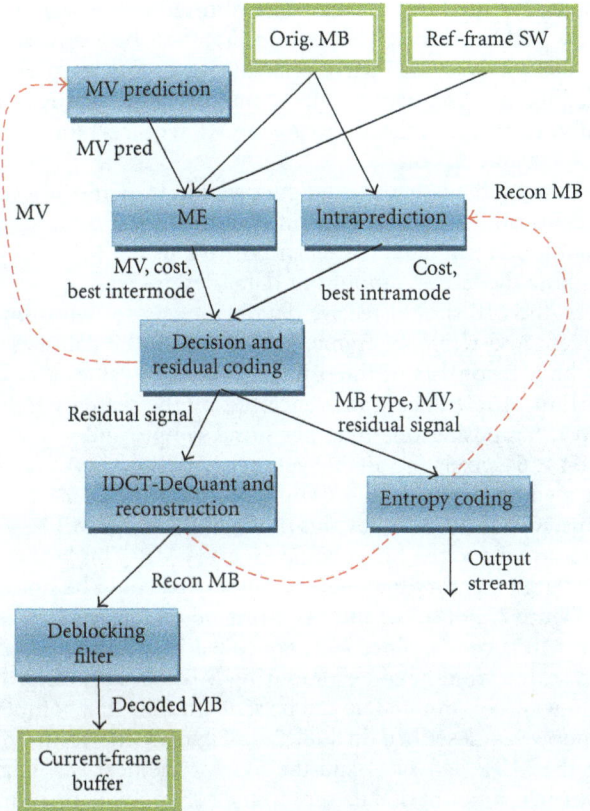

FIGURE 6: Encoder partition diagram.

picture buffer are not described. We preferred to focus on the encoder data flow in order to highlight the chances for module parallelisation. Moreover, the buffering mechanisms strongly depend on the architecture design implementation.

The here described partitioning seems to both fulfil the data dependency constraints and exploit the few opportunities of parallel execution available in a H.264 encoder. Moreover, the computational weight of the encoder components is quite well distributed among the different cores. The only exception is the ME, which is the most time-consuming module. In our encoder we utilize the SLIMH264 ME algorithm [27]. SLIMH264 is divided into two different stages: the first phase is common to all the partitions and

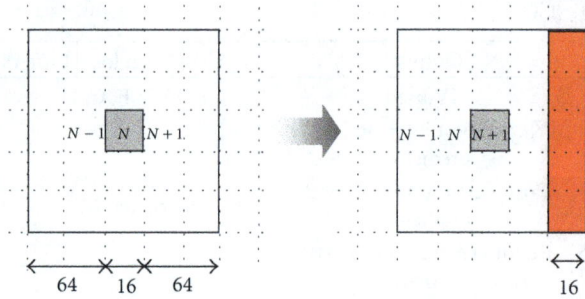

FIGURE 7: Search window update.

performs a fast search; the second step utilizes the coarse results coming from the first phase to refine the search for every MB partition. The second step can be executed in parallel for every MB partition. This leaves the designer the freedom to subsequently split the module to eventually avoid stalls in pipeline in the likely case the ME requires more cycles than intraprediction.

Among the issues the designer should take into account, there is still the memory bandwidth needed to feed the modules. From Table 1, we can notice that the ME module requests the largest amount of data. Besides coding parameters, the ME should receive data belonging to two frames: current and reference frame. For each MB, the data passed to the ME consists of the original MB luma values and the portion of reference frame enclosed by the search window (SW). Supposing one byte per luma sample and a SW set to 64×64 pixels (a suitable value for HD formats), we will get a width of $(64 + 16 + 64)$ pixels leading to 20736 bytes. Thus, we had 20736 bytes plus the original-image MB 16×16 bytes to send to the ME module for every MB. This leads to a very large memory bandwidth. Anyway, as could be noticed in Figure 7, not all of the SW must be resent every time a new MB is coded. Since MBs are coded in raster-scan order and search window of neighbouring MB overlaps, just a 16-byte-wide column update can be sent after the first complete window, as described in [28, 29]. Figure 7 shows the SW for the MB_N (left side) and the SW for the next MB (right side). The amount of data sent to the ME module for coding MB_{N+1} is shown as a red rectangle. When reaching the end of the row, MB_{N+1} does not need the update because this will be over the image border. Nevertheless, a SW update is written to the array, and it will be part of the SW of the first MB in the next row.

4.2. Modules Optimization. The H.264 encoder modules work on a block basis. Even though the basic block of the coding process is the macroblock, consisting of 16×16 pixel elements, the basic block of each module's computation can vary from 4×4 to 16×16. A number of experiments carried out at STMicroelectronics's Advanced System Technology Laboratories showed that, addressing HD resolutions, it is possible to disable interprediction modes involving the 8×8 blocks subpartitions without significant effects on video-quality and -coding efficiency. The same experiments also showed that fidelity range extensions are needed to improve

video quality at high resolutions, as one could expect. For this reason, we choose both to disable ME on partitions 4×8, 8×4, and 4×4 and to add intra 8×8 and transform 8×8. In this scenario, most of the encoder modules work on 8×8 blocks of 8-bit samples. The 4×4 blocks are still used in Intra-4×4 and DCT/Q/IQ/IDCT 4×4. The Intra-16×16 prediction works on the whole MB, whereas the correspondent transformations just iterate the 4×4 procedures.

Usually, inside each module the computations require 16-bit precision for intermediate results. Thus, a typical situation is as follows:

(i) load 8-bit samples from memory;

(ii) switch to 16-bit precision and compute the results;

(iii) store the results to memory as 8-bit samples.

Some of the modules, or at least some parts of them, require a 32-bit precision. Among them, it is worth noting a few computations for pixels interpolation and the Quantization and Inverse-Quantization process.

In order to evaluate the different performance achievable with the three different ISAs, we have inserted the SIMD instructions in an already optimized ANSI C code which is used as reference to evaluate the achieved speedup. For a better understanding of the presented work, the comparison is not only carried out at global level, but for every H.264 functional unit.

In the following, the implementation detail of the sum of absolute difference (SAD) and the Hadamard filter will be shown for all the three addressed ISAs. Among all the several modules implementations, we have chosen to describe these particular operations for different reasons: the SAD is one of the most time-consuming operations in video-compression algorithms; the implementation of Hadamard filter is a good example for describing how an ANSI C implementation can be rewritten to best fit the available SIMD ISA. The access to data stored in memory will be discussed as well because it is a typical issue in optimizing video compression algorithms using SIMD instructions. A complete description of the encoder SIMD implementation on the ST240 processor can be found in [30].

4.2.1. SAD Operation. The sum of absolute differences is a key operation for a large variety of video-coding algorithms. The number of times this operation is executed during a coding process can vary depending on the encoder implementation and it strongly depends on the motion estimation module, that it is not covered by the H.264 standard definition. Anyway, independently of specific implementations, this operation is a key factor for the whole-encoder performance.

Here, we will show three different SAD implementations using SIMD instructions, and we will compare them with an optimized ANSI C code.

Given the essential role the SAD plays in video coding algorithms, some instruction sets include specific instructions to speed up such operation. Here, we will compare SIMD instruction sets having different size and different degree of specializations.

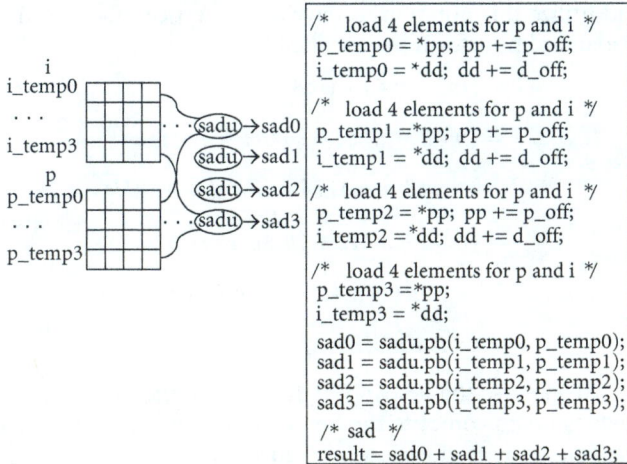

```
/*  load 4 elements for p and i  */
p_temp0 = *pp;  pp += p_off;
i_temp0 = *dd;  dd += d_off;

/*  load 4 elements for p and i  */
p_temp1 = *pp;  pp += p_off;
i_temp1 = *dd;  dd += d_off;

/*  load 4 elements for p and i  */
p_temp2 = *pp;  pp += p_off;
i_temp2 = *dd;  dd += d_off;

/*  load 4 elements for p and i  */
p_temp3 = *pp;
i_temp3 = *dd;
sad0 = sadu.pb(i_temp0, p_temp0);
sad1 = sadu.pb(i_temp1, p_temp1);
sad2 = sadu.pb(i_temp2, p_temp2);
sad3 = sadu.pb(i_temp3, p_temp3);
/*  sad  */
result = sad0 + sad1 + sad2 + sad3;
```

FIGURE 8: SAD implementation.

TABLE 2: SAD performance.

	Cycles	Operations	Load	Store
ANSI C version	36	134	8	0
SIMD version	14	30	8	0

Using the ST240 32-bit wide SIMD extension, the optimization of the SAD computation has been quite straightforward thanks to the SIMD instruction *sadu.pb*.

The SAD finds the "distance" between two 4×4 blocks, generally between a prediction block and the original image; given the two blocks in the left side of Figure 8, the pseudo-code computing the SAD can be viewed in the right side of the same figure. Besides loading the input data, it basically consists of four calls to the *sadu.pb* instruction.

The achieved speedup is shown in Table 2.

The xStream and P2012 architectures support 128-bit-wide vector registers, and they can perform 8-bit, 16-bit, or 32-bit arithmetic SIMD operations. Usually, SAD is performed using 8-bit precision, allowing for each SIMD calculation a capability to handle sixteen elements. Using vertical SIMD instructions, it is easy to achieve the absolute difference among several elements stored in two vectors, but the addition of the elements stored in a single vector is onerous because usually it requires several vertical SIMD inefficiently utilized. Both P2012 and xStream ISAs have horizontal addition of SIMD instructions, but with different capability. In xStream, it is allowed adding all the elements stored in the same vector, producing a scalar result. In P2012, VECx horizontal addition is limited to only add two adjacent elements inside a vector; in this way, four SIMD instructions must be used in order to achieve the scalar result of the SAD operation. This difference significantly impacts the encoder optimization. For example, when the SAD is calculated to evaluate the predictor cost in Intra 16×16, only two SIMDs are used with xStream against the six used with P2012. This is schematized in Figure 9.

Even if in P2012 ISA the lacking of a horizontal SIMD for addition partially wastes the obtained great gain, we still

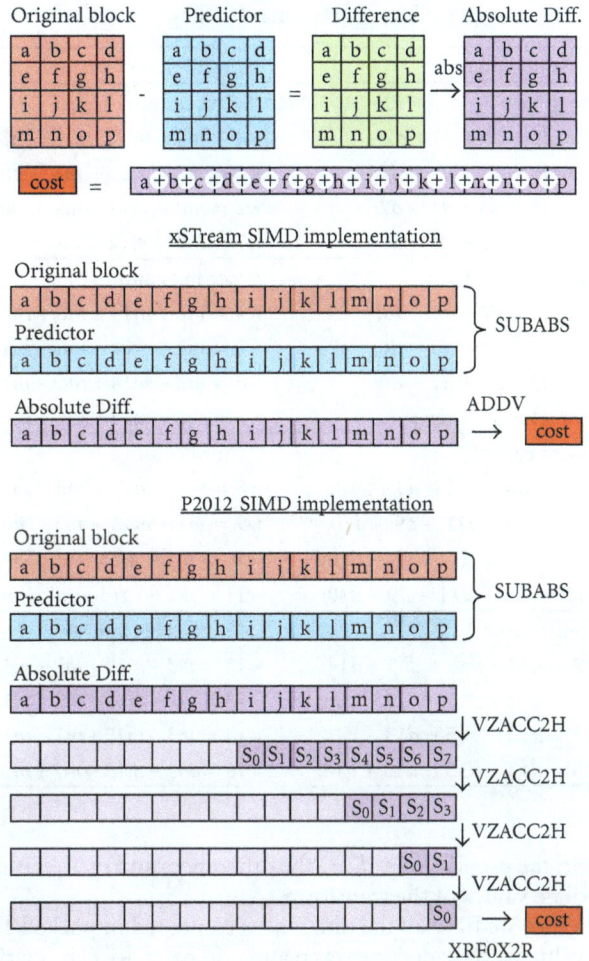

FIGURE 9: Predictor cost calculation.

complete the SAD operation using six VECx instructions and one scalar instruction, as shown in Figure 9, versus the 48 scalar instructions used in the ANSI C implementation (16 subtractions, 16 absolute values, and 16 additions).

4.2.2. Hadamard. We consider very interesting the Hadamard SIMD optimization because it involves a large number of instructions and can be considered a typical case study.

Although the Hadamard transform it is not currently used in the rest of the encoder, the intraprediction module utilizes such transform to find the best 16×16 intraprediction mode. The intramodule divides the predicted MB into sixteen 4×4 blocks. Each block is compared to the correspondent original-image's block, and sixteen differences are calculated. These sixteen values are filtered through the Hadamard transform before computing the SAD of the whole MB.

In the ST240 code, the optimization has started considering that Hadamard can be subdivided into two different phases: horizontal and vertical. The horizontal phase can be subdivided into 4 rows as well as the vertical phase into 4 columns, as shown in the portion of pseudocode in Table 3.

TABLE 3: Hadamard phases.

Horizontal phase	Vertical phase
/* first row */	/* first column */
$m0 = d0 + d3 + d1 + d2;$	$w0 = m0 + m12 + m4 + m8;$
$m1 = d0 + d3 - d1 - d2;$	$w1 = m0 + m12 - m4 - m8;$
$m2 = d0 - d3 + d1 - d2;$	$w2 = m0 - m12 + m4 - m8;$
$m3 = d0 - d3 - d1 + d2;$	$w3 = m0 - m12 - m4 + m8;$
/* second row */	/* second column */
$m4 = d4 + d7 + d5 + d6;$	$w4 = m2 + m14 + m6 + m10;$
$m5 = d4 + d7 - d5 - d6;$	$w5 = m2 + m14 - m6 - m10;$
$m6 = d4 - d7 + d5 - d6;$	$w6 = m2 - m14 + m6 - m10;$
$m7 = d4 - d7 - d5 + d6;$	$w7 = m2 - m14 - m6 + m10;$
/* third row */	/* third column */
$m8\ \ = d8 + d11 + d9 + d10;$	$w8 = m1 + m13 + m5 + m9;$
$m9\ \ = d8 + d11 - d9 - d10;$	$w9 = m1 + m13 - m5 - m9;$
$m10 = d8 - d11 + d9 - d10;$	$w10 = m1 - m13 + m5 - m9;$
$m11 = d8 - d11 - d9 + d10;$	$w11 = m1 - m13 - m5 + m9;$
/* fourth row */	/* fourth column */
$m12 = d12 + d15 + d13 + d14;$	$w12 = m3 + m15 + m7 + m11;$
$m13 = d12 + d15 - d13 - d14;$	$w13 = m3 + m15 - m7 - m11;$
$m14 = d12 - d15 + d13 - d14;$	$w14 = m3 - m15 + m7 - m11;$
$m15 = d12 - d15 - d13 + d14;$	$w15 = m3 - m15 - m7 + m11;$

In the pseudocode, d_i are the differences and m_i the intermediate values of the transform.

Once we have all the differences contained in packed 16-bit values subdivided into even and odd pairs, we can rewrite the first row of the horizontal Hadamard transform as

$$m0 = (d0 + d1) + (d2 + d3),$$
$$m1 = (d0 - d1) - (d2 - d3),$$
$$m2 = (d0 + d1) - (d2 + d3),$$
$$m3 = (d0 - d1) + (d2 - d3). \tag{1}$$

In such a way, we can exploit the packed 16-bit addition and subtraction to obtain the high and low halves of the m_i coefficients. As can be noted, the low and high halves of $m0$ and $m2$ are the same, but while the $m0$'s value is achievable by adding its halves, to compute the value of $m2$ we have to subtract its high half from the lower one. Similar considerations can be applied to the odd elements $m1$ and $m3$.

Anyway, since the vertical phase of Hadamard is yet to come, there is no need to compute such values at this point. In fact, we can rewrite the m_i coefficient as functions of their own halves as follows:

$$m0 = m0L + m0H,$$
$$m1 = m1L - m1H,$$
$$m2 = m2L - m2H,$$
$$m3 = m3L + m3H. \tag{2}$$

and utilize this notation to rewrite the vertical phase of the Hadamard transform as described below

$$
\begin{aligned}
w0 &= (m0 + m4) + (m8 + m12) \\
&= (m0L + m4L) + (m0H + m4H) \\
&\quad + (m8L + m12L) + (m8H + m12H) \\
&= (m0L + m4L) + (m8L + m12L) \\
&\quad + (m0H + m4H) + (m8H + m12H) \\
&= w0L + w0H.
\end{aligned}
\tag{3}
$$

We can use the low and high halves of the intermediate coefficients to compute the low and high halves of the final coefficients w_i as illustrated in Figure 10.

The Hadamard optimization with ST240 SIMD is quite complex. Due to shortness of SIMD, the standard algorithm has been modified in order to better match the SIMD ISA features.

Using the xStream and P2012 architectures, we followed a different approach. Our goal was the exploitation of 128-bit-wide SIMD minimizing the data reordering. Considering that the Hadamard transform can be defined as

$$
H_n = \begin{bmatrix} H_{n-1} & H_{n-1} \\ H_{n-1} & -H_{n-1} \end{bmatrix}
$$
$$
H_0 = 1, \tag{4}
$$

the Hadamard matrices are composed of ± 1 and are a special case of discrete fourier transform (DFT). For this reason, the calculation can exploit the FFT algorithm, usually known as Fast Walsh-Hadamard transform [31].

The only issue is obtaining a good implementation of the FFT butterfly with SIMD, avoiding wasting all the gain achieved using fast algorithm with the data reordering needed to implement the calculation. Our approach consists of a modified butterfly that allows using always the same butterfly structure for every level, even if we have to reorder data between stages (Figure 11).

The output values coming from every butterfly can be calculated for 16 samples at a time using two SIMD instructions, one calculating the additions and one calculating the differences. In this way, we have the advantage of computing the output of each level using a simple SIMD implementation, at the cost of swapping intermediate results between different levels.

Even if xStream and P2012 share this implementation mechanism, we have measured different performance. In this case, the difference depends on the different types of data manipulation instructions. The xStream ISA having the third operand allowing the permutation of results inside vectors is more flexible and can implement the above algorithm with a reduced number of instructions respect to VECx P2012 ISA. Algorithm 1 shows the xStream SIMD implementation.

4.2.3. Memory Access Issues. As previous exposed, a key factor to achieve a good performance improvement with

```
w0_pck = add(add(m0_2_pck,m4_6_pck),add(m8_10_pck, m12_14_pck));
w1_pck = sub(sub(m0_2_pck, m4_6_pck),sub(m8_10_pck, m12_14_pck));
w2_pck = sub(add(m0_2_pck, m4_6_pck),add(m8_10_pck, m12_14_pck));
w3_pck = add(sub(m0_2_pck, m4_6_pck),sub(m8_10_pck, m12_14_pck));
```

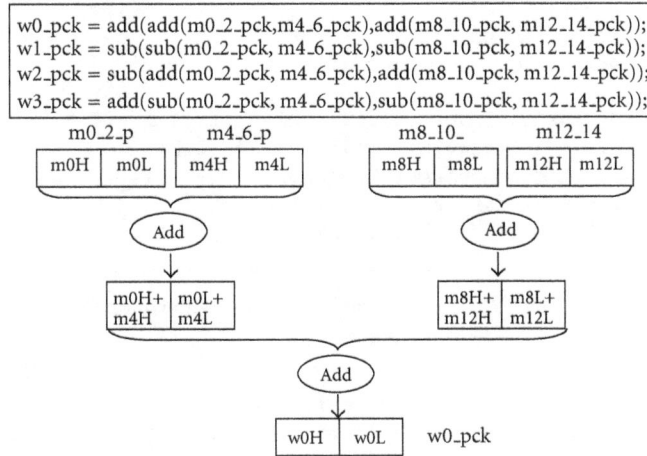

FIGURE 10: Hadamard vertical phase with ST240 SIMD.

```
w0_pck = add(add(m0_2_pck, m4_6_pck),add(m8_10_pck, m12_14_pck));
w1_pck = sub(sub(m0_2_pck, m4_6_pck),sub(m8_10_pck, m12_14_pck));
w2_pck = sub(add(m0_2_pck, m4_6_pck),add(m8_10_pck, m12_14_pck));
w3_pck = add(sub(m0_2_pck, m4_6_pck),sub(m8_10_pck, m12_14_pck));
```

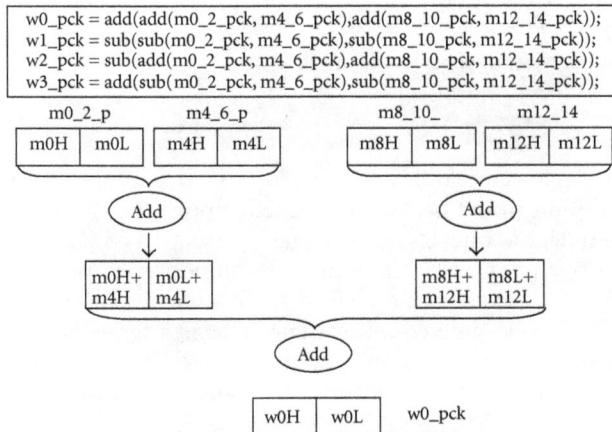

FIGURE 11: Hadamard modified butterfly.

SIMD optimization is the efficient handling of unaligned load operations. In general, programmers should structure the application data in order to avoid or minimize misaligned memory accesses. In video compression algorithm, the motion compensation is surely a case where it is not possible avoid unaligned memory accesses because it is impossible to predict motion vectors and consequently align data.

None of the three addressed architectures support unaligned load instructions. Therefore, it is important to efficiently use aligned accesses to load misaligned data from memory. The three ISAs support instructions to concatenate two vectors. This allows a solution consisting in two steps: first, we use two aligned load instructions for loading data in two vector registers, and, then, we concatenate and shift their elements in order to extract a single vector containing the needed data, as shown in Figure 12.

```
/*first level: one 16 samples butterfly*/
/*(s0 ÷ s7)+(s8 ÷ s15)*/
vaddh out_low = in_low, in_high
/*(s0 ÷ s7)−(s8 ÷ s15)*/
vsubh out_high = in_low, in_high

/*data reordering*/
/*0 1 2 3 8 9 10 11*/
vmrgbl in_low = out_low, out_high, perm
/*4 5 6 7 12 13 14 15*/
vmrgbu in_high = out_low, out_high, perm

/*second level: two 8 samples butterfly*/
vaddh out_low = in_low, in_high
vsubh out_high = in_low, in_high

/*data reordering*/
/*0 1 8 9 4 5 12 13*/
vmrge in_low = out_low, out_high
/*2 3 10 11 6 7 14 15*/
vmrgo in_high = out_low, out_high

/*third level: four 4 samples butterfly*/
vaddh out_low = in_low, in_high
vsubh out_high = in_low, in_high

/*data reordering*/
/*0 8 2 10 4 12 6 14*/
vmrgeh in_low = out_low, out_high
/*1 9 3 11 5 13 7 15*/
vmrgoh in_high = out_low, out_high

/*fourth level: eight 2 samples butterfly*/
vaddh out_low = in_low, in_high
vsubh out_high = in_low, in_high
```

ALGORITHM 1: Hadamard transform xSTream SIMD implementation.

```
uint32 AddressAt128;
vector_16b_sw Va, Vb, Vout;

AddressAt128b = ((uint32) (mref_ptr)) & (~0xF);
Offset = ((uint32) (mref_ptr)) & (0xF);
Va = ldq(AddressAt128, 0);
Vb = ldq(AddressAt128, 16);
Vout = wrot(Va, Vb, Offset);
```

ALGORITHM 2: Unaligned load SIMD implementation with concatenate instruction.

```
ui32_t PackCurr0 = *(orig_line);
ui32_t PackCurr1 = *(orig_line+1);
/* Pack to 128 bits */
TmpVectArray[0] = PackCurr0;
TmpVectArray[1] = PackCurr1;
Pack128In = ldqi(Pack128In, TmpVectArray,0);
/* Reorganize pixels */
Va = vmrgbeh(Va,Pack128In,VZero,permute0);
Vb = vmrgboh(Vb,Pack128In,VZero, permute1);
VPackCurr = vaddh(VPackCurr,Va,Vb,0);
```

ALGORITHM 3: Unaligned load SIMD implementation without concatenate instruction.

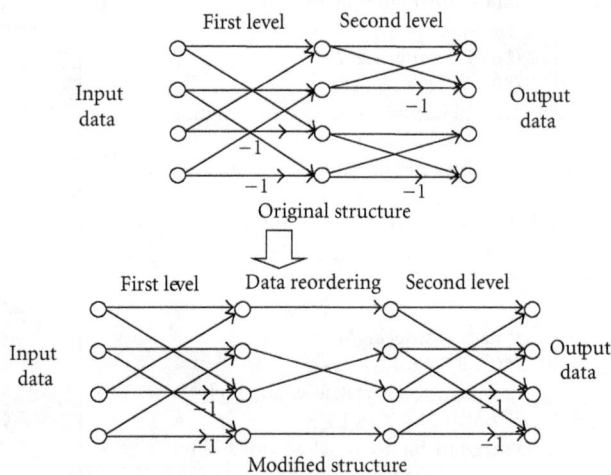

FIGURE 12: Unaligned load.

If an ISA does not define a SIMD performing this type of concatenate operation, then the unaligned load will be implemented with an extra cost due to the use of additional instructions for merging data between the two vectors.

Algorithm 2 shows the implementation of an unaligned load using xSTream. This solution can be compared to the same operation carried out without a concatenate instruction shown in Algorithm 3, in which we should add three instructions to reorganize the data for composing the required not-aligned vector.

It is very important that these concatenate instructions can take the offset argument not as a constant value but as a variable value; otherwise, modules such as motion compensation would not get any benefit from using them. For example, the Intel SSSE3 "palignr" instruction concatenates two operands and shift right the composite vector by an offset for extracting an aligned results, but the offset must be a compile-time constant value. This is a big issue for a module as motion compensation, in which it is impossible to know in advance the offset of a misaligned address.

5. Results

In the H.264 encoder, the most cycle-demanding modules have been optimized using SIMD instructions: motion estimation and compensation, DCT, Intraprediction, and so forth. The best way to compare different instruction sets in order to judge the effectiveness of both SIMD extensions and code optimizations is to measure the speedup obtained with the SIMD-based implementation versus the ANSI C version of the same source code. In order to separate the effect of SIMD performance improvement from ANSI C optimizations, we have inserted SIMD instructions in previously optimized ANSI C modules.

The results are provided in terms of average cycles spent to process one macroblock. The xSTream and P2012 architectures share the same modules subdivision. For the single-core DSP ST240, the subdivision is less fine, and related modules are joined together. In the reported tests, the presence of the ST240 processor is important because it allows comparing the single-processor elements of the multicore platforms to a single-core architecture. Tests are performed on a set of video sequences addressing different

TABLE 4: SIMD instructions for video coding.

Instruction description	Affected modules	Notes
Horizontal add: adds all the elements inside a vector register and produces a scalar result	ME, intraprediction	Speeds up SAD
Horizontal permute: rearranges elements inside a vector register	Intraprediction, DCT/Q/IQ/IDCT	Allows zig-zag scan and speeds up intra diagonal modes
Concatenate: concatenates two vector registers into an intermediate composite, shifts the composite to the right by a variable offset	Motion estimation and compensation	Allows software implementation of unaligned load
Promotion/demotion precision: an efficient support for promoting element precision while loading data from memory, and demoting the precision (with saturation) while storing data to memory	All the main modules	It will speed up the load and store operations for several modules
Absolute subtraction: for every element "a" in the first vector and every element "b" in the second vector performs the following operation: $|a - b|$	ME, intraprediction, deblocking filter	Speeds up SAD in conjunction with horizontal add; used in deblocking filter
Shift with round: performs the following operation for every element "a" in the vector operand: $(a + 2^{n-1}) >> n$, where n is a scalar value	IDCT, deblocking filter, motion compensation	Speeds up 1/2 pixel interpolation
Average: for every element "a" in the first vector and every element "b" in the second vector performs the following operation: $(a + b + 1) >> 2$	Intraprediction, deblocking filter, motion compensation	Speeds up 1/4 pixel interpolation

TABLE 5: Cycles/MB spent in each module for each ISA.

	xSTream			P2012			ST240		
	ANSI C	SIMD	Gain factor	ANSI C	SIMD	Gain factor	ANSI C	SIMD	Gain factor
Luma motion compensation	4788	2257	2.1x	8286	3965	2.1x			
Croma motion compensation	3064	658	4.7x	3626	1282	2.8x	265559	200380	1.3x
Motion estimation	303769	84342	3.6x	603182	114776	5.3x			
Intra 4 × 4	24366	10076	2.4x	38234	15760	2.4x	32013	19182	1.6x
Intra 8 × 8	15396	4997	3.0x	26972	9455	2,9x			
DCT/Q/IQ/IDCT 4 × 4	14994	7616	2.0x	20473	9088	2.3x	32013	19182	1.7x
DCT/Q/IQ/IDCT 8 × 8	18660	3498	5.3x	24486	11636	2.1x			

resolutions, and average results are resumed in Table 5. The results in Table 5 and Figure 13 show that the ST240, exploiting the instruction level parallelism (ILP) with a 4-issue VLIW architecture, achieves the best performance for the ANSI C implementation. All the SIMD implementations improve performance for every encoder module, but the ST240 with the shortest SIMD size obtains the lowest speedup factor. P2012 and xSTream with their wider SIMD can better exploit the data-level parallelism. In terms of pure number of cycles spent to encode one macroblock, the xSTream ISA achieves the best performance.

It is worth analyzing in detail these results to understand how different instruction sets lead to different performance. The xSTream processor elements take advantage from the "horizontal add" instruction that allow an efficient computation of the SAD operations: it is evident in the ME module, where xSTream spends about 25% fewer cycles than P2012 (84,342 versus 114,776 cycles/MB). The higher speedup obtained by P2012 is mainly due to the less-efficient ANSI C code generated by the P2012 compiler. We already described as the ST240 can exploit a specific instruction for

the SAD operation. In fact, its result is not far from the architectures having 128-bit-wide vector registers (the 200, 380 cycles/MB also include motion compensation). From these results, we can state that the support for horizontal SIMD will not only give a great performance improvement for the SAD operation, but it significantly impacts the whole ME module.

As earlier said, data manipulation instructions are a key factor to fully exploit SIMD implementations because operations such as transposing matrices or data reordering become frequent in this type of optimizations. An experimental result confirming this consideration can be seen in the DCT/Q/IQ/IDCT 8 × 8 module, covering all the toolchain performing the residual coding and decoding. This module involves several data-reordering operations, ranging from matrix transposition to zig-zag reorder. Both ST240 and xSTream instruction sets support the permutation of elements inside a vector in a very efficient way, as described in Sections 3.1 and 3.2. The P2012 SIMD extension includes a series of instructions for interleaving and merging elements between two vector operands. The great speed up the

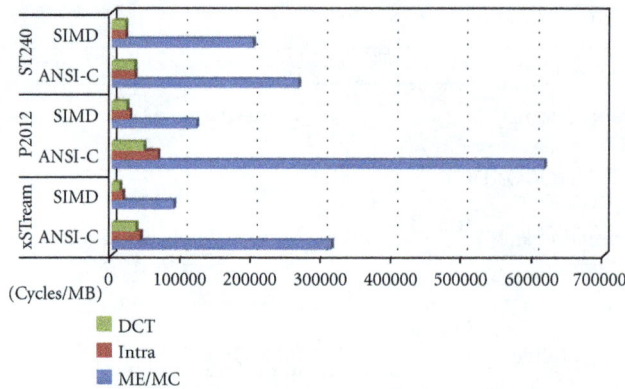

FIGURE 13: ISA comparison.

xSTream architecture gathers in comparison with P2012 is mainly due to the possibility to permute elements using a single instruction, in a sort of horizontal permute. The effect is emphasized in the 8×8 transform where the data reordering process is stressed more than in the 4×4 case. In our experience, we saw that if such instruction is available, then the zig-zag reordering can be effectively implemented with SIMD instructions; otherwise, we are forced to use the scalar implementation involving look-up tables to perform the reordering.

Intraprediction can exploit the horizontal permute instruction as well; the intraprediction modes involving diagonal directions require the permutation of elements inside the resulting vectors. For similar reasons, ST240 achieves great speedup factors in DCT and intramodules (resp., 1.7 and 1.6), considering that a 32-bit-wide SIMD can only perform two 16-bit-arithmetic operations.

There are other several SIMD instructions that in our opinion are to be considered as key instructions for optimizing video codec applications. Here, we assume an instruction set will already include SIMD for all the common arithmetic operations, compare, select, shift, and memory operations.

In previous sections, we already discussed about the impact of the unaligned memory access to the video codec performance. All the encoder modules are affected by the performance of the unaligned memory operations, but it becomes a keyfactor for motion estimation and compensation. An instruction concatenating two vectors and producing a vector at the desired offset is fundamental to implement an unaligned load instruction. As stated in Section 4.2.3, the capability to support variable offsets is a key factor for the instruction usability because the offset could not be known in advance.

Inside most of the modules, the computations require a 16-bit precision for intermediate results, but the input and output data contained into the noncompressed YUV images are 8-bit values. Thus, a typical operation at the beginning of a module is to load 8-bit input values and extend them to 16-bit precision. At the end, the output data precision is usually demoted down to 8-bit saturating the values before storing the results. Therefore, even if the support to 8-bit operations is not required, it would be very useful that an instruction set

will include SIMD instructions for promoting and demoting precision in a fast way. An optimal solution will also combine promotion with load operations and demotion with store instructions.

Usually, the video codec algorithms try to avoid the division operations because of its computational cost. When needed, divisors are power of two, and the division is substituted with a shift right with rounding as follows:

$$\frac{a}{2^n} \Longleftrightarrow (a + 2^{n-1}) \gg n. \tag{5}$$

Therefore, even if most instruction sets already include this type of instruction, it is important to remind its utility. Often, the shift right with rounding is used for averaging two or more values, as in the intraprediction and deblocking filter. In our implementation, one of the reasons the ST240 achieves a good speedup in the intraprediction module is the presence of an average SIMD instruction in the instruction set.

Table 4 summarizes our conclusions based on the presented work. The proposed instructions are described in the first column. For each instruction, the table indicates the H.264 modules that will be mainly affected by the introduction of the instruction, as well as a few notes about specific contributions to basic video coding operations.

6. Conclusions

This paper presents efficient implementations of the H.264/AVC encoder on three different ISAs. The optimization process exploits the SIMD extensions of the three architectures for improving the performance of the most time-consuming encoder modules. For each addressed architecture, experimental results are presented in order to both compare the different implementations and evaluate the speedup versus the optimized ANSI C code.

The paper discusses how SIMD size and different instruction sets can impact the achievable performance. Several issues affecting video-coding SIMD optimization are discussed, and authors' solutions are presented for all the architectures.

Most instruction sets have specific SIMD instructions for video coding. Even though these instructions can lead to great performance improvements, they could be useless for other application families. In this paper, we identify a set of generic SIMD instructions that can significantly improve the performance of video applications.

Besides presenting the SIMD optimization for the most time-demanding modules, the paper describes how a complex application as the H.264/AVC encoder can be partitioned to a multicore architecture.

Acknowledgments

The authors would like to thank STMicroelectronics's Advanced System Technology Laboratories for their support. This work is supported by the European Commission in the context of the FP7 HEAP project (#247615).

References

[1] "VC-1 Compressed Video Bitstream Format and Decoding Process," SMPTE 421M-2006, SMPTE Standard, 2006.

[2] T. Wiegand, G. J. Sullivan, G. Bjøntegaard, and A. Luthra, "Overview of the H.264/AVC video coding standard," *IEEE Transactions on Circuits and Systems for Video Technology*, vol. 13, no. 7, pp. 560–576, 2003.

[3] G. J. Sullivan, P. Topiwala, and A. Luthra, "The H.264/AVC Advanced Video Coding Standard: Overview and Introduction to the Fidelity Range Extensions," in *Applications of Digital Image Processing XXVII*, Proceedings of SPIE, August, 2004.

[4] D. Marpe, T. Wiegand, and S. Gordon, "H.264/MPEG4-AVC fidelity range extensions: tools, profiles, performance, and application areas," in *IEEE International Conference on Image Processing (ICIP '05)*, pp. 593–596, September 2005.

[5] H. Schwarz, D. Marpe, and T. Wiegand, "Overview of the scalable video coding extension of the H.264/AVC standard," *IEEE Transactions on Circuits and Systems for Video Technology*, vol. 17, no. 9, pp. 1103–1120, 2007.

[6] Joint Collaborative Team on Video Coding (JCT-VC), "WD4: Working Draft 4 of High-Efficiency Video Coding," 6th Meeting, Torino, Italy, July, 2011.

[7] J. Probell, "Architecture considerations for multi-format programmable video processors," *Journal of Signal Processing Systems*, vol. 50, no. 1, pp. 33–39, 2008.

[8] M. Koziri, D. Zacharis, I. Katsavounidis, and N. Bellas, "Implementation of the AVS video decoder on a heterogeneous dual-core SIMD processor," *IEEE Transactions on Consumer Electronics*, vol. 57, no. 2, pp. 673–681, 2011.

[9] M. Sayed, W. Badawy, and G. Jullien, "Towards an H.264/AVC HW/SW integrated solution: an efficient VBSME architecture," *IEEE Transactions on Circuits and Systems II*, vol. 55, no. 9, pp. 912–916, 2008.

[10] T. Rintaluoma and O. Silvén, "SIMD performance in software based mobile video coding," in *10th International Conference on Embedded Computer Systems: Architectures, Modeling and Simulation (IC-SAMOS '10)*, pp. 79–85, July 2010.

[11] H. Lv, L. Ma, and H. Liu, "Analysis and optimization of the UMHexagons algorithm in H.264 based on SIMD," in *2nd International Conference on Communication Systems, Networks and Applications (ICCSNA '10)*, pp. 239–244, July 2010.

[12] X. Zhou, E. Q. Li, and Y.-K. Chen, "Implementation of H.264 decoder on general-purpose processors with media instructions," in *Image and Video Communications and Processing*, Santa Clara, Calif, USA, January 2003.

[13] J. Lee, S. Moon, and W. Sung, "H.264 decoder optimization exploiting SIMD instructions," in *IEEE Asia-Pacific Conference on Circuits and Systems (APCCAS '04)*, pp. 1149–1152, December 2004.

[14] W. Lo, D. Lun, W. Siu, W. Wang, and J. Song, "Improved SIMD architecture for high performance video processors," *IEEE Transactions on Circuits and Systems for Video Technology*, vol. 21, no. 12, pp. 1769–1783, 2011.

[15] D. Talla, L. K. John, and D. Burger, "Bottlenecks in multimedia processing with SIMD style extensions and architectural enhancements," *IEEE Transactions on Computers*, vol. 52, no. 8, pp. 1015–1031, 2003.

[16] Z. Shen, H. He, Y. Zhang, and Y. Sun, "A Video Specific Instruction Set Architecture for ASIP design," *VLSI Design*, vol. 2007, Article ID 58431, 7 pages, 2007.

[17] M. Shafique, L. Bauer, and J. Henkel, "Optimizing the H.264/AVC video encoder application structure for reconfigurable and application-specific platforms," *Journal of Signal Processing Systems*, vol. 60, no. 2, pp. 183–210, 2010.

[18] P. Faraboschi, G. Brown, J. A. Fisher, G. Desoli, and F. Homewood, "Lx: a technology platform for customizable VLIW embedded processing," in *27th Annual International Symposium on Computer Architecture (ISCA '00)*, pp. 203–213, June 2000.

[19] J. Fisher, P. Faraboschi, and C. Young, "VLIW processors: from blue sky to best buy," *IEEE Solid-State Circuits Magazine*, vol. 1, no. 2, pp. 10–17, 2009.

[20] N. Coste, H. Garavel, H. Hermanns, F. Lang, R. Mateescu, and W. Serwe, "Ten Years of Performance Evaluation for Concurrent Systems using CADP," in *4th International Symposium on Leveraging Applications of Formal Methods, Verification and Validation ISoLA*, Heraklion, Greece, 2010.

[21] D. Pandini, G. Desoli, and A. Cremonesi, "Computing and design for software and silicon manufacturing," in *IFIP International Conference on Very Large Scale Integration (VLSI '07)*, pp. 122–127, October 2007.

[22] G. Desoli and E. Filippi, "An outlook on the evolution of mobile terminals: from monolithic to modular multi-radio, multi-application platforms," *IEEE Circuits and Systems Magazine*, vol. 6, no. 2, pp. 17–29, 2006.

[23] L. Benini, "P2012: a many-core platform for 10Gops/mm2 multimedia computing," in *21st IEEE International Symposium on Rapid System Prototyping*, Fairfax, Va, USA, June 2010.

[24] C. Silvano, W. Fornaciari, S. Crespi Reghizzi et al., "2PARMA: parallel paradigms and run-time management techniques for many-core architectures," in *IEEE Annual Symposium on VLSI*, pp. 494–499, July 2010.

[25] C. Mucci, L. Vanzolini, I. Mirimin et al., "Implementation of parallel LFSR-based applications on an adaptive DSP featuring a Pipelined Configurable Gate Array," in *Design, Automation and Test in Europe (DATE '08)*, pp. 1444–1449, March 2008.

[26] P. Paulin, "Programming challenges & solutions for multiprocessor SoCs: An industrial perspective," in *Design Automation Conference (DAC '11)*, June 2011.

[27] A. Kumar, D. Alfonso, L. Pezzoni, and G. Olmo, "A complexity scalable H.264/AVC encoder for mobile terminals," in *European Signal Processing Conference (EUSIPCO '08)*, Lausanne, Switzerland, August 2008.

[28] C. Y. Chen, C. T. Huang, Y. H. Chen, and L. G. Chen, "Level C+ data reuse scheme for motion estimation with corresponding coding orders," *IEEE Transactions on Circuits and Systems for Video Technology*, vol. 16, no. 4, pp. 553–558, 2006.

[29] B. Zatt, M. Shafique, F. Sampaio, L. Agostini, S. Bampi, and J. Henkel, "Run-Time Adaptive Energy-Aware Motion and Disparity Estimation in Multiview Video Coding," in *48th Design Automation Conference (DAC '11)*, pp. 1026–1031, San Diego, Calif, USA, June 2011.

[30] M. Bariani, I. Barbieri, D. Brizzolara, and M. Raggio, "H.264 implementation on SIMD VLIW cores," STreaming Day 2007, Genova, Italy.

[31] C. S. Lubobya, M. E. Dlodlo, G. de Jager, and K. L. Ferguson, "SIMD implementation of integer DCT and hadamard transforms in H.264/AVC encoder," in *Proceedings of the IEEE AFRICON*, pp. 1–5, September 2011.

15

Design Example of Useful Memory Latency for Developing a Hazard Preventive Pipeline High-Performance Embedded-Microprocessor

Ching-Hwa Cheng

Department of Electronic Engineering, Feng-Chia University, 100 Wen-Hwa Road, Taichung, Taiwan

Correspondence should be addressed to Ching-Hwa Cheng; chengch@fcu.edu.tw

Academic Editor: Yeong-Lin Lai

The existence of structural, control, and data hazards presents a major challenge in designing an advanced pipeline/superscalar microprocessor. An efficient memory hierarchy cache-RAM-Disk design greatly enhances the microprocessor's performance. However, there are complex relationships among the memory hierarchy and the functional units in the microprocessor. Most past architectural design simulations focus on the instruction hazard detection/prevention scheme from the viewpoint of function units. This paper emphasizes that additional inboard memory can be well utilized to handle the hazardous conditions. When the instruction meets hazardous issues, the memory latency can be utilized to prevent performance degradation due to the hazard prevention mechanism. By using the proposed technique, a better architectural design can be rapidly validated by an FPGA at the start of the design stage. In this paper, the simulation results prove that our proposed methodology has a better performance and less power consumption compared to the conventional hazard prevention technique.

1. Introduction

In current computer architecture, the multiple-instruction (pipeline, superscalar) microprocessors are proposed to improve the efficiency of a single-instruction microprocessor. There are usually four stages (instruction fetch, decode, execute, and writeback) adopted in a multiple-cycle processor. The *CPI* (cycle per instruction) value of the pipeline (multiple-instruction) microprocessor is several times larger than that of a single-instruction microprocessor. Generally, the pipeline architecture is combined with RISC (reduced instruction set computer) methodology to design high performance processors.

Pipeline microprocessor hazards occur when multiple instructions are executed. The pipeline architectural hazards that are introduced in [1, 2] make the program instructions unable to be parallely executed. In general, there are three types of hazards: structure, control, and data hazards. A *structural hazard* means that the hardware components

(resources) are insufficient to support the execution of the pipeline instructions in the same clock cycle. The frequently occurring case of the hardware components conflicting when sharing the single port memory means that they are unable to support the read/write operation at the same time. The second type of hazard is termed a *control hazard*, which arises from the present executed instruction's inability to make decisions because this instruction decision making should rely on the results from the next following executed instructions. An example is the branch instruction, which is unable to make a correct decision whether to jump or not during this instruction in the execution cycle. This is due to the most recent jump condition can not be obtained when a decision is made.

The third type is *data hazard* occurs when the current instruction's operands should refer to its earlier instruction's executing results, but the previous instruction final result is still not stable (the instruction is not working in the *writeback* stage,) as shown in Figure 1(a) where the reference

Design Example of Useful Memory Latency for Developing a Hazard Preventive Pipeline High-Performance Embedded-Microprocessor

161

(a) Hazard condition

(b) Hazard prevention by inserting NOP

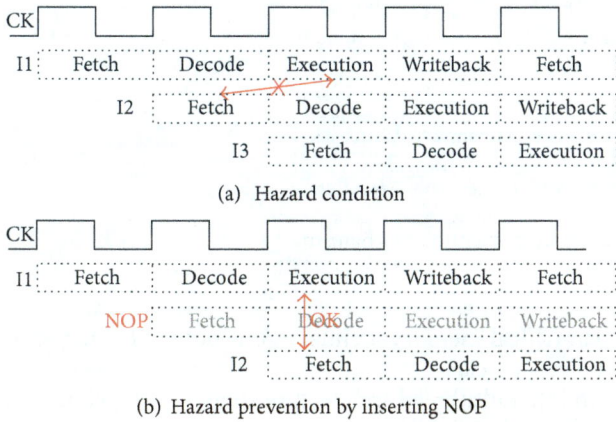

FIGURE 1: The conventional hazard prevention technique.

hazard occurred for instructions I1 and I2. The focus of most conventional designs is to analyze the hazard conditions and promote additional mechanism (insert NOP instruction) in order to resolve the hazard situations and obtain better pipeline performance, as shown in Figure 1(b).

Memory latency is not welcome because the access delay degrades the microprocessor's performance. An idea of how to manage the memory latency as a hazard prevention mechanism is shown in Figure 2. There is a half clock cycle memory latency for the fetched instruction which is loaded from the slow-speed main memory (SRAM, DRAM). Hence, the pipeline operation does not need to insert the NOP, and without the performance degradation penalty.

The different types of memory in the system board can help designers rapidly create a better pipeline hazard prevention architecture, such as a hazard prevention mechanism that utilizes nonuseful memory latency to reduce the hazard penalty. To design a better pipeline architecture, two factors should be taken into consideration simultaneously. First is the instruction's hazard conditions, and second is the memory access latency issue.

Most of the past architectural designs focus on the instruction hazard detection and prevention scheme from the aspect of pipeline function units. The pipeline-stall and forwarding techniques proposed in [1] are neglecting the use of useful memory latency. The possible reasons may stem from the following: one is that memory requires a particular design for better performance requirement in real microprocessors. This cannot be fully emulated by FPGA. Thus, most of the FPGA is used to verify the functional correctness of the fetch/decode/execution/writeback stages.

Freedom of design in different system's architecture is not available in a chip. Currently, FPGA is utilized to rapidly validate a feasible architectural design. The FPGA emulation process can help designers quickly adopt different memory architectures to reduce the hazard penalties during the system design phase. Better microprocessor performance adopts suitable cache-RAM-Disk memory volume in this hierarchy design. However, there are complex relationships (e.g., levels of cache, RAM access time, and volume) between the memory hierarchy and function units.

In this paper, we propose a superior pipeline architectural design obtained from the FPGA validation phase that does not merely use the FPGA to perform functional verification. As the memory latencies are dissimilar for different types of memory, the idea of an architectural design that applies useful memory latency can be rapidly validated by the FPGA. By choosing to adopt different types of FPGA board memory, we might find a better hazard detection/prevention mechanism.

When a designer attempts to utilize different types of memory in the FPGA board, the requirement is to compare the performance when utilizing the different memory latencies (internal register, flash, RAM, or ROM) within the core architecture.

We emphasize that the additional onboard memory can be well utilized to handle the hazards. The FPGA board not only helps the designer to validate the function units but also brings creative guidance to help the designer find better architecture and reduce hardware overhead to detect/prevent hazards.

One 16-bit X86-light pipeline RISC microprocessor (14 instructions) was developed to validate our idea for specific application (e.g. matrix multiplication). The design is Harvard architecture, where the data and the instruction are put in separate memories. In this paper, we focus on the hazard prevention realistic design using memory latency and verify our results using FPGA implementation for this microprocessor. For demonstrating the different paradigms of solving the hazard problems with/without using the FPGA onboard memory, two design approaches are adopted. Method-1 is the hazard detection and prevention mechanism design using additional hardware. Method-2 uses inboard memory latency to replace the hardware that was proposed in Method-1 and validates the results with a Xilinx FPGA demo board (XESS XSV-800). In the Method-2, the data memory (DM) and instruction memory (IM) SRAM is separate on board for pipeline processor.

In this paper, firstly, the data and control hazard detection and prevention techniques (forwarding, NOP insertion, and stall) for our architecture are introduced. Secondly, two validation approaches are used to verify the design architecture. For Method-1, the Instruction-Memory and Data-Memory are synthesized and embedded within the design architecture. Method-2 only synthesizes the core architecture into the FPGA and places the Instruction-Memory and Data-Memory in SRAM on the board. The test programs need to be loaded into the Instruction-Memory and synthesized with the design at the same time. For Method-2, the test program is loaded from IM and writes output to the Data-Memory during the validation process. Method-2 also supports flexible validation environments for quickly reevaluating the design architecture.

There are two contributions of this work. First, for the design phase, memory latency could be effectively utilized to avoid the hazard issues, for designing simpler and faster pipeline architecture; for example, the data hazard resolving mechanisms do not need to be embedded into the design. The second is not the same as conventional designing of the pipeline processor using internal registers as the processor's IM, DM memory. Using Method-2, the instruction memory

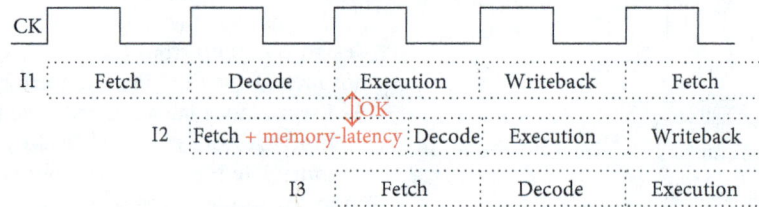

FIGURE 2: Using memory latency to design a hazard prevention mechanism.

access latency from on system board SRAM prevents the data hazard problems arising from pipeline operation. The functional testbench does not need to be synthesized with the design, so effortless verification methods can be used to rapidly validate the prototype advanced pipeline core architecture. A more flexible verification environment can be adopted for large amounts of varied test programs.

The aforementioned two approaches are successful in evaluating the designs. The synthesizable RISC architecture practically executes 35 MHz onboard, and the clock frequency of Method-2 is two times faster than that of Method-1.

As to the organization of this paper, Section 1 comprises the introduction. Section 2 is the design and hazards analysis of our RISC processor. The two FPGA verification methodologies are proposed in Section 3. Section 4 is the experimental results. This paper is concluded in Section 5.

2. The Pipeline Hazard and Memory Latency Surveys

In [3], the study investigates the relative memory latency, memory bandwidth, and branch predictability in determining the processor performance. The proposed basic machine model assumes a dynamically scheduled processor with a large number instruction window. This study claims that, if a system with unlimited memory bandwidth and perfect branch predictability, that memory latency is not a significant limit to performance. The simulation model with SPEC92 benchmarks is used to study the performance.

Reference [3] proves that the best existing branch prediction mechanism with very large table sizes also resulted in several times lower performance compared to perfect branch prediction method for many benchmarks. This means that perfect branch predict ability is the most import factor. This paper assumes that memory bandwidth is not usually a significant limit for the advanced technology. However, this assumption might not be achievable, as memory bandwidth is always a bottleneck in current harvard system architecture. There are also less currently advanced (multicore with multithread) designs with perfect branch predict ability while to tolerate high memory latency.

Reference [4] claims that the repeatable timing is more achievable than predictable timing. This research describes micro pipelining architecture and the memory hierarchy delivers repeatable timing can provide better performance compared to past techniques. The program threads are interleaved in a pipeline to eliminate pipeline hazards, and

a hierarchical memory architecture is outlined which hides memory latencies.

In [4], multithread architecture applies the pipeline operation from interleaving the memory access operation. The repeatable timing can speed up pipelining architecture, as the pipeline interleaving is within the pipeline processor and DRAM access. This specific architecture might not be suitable for generic system architecture. The designer applies the proposed technique that needs to consider the instruction dependence conditions.

The research [5] reviews the RISC microprocessor architecture, which presents a microthreading approach to RISC microarchitecture. This paper focus on the speculation that has high cost in silicon area and execution time, as a compiler can almost find some instructions in each loop which can be executed prior to the dependency is encountered. The proposed approach attempts to overcome the performance penalty from instruction control (branch, loop) statement and data missing problem. The proposed technique can tolerate high latency memory from avoiding the speculation in instruction execution. However, without well utilizing memory latency, only the compiler cannot obtain the best performance improvement for complicated programs.

3. The Demonstration RISC Microprocessor Architectures

The single instruction and the pipeline version demo architectures are written by Verilog HDL and validated using XCV-800 Xilinx FPGA [6].

3.1. The Single Instruction Architecture. Figure 3 shows the microarchitecture of the single instruction version. There are four stages used in this processor, for example, instruction fetch, decode, execute, and write back stages. The next instruction is fetched and should wait until the current instruction's results are written back to memory (register/data memory). The timing diagrams are shown in Figure 4. There are 4 clock cycles required for each instruction by this style, with each stage of the operation at the clock positive edge. There are partial instructions that do not need to execute the four steps shown in Table 5. The Verilog HDL codes are used to design the hardware function block, and *Xilinx* FPGA simulation/synthesis environment is used to fulfill the experiments.

The single cycle architecture shown in Figure 3 executes the opcode on each clock cycle positive edge. A simple description of the function unit is as follows: instruction

Design Example of Useful Memory Latency for Developing a Hazard Preventive Pipeline High-Performance Embedded-Microprocessor

163

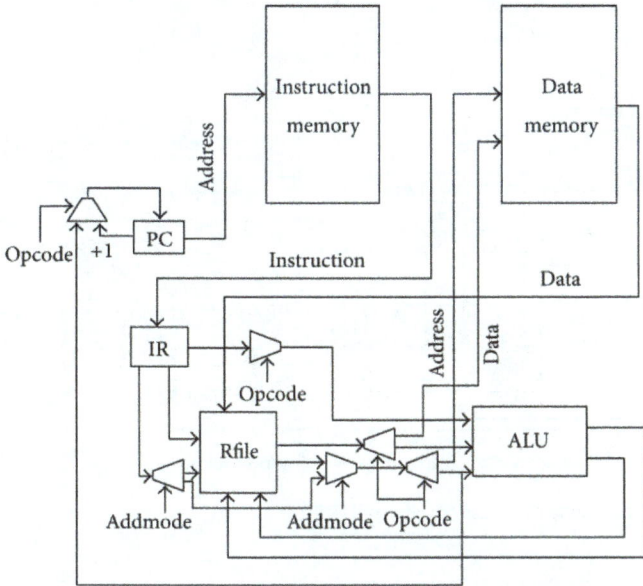

FIGURE 3: The single clock cycle architecture.

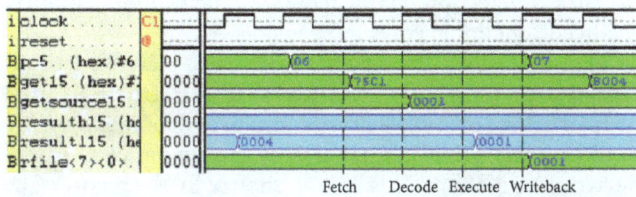

FIGURE 4: The function units are activated by each positive clock edge simultaneously.

memory: store the execution instructions (programs), data memory: store the execution results, PC: program counter, IR: instruction register, rfile: register file, sixteen 16 bit register, and ALU: arithmetic-logic unit. The opcode type-1: instruction of JGE, JMP. The opcode type-2: instruction of AND, OR, NOT, XOR, ADD, SUB, MUL, MOV, CMP. The opcode type-3: instruction of STA, LDA.

Each instruction execution is divided into four phases. Fetch: based on the PC value; fetch the instruction from the instruction memory then put it into the Instruction Register (IR). Decode: the MUX selects the proper mathematic/logic operand of ALU. In addition, the operators are offered from different (immediate/direct/indirect) addressing modes. Execute: instruction execution. Writeback: results are written back to the register file.

3.2. The Pipeline Architecture (Method-1). Our pipeline architecture is shown in Figure 5. The instruction/data memory is synthesized using the FPGA internal logic element, and the computed results are obtained in every clock cycle. There are four stages (fetch, decode, execute, and writeback) and three addressing modes (direct, indirect, and immediate) proposed for this design. The results can be obtained after each clock cycle in this four-layer pipeline architecture. The data hazards of pipeline processors are generated from data dependence

using the same registers, for example, RAW (read after write), WAW (write after write). The control hazard is made from the branch instruction, which decides to fetch a false next instruction during the pipeline operation. These hazards have been discussed in previous studies [1, 2].

3.3. The Hazard Analysis of Pipeline Architecture. Each instruction operation requires 4 clock cycles for Method-1. The incorrect pipeline operation occurs when the next instruction is executed in the following clock cycle. These conditions termed the structure, control, and data hazards are occurred.

The data hazard was raised from one instruction decoding operation using the same registers that corrupted the previous instructions during the execution (or writeback) stage, for example, the read after write (RAW) data hazard occurring in the pipeline architecture; for example, [SUB r1, r0] follow [ADD r0, #10] and the RAW hazard occurring on the register of r0. The write after write (WAW) hazard causes the register overwrite situation; for example, [STA m[r0], r0] follow [MOV r0, #10]; the register r0 becomes WAW hazard. Thus, the hazard detection and correct circuits were added to resolve these hazard problems in Figure 5.

For example, the data hazard occurs when two instructions are executed serially; for example, [SUB r1, r0] follow [ADD r0, #10]; the read after write hazard (RAW) occurs on register of r0. The renew value of r0 = r0+10 was not obtained until the ADD instruction *writeback* stage. The renew value is not ready to update the source register r0 of SUB instruction during the *decode* stage. We do not list the other types of hazards in detail such as WAW. The following Figure 6 shows the data hazard occurring on the two instructions of *add r0, #10* and *add r10, #10*.

The instruction format is "opcode, target operand, source operand". There are several types of occurrences that arise from this data hazard issue. We categorize these details in Table 1. The means hazard occurs when the target instruction register (target operand) at the *execution* stage combines with any third row instruction's source register (source target) in the *decode* stage. We just list the simple one here; the other detailed rules are shown in Appendix B. Several types of data hazards are categorized in Appendix C. To resolve this hazard issue, when detecting the aforementioned code sequence, the instruction execution cycle does need to wait for completion (the writeback cycle). The ALU quickly passes the computed results to the next instruction (as the direct input for next instruction). This method is called *forward*.

4. Memory Latency Utilized by Pipeline Architecture Method-2

Figure 7 shows the pipeline architecture Method-2 changing the instruction/data memory to use the FPGA onboard memory (RAM). The different approaches use these two pipeline architectures for Method-2, and each instruction operation extends to 8 clock cycles. There is no existence of any hazards because the decode/execute/writeback operation in FPGA overlaps with the next instruction access (fetch)

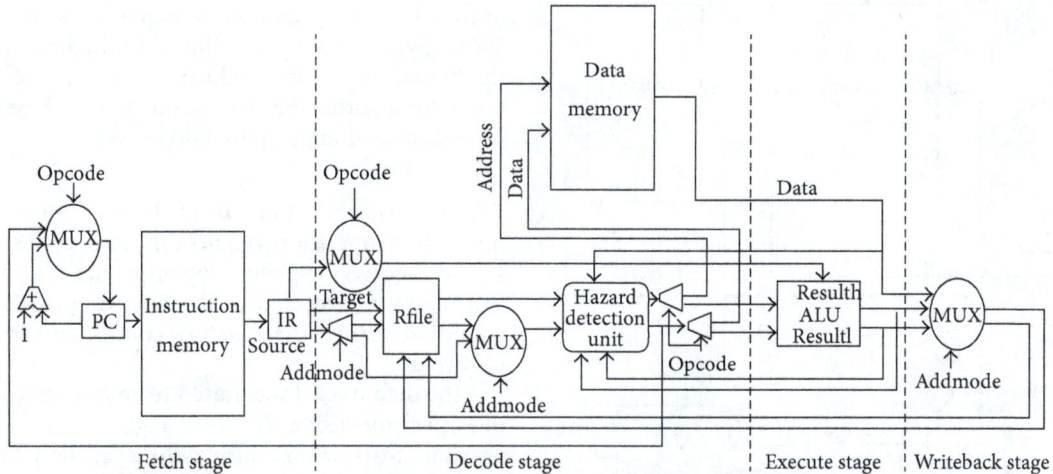

FIGURE 5: Our X86-light RISC pipeline processor architecture (Method-1).

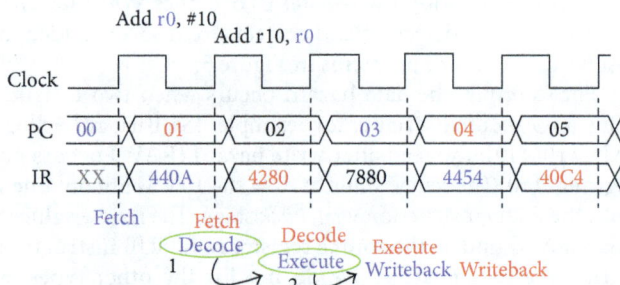

FIGURE 6: The data hazard example.

TABLE 1: Hazard-1. Type-1: RAW.

Dependent register	Instruction
EX.Register	AND, OR, NOT, XOR, ADD, SUB, MUL, MOV
ID.Register.Source	AND, OR, NOT, XOR, ADD, SUB, MUL, MOV

latency from IM (in the demo board). A simpler operating clock cycle is shown in Figure 8.

The memory load functional units require four clock cycles, as shown in Figure 8. We use the FPGA experimental board (*Xilinx XCV 800*) to demonstrate our idea. We focus on using the RAM of this board. There are 4 clock cycles required for SRAM read/write operation of *XESS XSV-800* board. The RAM read/write operation requires four clock cycles, and the detailed timing diagram is demonstrated in Figure 9.

The memory read/write requires four clock cycles. This gives us the chance to arrange 8 cycles to execute one instruction. There is one do-nothing cycle required to be inserted into the cycle. In total, eight-clock cycle is needed to execute an instruction. The insert do-nothing cycle is used to align the operation of the pipeline instructions. This methodology also has more freedom for hazard prevention. The two-level pipeline has better performance and less

hazard process hardware. The side with simple architecture (no hazard detection circuit required) allows more complex advanced functional cores to be inside.

Figure 10 shows the pipeline clock cycle plan of Method-2. There are 4 clock cycles required to read instructions from the instruction memory or write results to data memory. We should mention that the signal lines should be ready before the write operation. This is not shown in the figure. The latency can be utilized to cover the timing intervals of one instruction fetch stage with decode/execute/writeback/do-nothing stages of previous instructions, for example, instructions I2 and I3 in Figure 11. There are no hazards that occur when we move the instruction and data memory outside FPGA to the random access memory on board. The simpler redesign two-layer pipeline architecture is shown in Figure 7. There is no hazard detection/correct circuitry and each output can be obtained at every four clock cycles. If the access delay is not utilized, due to each stage extending to four clock cycles to fit the slowest fetch stage, each instruction operation should expand to 16 clock cycles (4 clock cycles for each stage). However, the hazard problems can be avoided for the four-layer pipeline architecture.

Both Method-1 and Method-2 architectures can be adopted and, as we stated, greater flexibility will be available to extend using the onboard RAM. When we need to match the memory read/write cycle, there is one alternation, as shown in Figure 12. All function units extend the operated cycle to four, thus the total clock cycle for executing an instruction increases to 16 clock cycles. In this architecture, the resulting output occurs every four clock cycles. A greater amount of architecture is able to be adopted when there is a sufficient use of different types of onboard memory, such as Flash and ROM. This helps the designer to have more freedom in choosing different design styles.

The hazard problem consideration of this architecture is simpler than the no-RAM version, due to each stage having four cycles. Each functional unit in the local element within FPGA only requires one clock cycle to execute the operation.

Design Example of Useful Memory Latency for Developing a Hazard Preventive Pipeline High-Performance
Embedded-Microprocessor

165

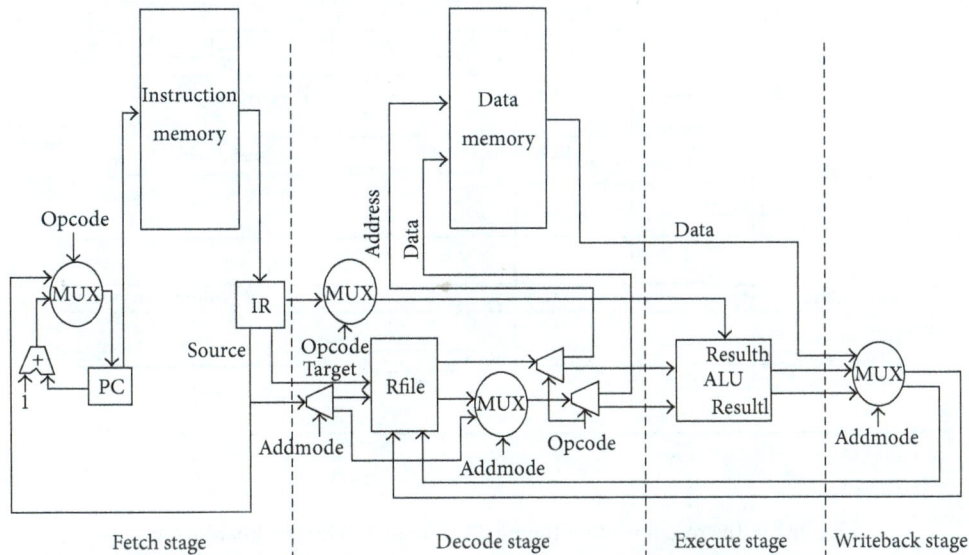

FIGURE 7: A simpler pipeline architecture without hazard detection/correction units (Mthod-2).

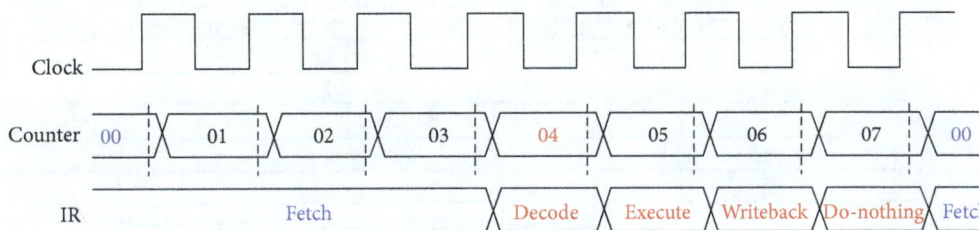

FIGURE 8: The pipeline operation of method-2.

TABLE 2: There are more benefits from pipeline Method-2.

Comparison items	Method-1	Method-2
Gate count	69,004	47,100 (−31.7%)
Maximum frequency	18.766 MHz	30.547 MHz (+62.8%)
Maximum net delay	18.140 ns	16.813 ns (−7.3%)

TABLE 3: The design specifications' comparisons.

Type	Properity		
	Gate count	Frequency	Max net delay
Single	48656	26.449 MHz	14.952 ns
16-cycle	68926 (+41.7%)	19.173 MHz (−27.5%)	18.674 ns (+24.9%)
8-cycle (Method-2)	47058 (+3.2%)	29.951 MHz (+1.3%)	15.607 ns (+4.3%)

TABLE 4: The performance/area comparisons of the synthesis results.

Type	Clock		
	19.00 MHz	26.00 MHz	29.00 MHz
Single	213.40 mW	282.14 mW	311.60 mW
16-cycle	258.21 mW	343.46 mW	380.00 mW
8-cycle (Method-2)	210.68 mW	278.42 mW	307.45 mW

5. The Experimental Result Analysis

In a conventional design concept, an FPGA chip is only used to validate the functional correctness of instruction operations in the fetch/decode/execution/writeback stages. In this research, the FPGA board not only helps the designer to validate the function units but also brings a creative contribution to help find new hazard detection/prevention strategies that have less hardware overhead in comparison to those in the past research. The performance evaluation of a prototype pipeline design needs to utilize different types of memory in the FPGA board as much as possible and to derive a better hazard detection and prevention mechanism.

There are four pipes for Method-1 and two pipes for Method-2. Thus, Method-2 has less hardware overhead. Table 2 shows that the synthesis results from Method-2 are

The remaining three cycles (3/4) provide more flexibility to solve the hazard issues than the tight conditions case (each stage assigns one cycle.) Figure 13 shows the data dependence being resolved by the forwarding operation.

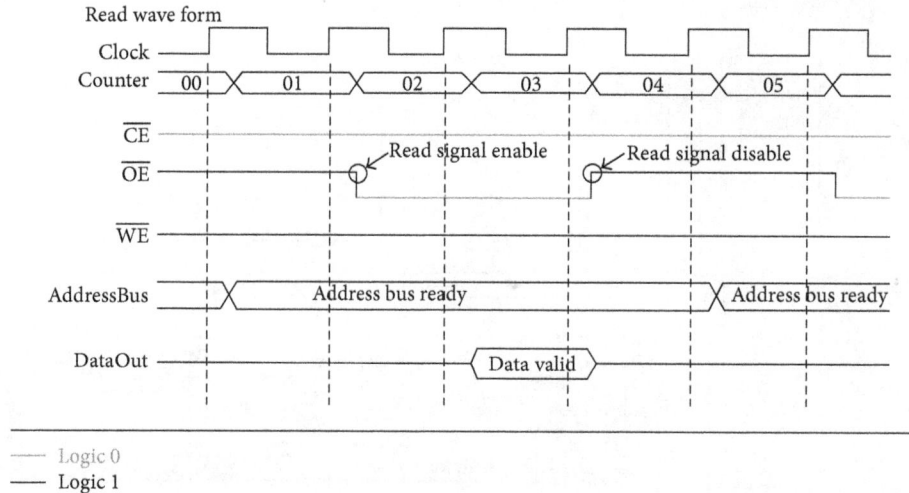

FIGURE 9: The read operation timing diagram of *XCV800* on board RAM.

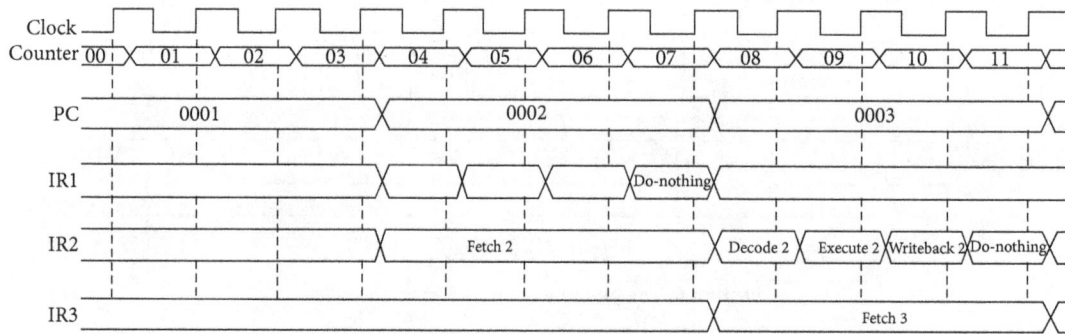

FIGURE 10: The clock cycle timing Method-2.

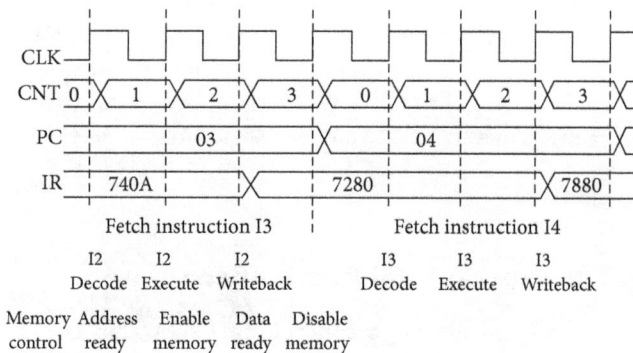

FIGURE 11: The three instruction execution cycles of Method-2.

The SINGLE and 8-cycle version are near equivalent; this represents the pipeline architecture performance is major limited by IM and DM access time. There are several benchmark test bench programs are used for the proposed design. In Table 4, the measure results are obtained from Bubble-Sort for large volume of data. Table 4 measures the power consumption. There is a less power consumption in the 8-cycle pipeline architecture under the three-clock frequency.

6. Discussions

In this paper, the proposed pipeline architecture utilizes the memory access latency to improve the performance when hazards occurred. As memory access speed is dissimilar for the different types of memory, well utilize the memory access latency cycles, the pipeline operation can be speed up. The proposed methods use the assumption of instruction and data can be obtained (hit situation) from memory. This means that the instruction and data can be found from the chip's internal register or onboard memory. These memories are enough to store the required instruction and data during program execution.

better than those from Method-1. Method-2 works frequency is 62.8% higher than Method-1.

Table 3 shows the comparisons of three architectures, and single instruction, 16-cycle pipeline (Figure 12), and 8-cycle pipeline (Method-2, Figure 10). All of the specifications are obtained from the synthesized reports by FPGA tool. For equal comparisons, it needs to be mentioned that the three types of instruction/data memory all use the onboard RAM.

Design Example of Useful Memory Latency for Developing a Hazard Preventive Pipeline High-Performance
Embedded-Microprocessor

167

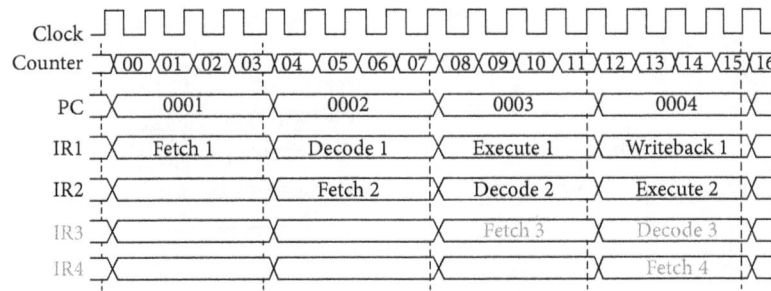

FIGURE 12: The alternative pipeline architecture.

TABLE 5: The instruction set format.

(a) Direct/Indirect Address Mode (bit location)

OPCODE		ADDMODE		TARGET		SOURCE		NOTHING	
15	12	11	10	9	6	5	2	1	0

(b) Immediately Address Mode (bit location)

OPCODE		ADDMODE		TARGET		IMMEDICATE	
15	12	11	10	9	6	5	0

(c)

Instruction	Function	Direct address mode	Immediately address mode	Indirect address mode	Comments
AND	AND operation	AND R0, R1	AND R0, #10	AND R0, @R1	R0 = R0 · R1
OR	OR operation	OR R0, R1	OR R0, #10	OR R0, @R1	R0 = R0 + R1
NOT	NOT operation	NOT R0			R0 = R0'
XOR	Exclusive OR	XOR R0, R1	XOR R0, #10	XOR R0, @R1	R0 = R0 ⊕ R1
ADD	Addition	ADD R0, R1	ADD R1, #10	ADD R0, @R1	R0 = R0 + R1
SUB	Subtract	SUB R0, R1	SUB R0, #10	SUB R0, @R1	R0 = R0 − R1
MUL	Multiply	MUL R0, R1			{R0, R1} = R0 ∗ R1
MOV	Move	MOV R0, R1	MOV R0, #10	MOV R0, @R1	R0 ← R1
JMP	Jump		JMP #10		PC value change
JGE	Condition jump		JGE #10		PC value change
CMP	Comparison	CMP R0, R1			
NOP	No operation				No operation
STA	Store to memory	STA M[R0], R1	STA M[0], R1		M[R0] ← R1
LDA	Load from memory	LDA R0, M[R1]	LDA R0, M[0]		R0 ← M[R1]

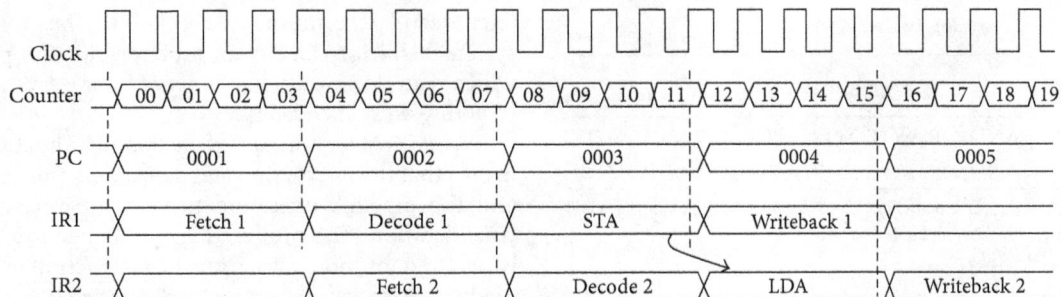

FIGURE 13: It is not complicated to manipulate the hazard for each stage with four clock cycles.

TABLE 6: The instruction action table.

Instruction	Operation			
	Fetch	Decode	Execution	Writeback
JMP	A	A	—	—
CMP	A	—	A	—
JGE	A	A	—	—
NOP	A	—	—	—
STA	A	A	—	—

TABLE 7: Hazard-1.

(a) Type-1: RAW

Dependent register	Instruction
EX.Register	AND, OR, NOT, XOR, ADD, SUB, MUL, MOV
ID.Register. Source	AND, OR, NOT, XOR, ADD, SUB, MUL, MOV

The second type of RAW hazard; for example CMP r1, r0 follow ADD r0, #10, RAW occurred onr0.

(b) Type-2: RAW

Dependent register	Instruction
EX.Register	AND, OR, NOT, XOR, ADD, SUB, MUL, MOV, LDA
ID.Register. Source	CMP

The second type of RAW and WAW hazard; for example, STA m[0], r0 follow ADD r0, #10, RAW occurred on r0; STA m[r0], r1 follow ADD r0, #10, WAW and occurred on r0.

(c) Type-3: RAW, WAW

Dependent register	Instruction
EX.Register	AND, OR, NOT, XOR, ADD, SUB, MUL, MOV
ID.Register. Source	STA, LDA

(d)

Dependent register	Instruction
EX.Register	LDA
ID.Register. Source	AND, OR, NOT, XOR, ADD, SUB, MUL, MOV

The third type RAW data hazard will occur as the following example ADD r1, r0 follow LDA r0, m[10] occurred on the r0.

(e) Type-4: RAW

Dependent register	Instruction
WB.Register	AND, OR, NOT, XOR, ADD, SUB, MUL, MOV
EX.Register	AND, OR, NOT, XOR, ADD, SUB, MUL, MOV

The third type RAW data hazard will occur as the following example ADD r1, r0 follow SUB r1, r0, RAW occurred on the r0.

The designer applies the proposed concept to obtain a better system architecture in preplan stage. Form well arranging this memory latency situation with pipeline instruction in

TABLE 8: Hazard-2.

(a) Type-1: RAW

Dependent register	Instruction
WB.Register	AND, OR, NOT, XOR, ADD, SUB, MUL, MOV
ID.Register. Source	AND, OR, NOT, XOR, ADD, SUB, MUL, MOV

Ex: ADD r0, #10 → SUB r1, r2 → AND r3, r0, RAW at r0.

(b) Type-2: RAW

Dependent register	Instruction
WB.Register	AND, OR, NOT, XOR, ADD, SUB, MUL, MOV, LDA
ID.Register.Source	CMP

Ex: ADD r0, #10 → SUB r1, r2 → CMP r3, r0, RAW at r0.

(c) Type-3: RAW, WAR

Dependent register	Instruction
WB.Register	AND, OR, NOT, XOR, ADD, SUB, MUL, MOV
ID.Register. Source	STA, LDA

Ex: ADD r0, #10 → SUB r1, r2 → STA m[0], r0, RAW at r0.

(d) ADD r0, #10 → SUB r1, r2 → STA m[r0], r2, WAW at r0

Dependent register	Instruction
WB.Register	LDA
ID.Register. Source	AND, OR, NOT, XOR, ADD, SUB, MUL, MOV

Ex: LDA r0, m[10] → SUB r1, r2 → ADD r2, r0, RAW at r0.

TABLE 9: Hazard-3. Type-1: RAW, WAW.

Dependent register	Instruction
EX.Register	LDA
ID.Register.Source	STA

Ex: LDA r0, m[1] → STA m[2], r0, RAW at r0.
LDA r0, m[1] → STA m[r0], r1, WAW at r0.

fetch/decode/execution/writeback state. The proposed nature system behavior can obtain a better pipeline performance, which can be easily implemented in a system memory structure with fixed latency.

When instruction and data missing situations occur, it means that the instruction and data access time might length, and the pipeline operation need to be varied under this consideration. The proposed methods need to be modified by waiting for more cycles for the instruction and data enter ready state under instruction and data missing situations.

In the future study, a reliable system need to consider the instruction and data missing situation. Hence, the proposed Method-1,2 need to be modified for inserting different

Design Example of Useful Memory Latency for Developing a Hazard Preventive Pipeline High-Performance Embedded-Microprocessor

169

wait cycles for instruction and data memory missing situations. A flexible system architecture requires to include the instruction and data hit-miss conditions for various memory architectures. However, such a pipeline architecture is hard to design by including different memory waiting cycles.

7. Conclusion

We find that a good cycle timing plan is the most important issue for designing a pipeline CPU. The processor performance depends on how well the clock cycle, the control, and the data flow are managed. Also, the design style has a good chance to be improved if one does consider that the memory latency can be utilized for hazard prevention. When the microprocessor can utilize different types of memory (internal register, flash, RAM, or ROM) in the system board, this gives flexibility and helps to achieve a system architecture with better performance. The hard to use onboard memory will be regularly ignored during the prototype verification phase. The designer should not forget the onboard FPGA memory, although it differs from the real CPU memory, and it is also inconvenient to use it (to coordinate the read/write operation with kernel function units). This might be the best chance to reevaluate the preliminary design during the verification phase because one might find another better hazard free structure from memory latency. In our experience, the designer can obtain a greater number of different architectures by spending time to try to use onboard memory. The functional test program should take the memory access delay into consideration when the execution programs are moved to the outside memory in the second method. The design that applies memory latency for hazard prevention has a better performance with less power consumption than that of the conventional design.

Appendices

A.

There are 14 instructions supported in this design; the details of the fourteen instruction formats are shown in Table 5.

The partial instructions might not operate at all functional stages as shown in Table 6. The symbol **A** means the instruction has an action on this stage, and — means the instruction has no operation in this stage.

B.

The hazard condition could regulate listing as the following rules. The explanation of the Hazard-1 condition is the source operand (in instruction *decode* stage) using same register with the previous instruction target register (operand) in the *execution* stage. Hazard-2 and Hazard-3 are described as the former description. However, the hazard condition is different for pipeline architecture Method-1 and Method-2. All three hazard conditions will occur only with pipeline Method-1, but pipeline Method-2 only has Hazard-1 and Hazard-2 (no Hazard-1) occurring.

Hazard-1.

if ((ID.RegisterSource = EX.Register) or (EX.Register = WB.Register)).

Hardware Solution: hazard detection unit-Forward.

Hazard-2.

if ((ID.RegisterSource = WB.Register)).

Hardware Solution: hazard detection unit-Forward.

Hazard-3.

if ((ID.Instruction. Sta = EX.Inxtruction. Lda) and (ID.RegisterSource = EX.Register)).

Software Solution: insert NOP Instruction.

C.

We categorize the data hazard conditions to interpret the data utilization error occurring at different stages. Condition 1 is control/data hazard occurring on Tables 7, 8, and 9.

References

[1] D. A. Patterson and J. L. Hennessy, *Computer Organization and Design*, M.K. Publishers, 1998.

[2] J. L. Hennessy and D. A. Patterson, *Computer Architecture—A Quantitative Approach*, M.K. Publishers, 2003.

[3] N. P. Jouppi and R. Parthasarathy, "The relative importance of memory latency, bandwidth, and branch limits to performance," in *Proceedings of the Workshop on Mixing Logic and DRAM: Chips That Compute and Remember*, 1997.

[4] S. A. Edwards, S. Kim, E. A. Lee, I. Liu, H. D. Patel, and M. Schoeberl, "A disruptive computer design idea: architectures with repeatable timing," in *Proceedings of the IEEE International Conference on Computer Design (ICCD '09)*, pp. 54–59, Lake Tahoe, Calif, USA, October 2009.

[5] C. Jesshope and B. Luo, "Micro-threading: a new approach to future RISC computer," in *Proceedings of the 5th Australasian Architecture Conference*, 2000.

[6] http://www.xess.com.

16

Homogeneous and Heterogeneous MPSoC Architectures with Network-On-Chip Connectivity for Low-Power and Real-Time Multimedia Signal Processing

Sergio Saponara and Luca Fanucci

Department of Information Engineering, University of Pisa, Via G. Caruso 16, 56122 Pisa, Italy

Correspondence should be addressed to Sergio Saponara, sergio.saponara@iet.unipi.it

Academic Editor: Marcelo Lubaszewski

Two multiprocessor system-on-chip (MPSoC) architectures are proposed and compared in the paper with reference to audio and video processing applications. One architecture exploits a homogeneous topology; it consists of 8 identical tiles, each made of a 32-bit RISC core enhanced by a 64-bit DSP coprocessor with local memory. The other MPSoC architecture exploits a heterogeneous-tile topology with on-chip distributed memory resources; the tiles act as application specific processors supporting a different class of algorithms. In both architectures, the multiple tiles are interconnected by a network-on-chip (NoC) infrastructure, through network interfaces and routers, which allows parallel operations of the multiple tiles. The functional performances and the implementation complexity of the NoC-based MPSoC architectures are assessed by synthesis results in submicron CMOS technology. Among the large set of supported algorithms, two case studies are considered: the real-time implementation of an H.264/MPEG AVC video codec and of a low-distortion digital audio amplifier. The heterogeneous architecture ensures a higher power efficiency and a smaller area occupation and is more suited for low-power multimedia processing, such as in mobile devices. The homogeneous scheme allows for a higher flexibility and easier system scalability and is more suited for general-purpose DSP tasks in power-supplied devices.

1. Introduction

Telecommunications, consumer, and infotainment applications are characterized by a growing interest in the real-time and low-power implementation of DSP techniques, for images, videos, and audio, to improve system performances in terms of coding efficiency, visualization quality, and audio reproduction fidelity. At the state-of-the-art, several algorithms have been proposed [1–13] which represent mature solutions for video and audio coding, noise and artifact suppression, high-dynamic-range signal management, interpolation, video motion estimation, and low-distortion digital audio amplification. The implementation of such algorithms in real-time and with low-power consumption is a challenging open issue. To face the high computational power required by real-time multimedia processing applications multicore architectures are needed [14, 15]. At the state-of-the-art, in most cases, the proposed algorithms still refer to off-line software implementation on a programmable platform. Even in case real-time processing is obtained by using graphic-specific processing units (GPUs) [16–20], the relevant power consumption is unsuitable for embedded or mobile systems. A GPU has a power consumption ranging from tens to hundreds Watts, depending on the workload [21]. Dedicated integrated circuits (ICs) have been proposed in literature [12, 22–30] whose power consumption is limited to hundreds mW; however, they are dedicated to a specific algorithm, for example, motion estimation for interframe video coding in [28] or dynamic range compression for display of mobile devices in [27] or audio oversampling and noise shaping in [12, 30]. Instead, a programmable solution covering multiple tasks is needed. The availability of low-power and integrated solutions, offering enough programmability to support different classes of audio and image/video processing algorithms, is a strategic target to be achieved to foster the adoption of such techniques in new

Homogeneous and Heterogeneous MPSoC Architectures with Network-On-Chip Connectivity for Low-Power and
Real-Time Multimedia Signal Processing

171

application fields. The problem is not only the design of the computing architecture but also of the communication infrastructure due to the nature of multimedia algorithms, dominated by data transfer and storage costs [31].

To address the above issues, this work presents 2 programmable and scalable MPSoC architectures for multimedia signal processing:

(i) One is based on the use of an array of multiple heterogeneous tiles, each acting as an application-specific instruction set processor (ASIP) covering a DSP algorithmic class for 1D (audio) and/or 2D (image and video frames) signals;

(ii) The other is based on an array of homogeneous tiles, each composed of a 32-bit RISC core plus a 64-bit VLIW computing unit.

In both cases, a network-on-chip (NoC) with Spidergon topology is adopted as communication infrastructure to overcome the limits of classic circuit-switched bus. The paper is structured as follows. Section 2 reviews state-of-the-art computing architectures for real-time multimedia processing. Sections 3 and 4 describe the homogeneous-tile and the heterogeneous-tile MPSoCs, respectively, for audio, image, and video processing. Section 5 discusses the design of the NoC infrastructure for both architectures. Functional characterization of the MPSoC architectures with reference to two case studies is reported in Section 6: an H.264/MPEG4 AVC video coder and a low-distortion reproduction audio system. Implementation results in submicron CMOS technology for the MPSoC architectures are discussed in Section 7. Conclusions are drawn in Section 8.

2. Computing Architectures for Multimedia Processing

The computing architectures proposed in literature to achieve multimedia real-time processing can be clustered into 3 main categories.

(1) Hardware platforms achieving real time for handheld/mobile devices at power costs lower than 1 Watt, but limited to a specific algorithm, for example, [22–25, 27–29, 32, 33].

(2) Software implementation on general purpose processors with clock frequencies in the GHz domain and with a single-core, for example, Atom in [26], or multicores, for example, Core2 in [26, 34], depending on the computational load. The power consumption of such solutions is up to tens of Watts. The real-time processing of multimedia algorithms is typically limited to simple tasks or to small image/video formats.

(3) Software implementation on massively parallel processors, such as GPUs [16–20] with computational throughput of billions of floating point operations per second (GFLOPS) ensuring real-time multimedia processing but with power cost up to hundreds of Watts.

From a technology point of view, although the performances achievable with FPGA (field programmable gate array) are continuously increasing, the realization of dedicated or programmable platforms on submicron CMOS technology (with standard cells or full-custom or hybrid design flows) is still the approach ensuring the best trade-off between power consumption and computational/storage capabilities. As example the Virtex 6 or the StratixIV FPGA families include devices with several Mbits of RAM on chip, several hundreds of DSP blocks with multiply and accumulate (MAC) capabilities at hundreds of MHz, I/O transceivers running up to several Gbits/s. Both configurable hardware logic and hard or soft microprocessor cores are available in today FPGAs. However, the overall complexity of such FPGAs in 45 nm silicon technology is in the order of billions of transistors, occupying an area of several hundreds of mm^2, and the power consumption is up to tens of Watts. As example, from the analysis carried out in [35], the Virtex6 LX760 FPGA device offers a DSP computational capability of roughly 380 GFLOPS when implementing a fast Fourier transform (FFT) algorithm but for a power cost higher than 50 W. From [35] it emerges how the LX760 FPGA has worse performances, measured as the ratio between computational capability and power consumption (GFLOPS/W) or area (GFLOPS/mm^2), by one order of magnitude versus a dedicated ASIC design and similar performances to a GPU (GTX480 from NVidia) realized in a 45 nm CMOS technology. Moreover, the market of multimedia and consumer platforms is a large volume one and hence is suited for CMOS designs characterized by higher nonrecurrent costs but lower recurrent costs versus FPGAs.

Although a comparison of hardware/software platforms with different architectures, instruction sets, and computing performances is difficult to implement, Figure 1 provides a visual representation of the trade-off between die size (due to circuit complexity spent for increased parallelism and hence increased performance), power consumption, and computational capabilities in state-of-art single-core and multicore systems. Reported data refer to platforms realized in 45 nm silicon technology nodes and to operations on 32-bit data (integer or single-precision floating-point).

Figure 1 highlights that considering as performance metric the ratio GOPS/mm^2 there are two different trend-lines: one for platforms optimized for general purpose computing applications (e.g., Atom, Core2) and one for platforms optimized for DSP applications. When the level of parallelism increases (e.g., from 1 in the SP_CELL to many cores in the CELL-BE or Tile64 or GTX480), the difference between the two trend-lines is reduced since at high parallelization levels the performance bottleneck is represented by communication and the computational power increases slowly versus the number of cores.

The target of our work is designing programmable multi core architectures with power consumption limited to few Watts, for example, <3Watts, a die size around 50 mm^2, and enough computing performances to ensure real-time processing of CD and DVD-quality audio (up to 96 kHz, 16 bit/sample) and high definition (HD) images and videos with frame resolutions up to 1080×720 or 1024×768.

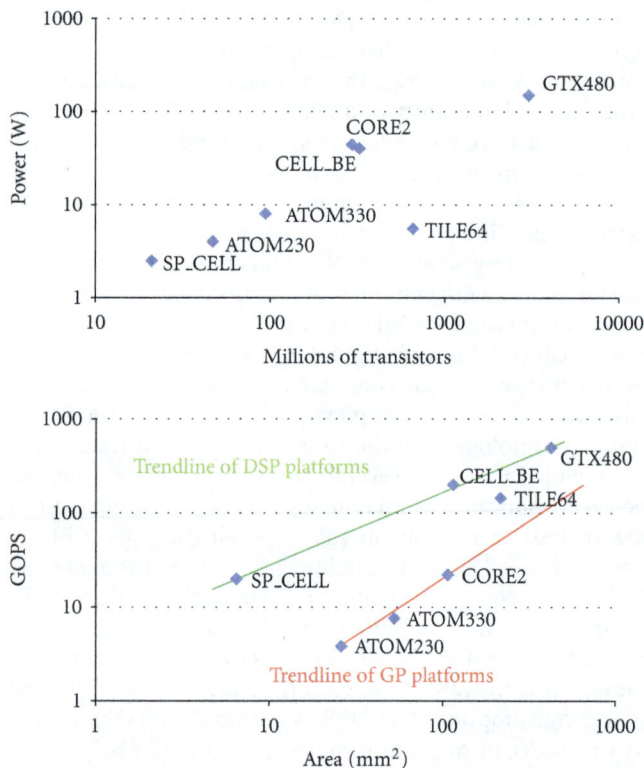

FIGURE 1: Die size, number of transistors, computational power (GOPS), and power consumption of state-of-art single-core, ATOM 230 and 1 synergistic processor (SP) of the Cell BE, and multi core systems, ATOM330, Fermi GPU GTX480, Core2, Cell BE, and Tile64.

The display and camera resolution of smart phones and handheld multimedia devices is typically lower, for example, 30 Hz 352 × 288 CIF format or 640 × 480 VGA format [36] and hence the functional assessment of the MPSoC architecture in Section 6 has been carried out mainly with CIF and VGA test input videos. Similarly, the most common audio sampling rates in handheld devices are 44.1 and 48 kHz [36], used for the tests in Section 6.

As example of pure software implementations of a scale invariance feature transform (SIFT) algorithm for image feature extraction [26] and of a 3D DCT/IDCT video codec have been realized on the Intel Core 2 processor. In [26], an Intel Core 2 6600@2.4 GHz implements the SIFT algorithm with a processing-rate of 16 Hz for VGA format. In [34], an Intel Core 2 6300@1.86 GHz achieves real-time processing of 3D DCT/IDCT codec for 24 Hz VGA format. Therefore, software implementations of multimedia processing tasks are possible although limited to suboptimal techniques. since a 3D DCT/IDCT codec has poor coding performance, in terms of compression efficiency, when compared to the H.26x/MPEGx motion-compensated hybrid scheme considered in Section 6. A Core 2 processor, realized in 65 nm CMOS technology, integrates 2 64-bit CPU cores and up to 4 Mbytes of L2 cache memory. It has a die size of roughly 143 mm^2 and a transistor count of about 300 millions; the power consumption is up to 65 Watts, unsuited for battery powered, mobile or handheld terminals. Realizations of the same Core 2 processor in 45 nm CMOS technology, with a clock frequency still of 2.4 GHz and a L2 cache of 6 Mbytes, occupy an area of 107 mm^2 with a power consumption up to 44 W.

A lower power consumption, in the order of several Watts, can be achieved by targeting single-core CPUs such as the Intel ATOM 230@1.6 GHz used in [26] to implement the already-mentioned SIFT algorithm. The Intel ATOM processor, realized in 45 nm CMOS technology, integrates a single 32-bit CPU core with 512 kbytes of L2 cache. It has a die size of 26 mm^2, a transistor count of 47 millions and a power consumption of few Watts, less than 4 Watts. The lower size and power cost of ATOM versus Core2 is paid in terms of frame-rate performances. Indeed, in [26], only 5 images per second can be processed with the single-core ATOM processor which is not enough for real-time video representations.

As a matter of fact the computational capability of the Atom core at 1.6 GHz, measured with Dhrystone benchmarks, is around 3,8 billions of operations per second (GOPS) while that of a dual-core Core2 processor at 2.4 GHz is roughly 5.5 times higher.

Dual-core versions of the 32-bit ATOM processor have been realized in 45 nm CMOS technology, such as the Intel ATOM 330@1.6 GHz, with an overall L2 cache of 1 Mbyte: the computational capabilities are doubled versus the single-core ATOM230 processor but at the expenses of increased area, transistor count and power consumption by a factor of 2: 52 mm^2, 94 millions of transistor and less than 8 Watts, respectively.

Multi core platforms for embedded multimedia applications, based on ARM processors, have been presented in literature as MALI family: as example the MALI-200 has 2 ARM cores with AMBA AXI 3 interface; realized in 65 nm CMOS technology it occupies 4 mm^2 and has a computational capability of 275 Millions of pixels/s at 275 MHz. Such computational capability is not enough to support complex multimedia algorithms, such as a complete H.264 encoder at high definition; indeed for such applications dedicated coprocessors, such as the MALI VE6 video engine, has to be added [37].

Real-time performance with complex image/video processing algorithms, such as a complete H.264/MPEG4 AVC video coder, can be achieved with GPUs, for example, the cell broadband engine (BE) from IBM [16, 17, 19, 20, 38] or the FERMI architecture from NVIDIA [18, 35]. The cell BE has a capability of more than 200 GFLOPS with a 3.2 GHz clock when realized in 90 nm CMOS silicon-on-insulator (SOI) technology. In 65 nm and in 45 nm CMOS SOI, the clock frequency can rise up to 6 GHz and the peak computational throughput is roughly 380 GFLOPS [20, 38]. The cell BE integrates on-chip one 64-bit power processor element (PPE) core in charge of operation scheduling and data flow control, and 8 synergistic processors (SPs) dedicated to DSP computing and working according to an SIMD (single instruction multiple data) scheme. The cell BE has 256 kbytes of local memory for each SP, 32 kbytes of L1 data and instruction caches for the PPE, and 512 kbytes L2

Homogeneous and Heterogeneous MPSoC Architectures with Network-On-Chip Connectivity for Low-Power and
Real-Time Multimedia Signal Processing

173

shared cache. Each SP core has a transistor count of 21 millions, an area of 14.77 mm^2 and a power cost of 5 Watts in 90 nm CMOS technology [19]. Each SP has an area of 11.08 mm^2 and 6.47 mm^2 in 65 nm and 45 nm technology respectively. The multi core cell BE has an area of 235 mm^2 in 90 nm CMOS technology, 174 mm^2 in 65 nm CMOS, and 115 mm^2 in 45 nm CMOS SOI. Therefore, in submicron CMOS technology by scaling the same architecture from 90 nm to 65 nm and then to 45 nm technology node, the overall processor area scales with a factor roughly $\lambda2/\lambda1$ being $\lambda2$ and $\lambda1$ the channel length of the MOS devices in the two technologies; in the past, see [39, 40], as example, with larger technology nodes above 100 nm such ratio was $(\lambda2/\lambda1)^\theta$ being θ a fitting parameter between 1 and 2. The power cost of the Cell BE at 3.2 GHz is up to 80 W in 90 nm CMOS technology while at the same clock frequency in 45 nm CMOS technology is up to 40 W.

A high parallel multi core processor, called AsAP, has been proposed in [41]: it consists of 36 basic processors (32-bit RISC cores with 9 stage pipeline, 40-bit multiply and accumulate unit and 8Kbit of local memory) connected through a nearest-neighbor topology and running at 500 MHz. The area of the whole chip is 32 mm^2 (5.65 mm $\times$ 5.68 mm) and the power consumption is 1.15 W in typical conditions. Although the area and power consumption of AsAP meet the target requirements of our application, the offered computational power is enough only for real-time JPEG coder, that is, encoding of still images not H.264/AVC video coding. This is mainly due to the limited DSP capabilities of the basic AsAP tile. In [42], an evolution of the AsAP architecture, nicknamed AsAP2, has been proposed in 65 nm CMOS technology composed of 164 basic processors running at 1.2 GHz with a computational capability of 196.8 billions of multiply and accumulate operations per second (GMAC/s). AsAP2 also integrates 3 coprocessors dedicated to computing intensive DSP tasks such as FFT, video motion estimation, and viterbi decoder, and 3 SRAM blocks each of 16 kbytes. The overall complexity of AsAP2 rises to 55 millions of transistors for an area of roughly 40 mm^2. The computational power is enough to support 1080 HD multimedia applications at 30 frames/s with a power consumption of roughly 10.5 W.

A tentative realization of an MPSoC based on heterogeneous processors has been done in [43], which however is limited to a specific application domain (only the enhancement of image and video signals) while the aim of this work is more general: targeting real-time multimedia (image, video and also audio) tasks. GPUs are now evolving as massively parallel platforms for graphics but also for computing-intensive general purpose algorithms with high degree of parallelism. The Fermi architecture is composed of 512 CUDA (compute unified device architecture) processing cores hierarchically organized in 16 streaming multiprocessors each with 64 kbyte L1 cache and 128 kbyte local register file, and sharing a common 768 kbyte L2 cache. The total amount of on-chip memory is roughly 4 Mbytes. The Fermi GPU architecture has also PCIe host interface and six 64-bit DDR (double data rate) DRAM interfaces. Each CUDA core is capable of both integer and floating point operations with 64-bit results. Each of the 16 streaming processors has also 4 40-bit special function units for fast approximation of nonlinear operators (square root, sin, cos, exp, log functions) and 16 load/store units. The Fermi NVIDIA GPU leads to high computational power, up to 1500 GFLOPS in single-precision, orders of magnitude higher than Core 2 or Atom general-purpose processors. Although real-time processing is not an issue for such GPUs, their area and power consumption are suited only for desktop applications and workstations, not for handheld or mobile devices. As an example from the analysis in [35], the GTX480 Fermi NVIDIA, realized in 45 nm CMOS technology, has a transistor count of 3 billions, an area occupation for the core of 422 mm^2, and the power consumption higher than 150 W when implementing algorithms with a computational workload higher than 500 GFLOPs. Scaled versions of CUDA-based GPU exist: for example, the NVIDIA Quadro 4000 GPU has 256 CUDA cores, 4 64-bit DDR DRAM interface, and a maximum power consumption of 100 W. The massively parallel approach has been followed also by TILERA: its Tile 64 processor proposed in [44] realizes in CMOS 90 nm an array of 64 basic processors with 64-bit instruction set and 32-bit integer ALU communicating through a 2D Mesh on-chip interconnect. The Tile 64 reaches a computational throughput of 144 integer GOPS (32-bit operations) and 384 integer GOPS (8-bit operations) with 750 MHz clock. Its complexity amounts to 615 millions of transistors; the area is 433 mm^2, 215 mm^2 in 45 nm, and the power consumption at full load is roughly 11 W.

To reduce the power consumption of general purpose CPUs or high-parallel GPUs, several configurable hardware designs have been proposed in literature. Such designs have a power consumption below 1 Watt but achieve real-time processing only for a specific task and for low/mid-size image and video formats. Most of the proposed solutions use an RISC-like processing core with local instruction/data cache, for operation and data control flow, enhanced by external DRAM controller and hardware accelerators for the most demanding tasks. However, these hardware designs are customized for a specific algorithm: for example, dynamic image compression in [27], feature extraction in [26], 3D rendering in [23], video or audio coding in [12, 25]. If a mobile device needs a number of these functions several ICs should be used and mounted on a PCB board. Moreover, the communication between the RISC core and the coprocessors is based on classic bus architectures, for example, AMBA AHB bus. This limits the scalability of the solution and represents a bottleneck in case of multiple algorithms to be executed in parallel.

To overcome the limits of the state-of-the-art the following sections present two MPSoC architectures to achieve real-time processing for multiple algorithmic classes, running in parallel, at a reasonable power cost. The target is a power consumption of few Watts when implementing complex systems such as a complete H.264/MPEG AVC encoder in real time. To this aim, the following ideas will be exploited

(i) NoC as on-chip communication infrastructure, easing architecture scalability and management of

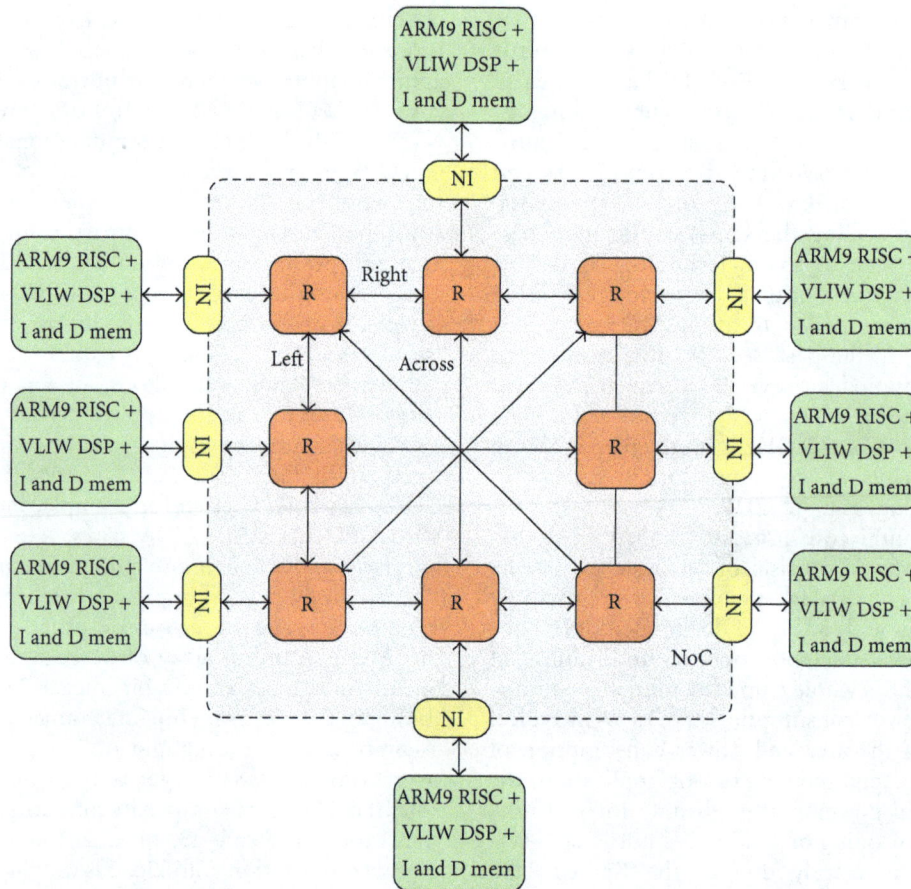

FIGURE 2: Block diagram of the Homogeneous NoC-based MPSoC architecture.

computing tasks with irregular or massively parallel data flow thanks to the packet-switched communication scheme.

(ii) Use of multiple tiles, with heterogeneous and homogeneous approaches, working in parallel to provide the computational power required by real-time multimedia tasks.

3. Homogeneous NoC-Based MPSoC

The proposed homogeneous MPSoC architecture is the one developed in the SHAPES (scalable software hardware architecture platform for embedded systems) European project involving among the partners University of Pisa, STMicroelectronics and ATMEL. In SHAPES, eight identical tiles building a scalable MPSoC are interconnected by means of a packet-switched on-chip network, see Figure 2. The NoC uses eight 4-port Routers (R in Figure 2): three router ports are dedicated to *across*, *right*, and *left* router-to-router connections and one port is dedicated to the connections between the NoC and the tiles. Conversion of protocol, data size, and clock frequency between the NoC and each tile is managed by eight Network Interfaces (NIs). A SHAPES tile, see Figure 3, contains the following:

(i) A 32-bit RISC processor based on the ARM926 core with dedicated instruction and data caches, 64 kbits overall, and a computational power of 1.1 MIPS/MHz evaluated with Dhrystone 2.1 (i.e., 275 MIPS when the tile is clocked at 250 MHz, see CMOS implementation results in Section 7);

(ii) A VLIW floating-point DSP based on the 64-bit mAgicV architecture by ATMEL featuring in submicron CMOS technology a complexity of 915 kgates and roughly 2 Mbits of program and data memory;

(iii) A distributed network processor (DNP) interface for extra tile communication which is connected to the NoC through a network interface; the DNP has also resources to be interconnected with off-chip host processors; there are also low-speed peripherals (UARTS, timers, interrupt controller) plus an interface towards external RAM.

The mAgicV core [45, 46] is capable of 10 arithmetic operations (multiply, add, subtract) per cycle and operates on IEEE 754 40-bit extended precision floating-point and 32-bit integer numeric format for numerical computations. Internal memory accesses are supported by a 16-bit MAGU (multiple address generation unit) with 4 different addresses that can be generated at each cycle. Hence the VLIW DSP

Homogeneous and Heterogeneous MPSoC Architectures with Network-On-Chip Connectivity for Low-Power and
Real-Time Multimedia Signal Processing

175

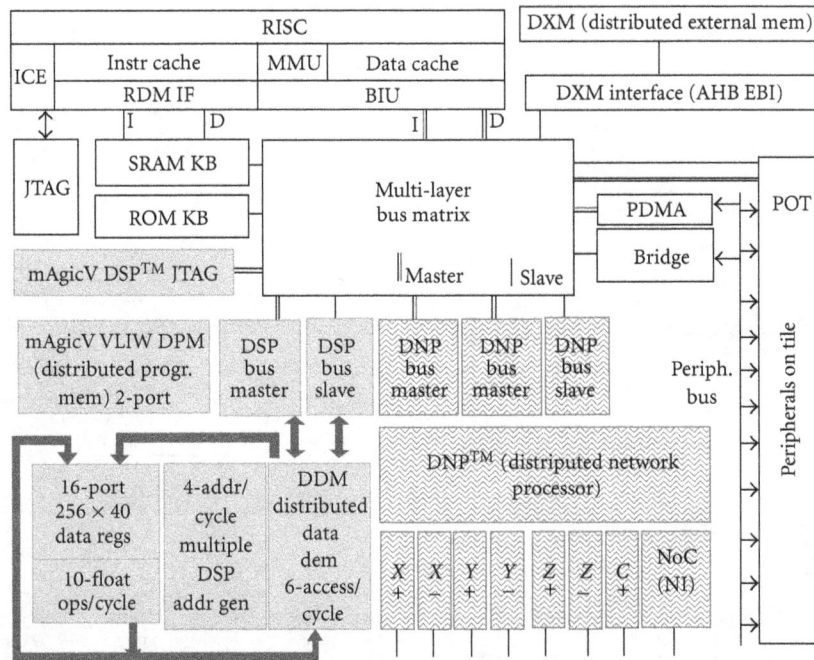

FIGURE 3: Content of a typical SHAPES tile [45].

is suited to process both 1D (audio) and 2D (images, video frames) data. All the above blocks are intratile interconnected through a 32-bit AXI bus. As detailed in Section 7, the 8-tile homogeneous MPSoC processor can be clocked at 250 MHz in 45 nm and 65 nm CMOS technologies. At such frequency it is capable of a computational power of 2.5 GFLOPS for each core, that is, up to 20 GFLOPS for the whole 8-tile MPSoC. The platform has a total on-chip memory of about 16 Mbits equally distributed among the tiles.

4. Heterogeneous NoC-Based MPSoC

The heterogeneous MPSoC architecture is sketched in Figure 4. It is composed of one general purpose core acting as control unit, one tile integrating external memory (DRAM, FLASH) controllers plus seven tiles each acting as an ASIP for a different class of algorithms: some units (in red) are dedicated to image/video processing such as the motion estimation (ME), some implements configurable 1D and 2D functions (in gray) such as frequency transforms or filters, one is dedicated to audio processing for low distortion digital input audio amplification (DIAA unit in blu in Figure 4). There are also seven shared SRAM memories, and in addition each computing tile has local memory resources. All tiles and memories are interconnected through an NoC which uses different routers from the architecture in Figure 2. In the heterogeneous MPSoC, new 5-port routers are used: three router ports are dedicated to *across*, *right*, and *left* router-to-router connections and two ports are dedicated to the connections between the NoC and the computing or memory tiles. The tile labeled *CPU* in Figure 4 is a 32-bit SPARC V8 core and it is in charge of instruction and data flow management. This tile has a 5-stage integer pipeline

plus hardware support of MAC operations and 8 kbytes of instruction cache and 8 kbytes of data cache. The *CPU* tile has also the following low-speed local peripherals: timers, interrupt controller, memory controller for connection to external E2PROM, UART, GPIO, and JTAG interface. When compared to other SPARC V8 cores available in literature such as LEON2 and LEON3 the proposed CPU integrates also an IEEE-754 compliant floating point coprocessor, missing in the LEON2, while versus the LEON3 [47] it is characterized by a reduced complexity since a 5-stage pipeline is adopted instead of a 7-stage ones. The reason is that our CPU is part of an heterogeneous core with dedicated DSP coprocessors and hence its role will be more devoted to control tasks rather than on computing intensive applications. The computational power of the designed CPU amounts to 1 Dhrystone MIPS/MHz, that is, about 250 MIPS when the tile is clocked at 250 MHz (see CMOS implementation results in Section 7). The complexity of the *CPU* tile is 70 k logic gates. The tile named *Ext Mem Ctrl* works as an external memory controller (DDR-DRAM or ROM/EEPROM). The tile has DMA (direct frame memory access) functionalities and 1 Mbit local buffer. After describing the general purpose tiles, hereafter, the architectural block diagram of the main tiles for signal processing tasks is discussed. By exploiting a design reuse approach these tiles are implemented starting from the RTL descriptions of IP cells, we proposed as stand-alone processors in past works [28, 32, 39, 48].

Two tiles named *ME* are dedicated to motion estimation processing, and they work on 2D image blocks. Each ME application-specific tile, see Figure 5, consists of a 2D hardware search engine plus local buffer memories, a 32-bit AHB interface towards the NoC and an ME controller to

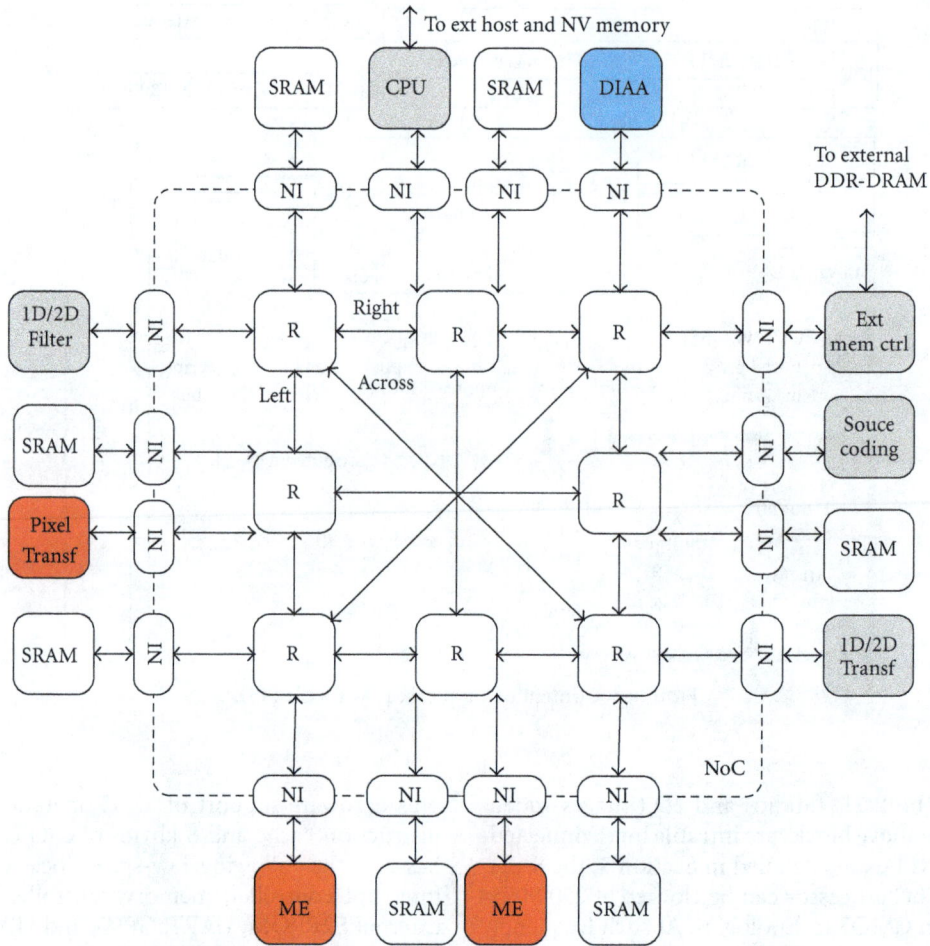

FIGURE 4: NoC-based Heterogeneous MPSoC: tiles for audio and image/video tasks are in gray; tiles dedicated to audio tasks are in blue, tiles dedicated to image/video tasks are in red. Memories and NoC units are in white.

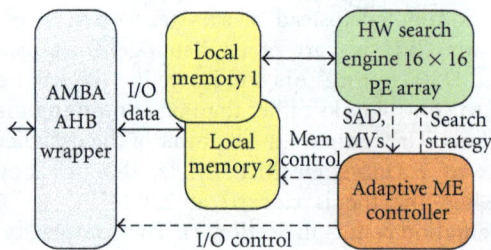

FIGURE 5: Architectural block diagram of the *ME* tile.

support fast search strategies. The search engine is a regular array of 256 processing elements each implementing at pixel-level absolute difference (AD) operations plus an adder tree, a unit for minimum detection calculation, and a unit used to compare block matching results with programmable thresholds (in case of fast ME search with early stop criteria). The search engine supports programmable search size and implements sum of absolute difference (SAD) and motion vector (MV) field calculation; it is derived from a parametric 2D search engine we previously proposed in [39]. In this work, the ME core has been configured for a 256 PE array

capable of 256 AD operations per clock cycle, that is, 64 billions of AD operations per second (GAD/s) when the tile is clocked at 250 MHz. The use of both tiles allow for a computational capability of 128 GAD/s. The local memory resources for each ME tile have a size of 40 kbits, enough to (i) store a 16×16-pixel block and its search area with ± 16 pixel displacement in horizontal and vertical directions and (ii) ensuring the prefetch of the next image block and its search area. The logic complexity of the two ME tile amounts to 230 k logic gates.

The architecture in Figure 5 implements a Full Search, but thanks to a programmable search area and to the support of early termination criteria it also realizes fast ME algorithms, such as the predictive ME in [28, 49]. In case of fast ME algorithms, the context-aware control strategies are elaborated by the configurable ME controller reported in Figure 5, which adapts the search strategy analyzing SAD and MV results.

The 1D/2D *Transf* tile supports frequency transforms, such as DCT/IDCT and FFT/IFFT, applicable to array or matrix data structures and hence to 1D audio signals or 2D image and video frames. Twiddle coefficients are stored in ROM memories and hence the type of transform can

Homogeneous and Heterogeneous MPSoC Architectures with Network-On-Chip Connectivity for Low-Power and
Real-Time Multimedia Signal Processing

177

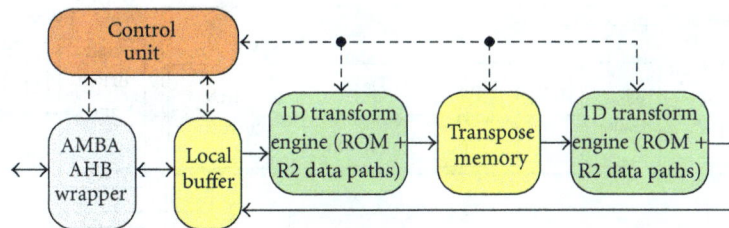

FIGURE 6: Architectural block diagram of the 1D/2D *Transf* tile.

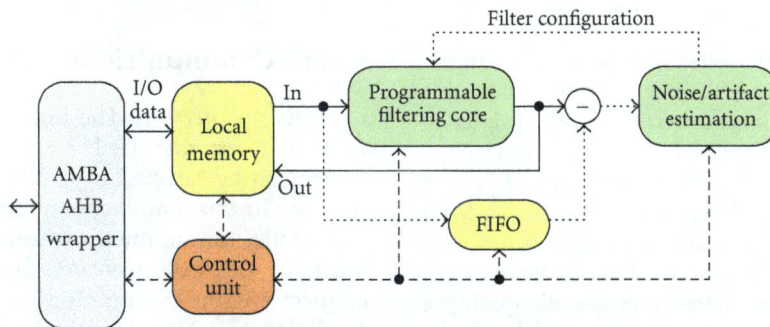

FIGURE 7: Architectural block diagram of the *Filt* tile.

be changed by selecting the proper coefficient memory. By exploiting the separability technique of 2D transforms, the *Transf* tile is composed of the cascade of 1D transform engines, see Figure 6. The second stage is bypassed in case of audio signal. The elementary data structure is an array of 8 samples. Image and video frames are decomposed in blocks of 8 × 8 pixels, elaborated row-wise in the 1st stage while the second engine works on the transposed results. Each engine is composed by 4 radix-2 (R2) butterflies, implementing MAC operations between input samples and twiddle coefficients with block floating point arithmetic. A processing capability up to 250 illions of transform operations per second is possible by clocking the tile at 250 MHz. The circuit complexity is less than 50 k logic gates. This tile also adopts a local memory buffer of 32 kbits: 16 kbits to prefetch up to 1024 samples (corresponding to sixteen 8 × 8 blocks image blocks), and 16 kbits to store the results for 1024 samples. The memory occupancy of each sample is 16 bits.

The 1D/2D *Filter* tile, see Figure 7, is an ASIP dedicated to filtering of 1D and 2D data structures; it supports linear (e.g., FIR, IIR) or nonlinear (e.g., rational filters) operators working at array level for audio signals and image block level for image/video frames. The filter tile has a 32-bit AHB interface towards the NoC and a throughput of 1 sample/clock cycle, that is, 250 Msample/s can be processed in real-time with a 250 MHz clock. The total local memory plus FIFO amounts to 35 kbits while the circuit complexity is around 45 k logic gates. The 1D/2D *Filter* is made up of the following building blocks:

(1) a filtering core configurable to implement linear or rational nonlinear filters;

(2) a unit for noise/artifact estimation and filter tuning;

(3) memory resources for data flow management;

(4) a control unit providing all relevant control signals.

The tile named *Pixel Transf* is dedicated to special functions widely used in image and video processing: pixel transformations such as gamma correction and contrast enhancement, color domain conversion (supporting RGB, YUV, YCrCb), frame size conversion, log-linear and linear-log domain conversion, and clipping. The tile named *SourceCoding* is in charge of source coding techniques such as variable-length coding or context-adaptive binary arithmetic coding (CABAC). It works at bit stream level, it can be applied to both 1D and 2D data structures. It has a complexity of 30 k logic gates plus 30 kbits of local memory. The core of this tile is the coding engine we proposed in [29].

The DIAA tile is an ASIP implementing the audio DSP tasks needed for a high-quality reproduction of digital input audio streams (e.g., 16-bit PCM digital tracks stored in a CD or DVD or memory card) through a switching power output stage. With reference to scheme of a complete digital input power amplifier in Figure 8, the tasks supported by the DIAA ASIP are the mainly oversampling, noise shaping, multi level PWM and dead time insertion. Other supported audio function such as SPDIF interfacing or volume control or signal decimation are not shown in Figure 8. The oversampling unit implements zero padding and subsequent digital filter interpolation, needed to increase the incoming input frequency (F_{IN} ranging from 44.1 kHz to 96 kHz) by a factor M (typically $M = 8$ or 16), thus easing the reject of distortions introduced by the quantization and sampling process from the base-band audio signal (which is typically

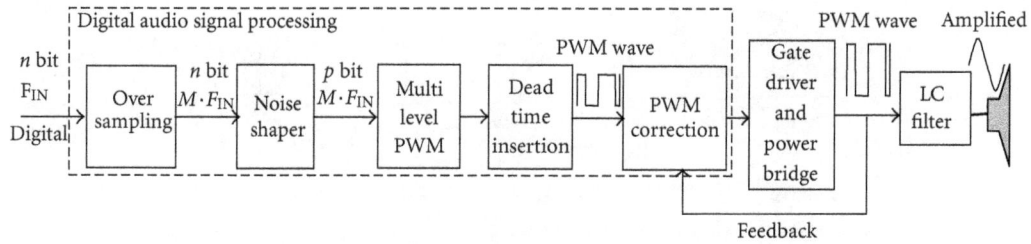

FIGURE 8: Architectural block diagram of a low distortion audio amplification processing chain.

below 15-16 kHz). The digital filter interpolation is implemented through a polyphase FIR filter with programmable coefficients. Converting the n-bit oversampled PCM signal to PWM leads to a minimum impulse time $T_{min} = 1/(M \cdot F_{IN} \cdot 2^n)$ s, for example, roughly 0.25 ns considering $M = 16$ and the 44.1 kHz 16-bit PCM signal of audio CDs. Such values are too low for commercial power transistors with rise and fall times, T_r and T_f, of tens of ns. To reduce such requirement while keeping unaltered source audio quality, a noise shaper is used. It reduces the used bits from n to p, while the added quantization noise can be spread outside the audio band using a K-th order FIR shaping filter. For stability reasons, the maximum supported order is $K = 5$ while p is at maximum 6 so that $T_{min} = 1/(M \cdot F_{IN} \cdot 2^p)$, that is, 22 ns for $M = 16$, $F_{IN} = 44.1$ kHz, compliant with rise and fall times of fast power MOS available in the market [50].

The multi level PWM and dead time insertion units convert the oversampled and noise-shaped PCM audio track to a PWM signal that can be binary-coded $(-1, +1)$ or ternary-coded $(-1, 0, +1)$. The latter technique achieves higher efficiency: when the audio signal is null there is no switching activity while in binary-coded PWM when the audio signal is null, the PWM signal is switching with a duty cycle of 50% between -1 and $+1$ symbols. Configurable guard time intervals can be digitally inserted (dead times) to compensate the turn-on and turn-off delays of real off-chip Power MOS and hence avoiding short-circuit current flowing in the power H bridge between supply and ground (e.g., if a low-side MOS already turned on while the high side MOS is still not turned off or vice versa).

The complexity of the DIAA tile amounts to roughly 30 Kgates. To be noted that the DIAA supports the DSP audio task needed for high-fidelity digital audio reproduction. Other audio DSP functionality, such as 1D filtering or FFT for noise removal, signal enhancement or equalization, insertion of polyphonic effects, among others, are supported by the other ASIP tiles, particularly the *1D/2D Filter* and *1D/2D Transf* ones.

All computing and memory tiles in Figure 4 have a 32-bit AHB interface towards the corresponding NI. The size of each memory unit is a parameter of the HDL description configurable at synthesis time. In this work, a size of 3 Mbits is considered for each of the 7 SRAM blocks in Figure 4 which is enough to store a whole VGA frame or half of a HD frame in a single memory block. In case of audio processing, a single memory can store up to 200 K samples at 16 bits, that is, 4.5 s for a single CD-quality channel at 44.1 kHz.

5. NoC Communication Infrastructure Design

5.1. NoC Architecture. The high computational power and memory access rate needed for real-time multimedia processing can be satisfied by exploiting parallel computation [51–56]. To this aim, a communication backbone allowing for parallel communication among the tiles is needed. For both the heterogeneous and homogeneous MPSoC architectures, the communication infrastructure is based on the design of a NoC (and its building macrocells, NI, and routers) with a Spidergon topology. The Spidergon approach includes the classic ring topology where each router is connected to its *Left* and *Right* neighbours, see Figures 2 and 4, but as enhancement an additional diagonal connections is introduced (each router is linked to the *Across* neighbour; see Figures 2 and 4) to minimize the number of hops to reach the destination. In the NoC in Figures 2 and 4, any packet goes across a maximum of three routers to reach its destination, since no more than two hops separate any two routers in the network. Each 4-port router in Figure 2 presents a port to the local NI, while in Figure 4 the 5-port routers are connected to two NIs: one NI for interfacing to the computing tile and one NI for interfacing to a memory block. With respect to a classic on-chip communication infrastructure based on bus hierarchy and circuit-switched connections, the NoC has two advantages: it is easier scaling the MPSoC architecture adding other computing or memory tiles; the packet-switched approach allows multiple tiles working in parallel and hence the capabilities (computational throughput and/or supported set of algorithms) for the whole homogeneous and heterogeneous MPSoCs are higher than the single tile.

The proposed NoC routers implement packet-switched wormhole routing with credit-based flow control, consequently they require small buffers and allow for a deep pipelined packet communication. Since the routing path is assigned by the NI at packet injection, the router does not need slow complex logic or look-up tables, but it simply extracts the forward information from the packet header (that has a fixed size due to the symmetry of the topology). Also the QoS (quality-of-service) information is encoded in the packet header at the injection point, and it is not explicitly linked to the path of a data flow through the network. The routers then perform a simple 2-step arbitration based on the priorities extracted from the headers, hence the QoS support in the router does not require any complex computation logic. Figure 9 illustrates the main blocks of the router, with reference to a 5-port configuration:

Homogeneous and Heterogeneous MPSoC Architectures with Network-On-Chip Connectivity for Low-Power and
Real-Time Multimedia Signal Processing

179

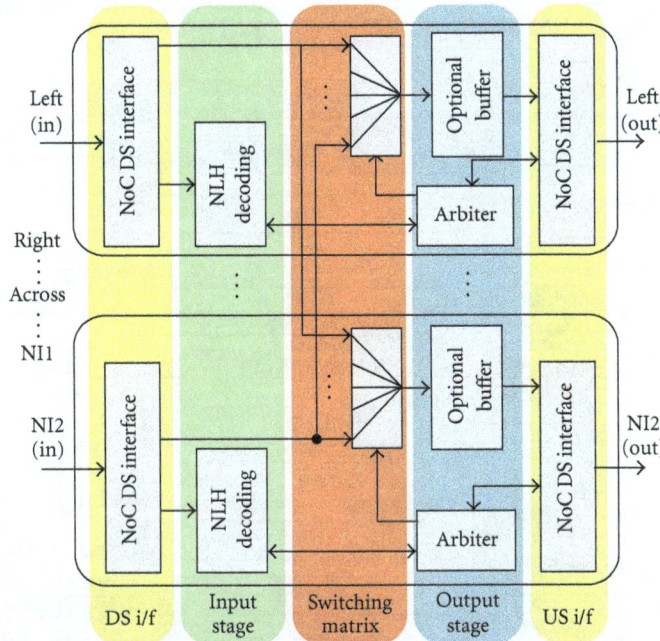

FIGURE 9: NoC router architecture.

the NoC DownStream (DS) interface, which is the input from the network and contains the small buffers for the incoming traffic; the Input stage, where the routing and QoS information are extracted from the packet header (network layer header, NLH in Figure 9); the switching matrix, which connects any router input port to any output port; the output stage, where multiple requests to access the same output are arbitrated, and a single input request is granted access (extra optional buffering can also be performed, if required by the application); and the NoC UpStream (US) interface, which is the output to the network. The architecture is modular, and the logic depicted schematically in Figure 9 for ports Left and NI2 is replicated on each port.

The router has a configurable number of pipeline stages in the data path, where registers can be removed and buffers are optional or by passable, thus resulting in a minimum crossing latency variable from 0 up to 2 clock cycles. Further details can be found in our recent patent [57]. The reduced crossing latency of the NI and of the router and their configurability are the main features of the proposed NoC design versus the state-of-the-art. Minimization of communication and processing latency is important at application level for interactive multimedia systems such as videoconferencing, videotelephony. The actual latency value may be increased by other factors, for example, if the packet does not win immediately an arbitration, or if the output port is temporarily stalled due to full buffers at the other side of the link. The NoC NIs in Figures 2 and 4 constitute the IP cores entry point to the communication backbone. Figure 10 shows the basic architecture of an NI. The upper part of the figure, with traffic from the IP core to the NoC, is the request path, while the lower part, from NoC to the IP bus, is the response path. In the NI, two main components can be identified: Shell and Kernel, from left to

right in Figure 10. The Shell is responsible for the handshake with the IP bus and for encoding/decoding the NoC packet header (network layer and transport layer headers, NLH and TLH, respectively, in Figure 10) thus translating IP transactions into NoC packets and vice versa. The Kernel, IP-protocol independent may contain the header and payload bisynchronous FIFOs where size and frequency conversion between the IP and the NoC domains are performed. The FIFOs are also exploited in case store & forward transmission is required, where a whole packet is stored before being sent: this feature can change an intermittent transmission into a bubble-free traffic, hence optimizing the use of the network. The Kernel also contains the Upstream and Downstream interfaces, where the credit-based flow control is managed.

The NI data pipeline stages are configurable, from 0 to 3. When all input/output registers and FIFOs are removed (zero pipeline stages), size or frequency conversion are not supported, nor the store & forward mechanism. The minimum NI crossing latency equals the number of pipeline stages instantiated, but its actual value may be increased by other factors (e.g., the synchronization delay if the frequency conversion support is enabled). The proposed NI has also optional support of advanced features such as handling of out-of-order transactions, detection of error transactions, secure memory access control. However, such features are not necessary in our application, and the NoC building blocks are configured to implement just the basic functionalities: conversion of protocols, data size, and frequency. To guarantee efficient parallel communication between cores and memory spaces, the store & forward mechanism is enabled.

5.2. NoC Packet Format, Latency, Bandwidth. The NoC packet format, carrying header and payload data, is

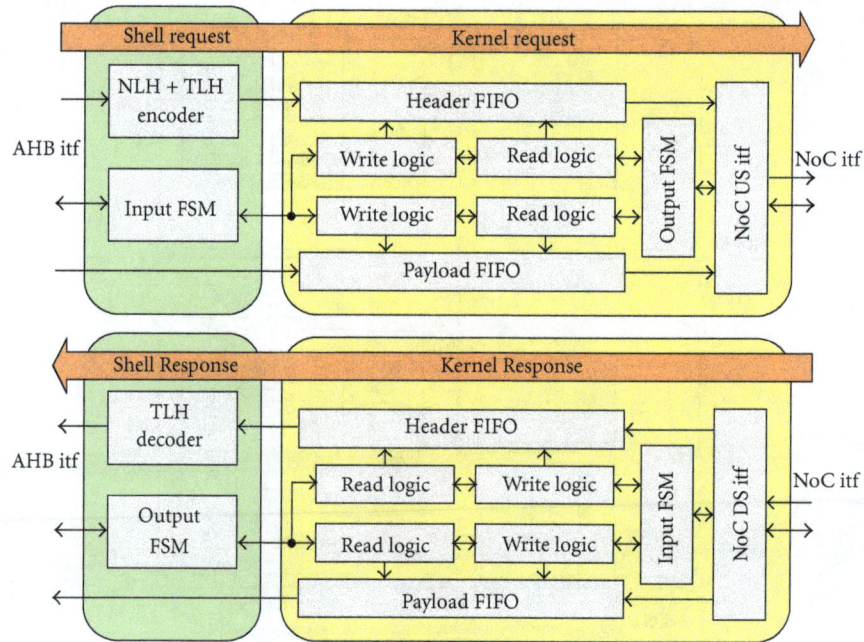

FIGURE 10: Main blocks in the NI architecture.

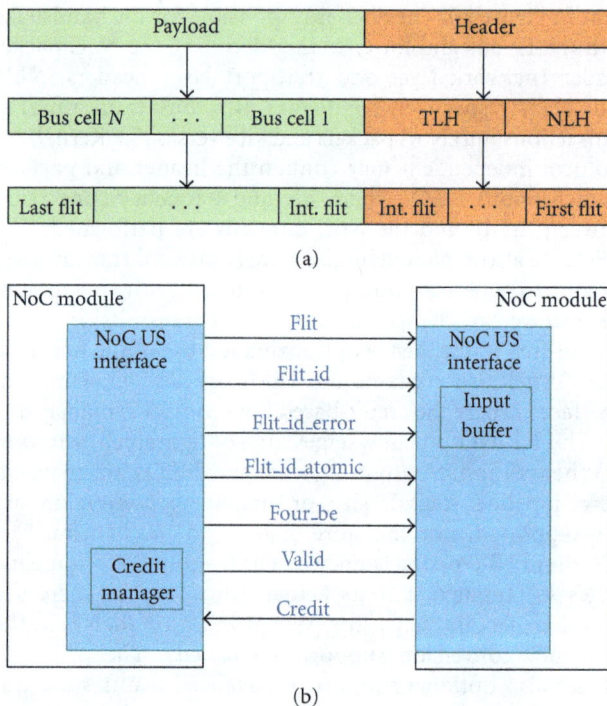

(a)

(b)

FIGURE 11: (a) Internal organization of a NoC packet, (b) physical link at the NoC side.

FIGURE 12: PSNR versus bit-rate for Stefan, a fast ME is used for homogeneous and heterogeneous MPSoC.

illustrated in Figure 11(a). The header field is composed of a network layer header (NLH) and a transport layer header (TLH). In the NLH, there are the packet routing and QoS information that are used by the NoC routers; the TLH, instead, contains IP protocol information and is necessary at the destination node, to translate the received NoC packet

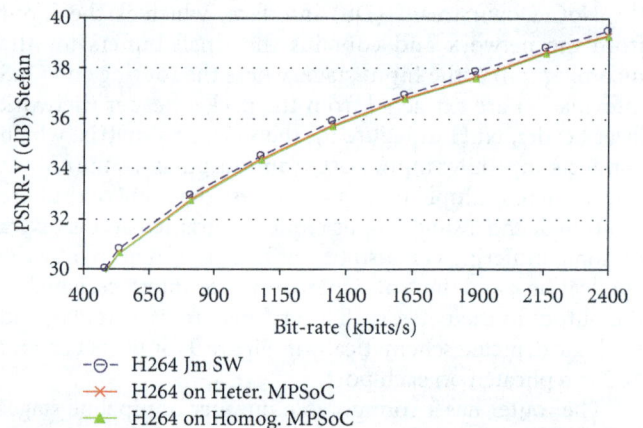

into a bus transaction. The payload of the NoC packet transports the payload data cells of the bus transaction. The NoC packet travels over the network in subunits called flits. Header and payload information need to travel in separate flits, and all flits of a packet are routed through the same path across the network. As far as the physical link is concerned, at the NoC interface side, there are the hardwired lines showed in Figure 12.

N-bit flits are used to transfer NoC packets, with N configurable at synthesis time. The flit_id field identifies first, intermediate, and last flits of a packet or if a flit is a single one. The optional K-bit four_be signal ($K = N/32$) marks meaningful 32-bit pieces of data within a flit and is used in end-to-end size conversion. The optional 2-bit flit_id error is used for signaling a slave side error or an interconnect error.

The optional flit_id atomic signal enables support of atomic operations: an NI can lock paths towards a Slave IP so that a Master IP can perform a generic number of consecutive operations without any interference from other masters. The credit and valid signals allows for credit-based flow control. A flit is sent only when there is room enough to receive it: neither retransmission nor flit dropping are allowed. This is done automatically by setting an initial number of credits in the U.S. interface (in its credit manager), equal to the size of the input buffer in the DS interface it communicates with. Since the U.S. interface sends flits only if the connected DS interface can accept them, there are no pending flits on the link wires.

In both homogenous and heterogeneous MPSoC designs, the AMBA AHB IP cores have a 32-bit data bus, while the NoC uses 128-bit flit size. Moreover, the different tiles might be running at different speeds because of different operating modes (see results in Section 7). For these reasons, the NIs are configured to support both size and frequency conversion, and thus they instantiate bisynchronous FIFOs in the Kernel. No extra pipeline stages are needed, as the maximum supported frequency is only 250 MHz.

The NoC routers are also configured to have a single pipeline stage (so, a minimum latency of one cycle), since it is enough to support the target NoC frequency. The global network latency can, therefore, be estimated as the sum of the routers crossing latency (i.e., a maximum of three cycles, since a maximum of three 1-cycle-latency routers are crossed to reach any destination in the network) plus the NIs latency, resulting from the size conversion operation and the synchronization delay required by frequency conversion. Any contention to access the link adds up to the global latency of the system. However, in the considered application, the network latency is critical only if the local frame memories become empty during operation. In both the heterogeneous and homogeneous MPSoC platforms, the NoC interconnect has been properly sized (128 bit NoC versus 32 bit of each connected IP) to support a data flow capable to feed the local memory resources and avoid them to go empty, thus resulting in a system which is not latency-sensitive.

The 128-bits NoC data size together with a 500 MHz NoC clock frequency (supported by the selected NoC building blocks configurations) can guarantee a nominal throughput of 128 bits × 500 MHz = 64 Gbps per link. This throughput is enough to support all target image/video and audio processing functionalities, and allows sustaining the maximum throughput theoretically generated by 8 cores working concurrently in Figures 2 or 4 through the 32 bit AMBA interface running at 250 MHz. Moreover, thanks to the NI store & forward support, the path in the network is engaged only by a bubble-free data flow, thus making full use of the available bandwidth. To exploit the high bandwidth available in the NoC interconnect it is necessary to execute long AMBA AHB transactions, since NoC packet length depends on the amount of data to be transferred. To clarify this point, suppose to execute "Store 1 byte" operations: this generates NoC packets composed of a header (whose overhead is approximately 80 bits per packet, and is transmitted in a separate flit) plus a 128-bit payload

flit where only a single byte is significant, thus resulting in a real throughput of only 2 Gbps. Executing "Store 16 bytes" operations, instead, produces NoC packets composed of a header flit and a payload flit containing 128 data bits, thus achieving a 32 Gbps throughput. It is, therefore, clear that for an efficient use of the high NoC bandwidth it is important to properly select the size of the operation executed in the AMBA AHB domain. In our application, the NoC QoS mechanism is not exploited, since the NoC bandwidth capability is much higher than the AHB one and thus there is no real need to give higher priority to critical data flows. All data flow injected in the NoC by the NIs have the same bandwidth reservation (the same priority), and the arbitration performed in the routers is a simple Least Recently Used. The proposed NoC sizing is done to avoid saturation in case of the maximum bandwidth request by the target applications, that is, when running video applications like H264 encoding of one multimedia stream with 30 frames/s HD videos.

6. MPSoC Functional Assessment

The heterogeneous architecture in Figure 4 is tailored for power and area efficient real-time implementation of 1D (audio) and 2D (image/video) processing functionalities: 250 MIPS of general-purpose computing capabilities are provided by the CPU tile, while the other heterogeneous tiles provide high-throughput DSP capabilities for several algorithms: 128 billions of AD operations (GAD/s) for ME plus 250 millions of operations per second (MOPS) for each of the 5 ASIPs for filtering, transform, source coding, audio processing, pixel-level image processing. Large memory resources are integrated on-chip since multimedia applications are data-dominated. Memory resources are hierarchically organized in a first level of local memories in the computing tiles, 1.5 Mbits, plus a second level of shared frame memories (21 Mbits SRAM) accessible via NoC, and a third level of off-chip DRAMs accessible via the *Ext Mem Ctrl* tile.

The homogeneous architecture in Figure 2 offers a higher degree of flexibility since the 8 tiles provide general-purpose capabilities (for each tile, 275 MIPS of a standard ARM CPU core plus 2.5 GOPS of a DSP core) useful not only for audio or image/video processing, but also for DSP applications in general. The memory hierarchy organization of the homogeneous MPSoC is different from the heterogeneous one, since it is based on 2 levels. At Level 1, large memory resources are integrated on-chip, 16 Mbits, distributed 2 Mbits for each tile. A dedicated external DDR-DRAM can be added for each tile in Figure 2. Table 1 summarizes the computation, memory, and on-chip communication capabilities of the two MPSoCs.

In both heterogeneous and homogeneous MPSoC, thanks to the NoC paradigm, the architectures can be scaled by modifying the configuration discussed in this work in terms of: number of used computing tiles, number and size of the memories, as well as size of the tile data bus and of the NoC links. This way, the desired trade-off between complexity and performance can be set. Being composed by identical tiles, the homogeneous architecture requires less

TABLE 1: Performance of the MPSoC architectures.

MPSoC	Memory	DRAM interface	Computation capabilities	NoC	Scalability
Homogen.	16 Mbits L1	1 for each tile	2200 MIPS-ARM + 20 GFLOPS DSP	8 NIs + 8 4-port Routers 32 bit IP bus at 250 MHz, 128 bit flit at 500 MHz	Easy for on-chip communication and tiles
Heterogen.	22.5 Mbits (L1 + L2)	1 shared	250 MIPS-SPARC + 1.25 GOPS + 128 G AD/s	16 NIs + 8 5-port Routers 32 bit IP bus at 250 MHz, 128 bit flit at 500 MHz	Easy for on-chip communication, not for tiles

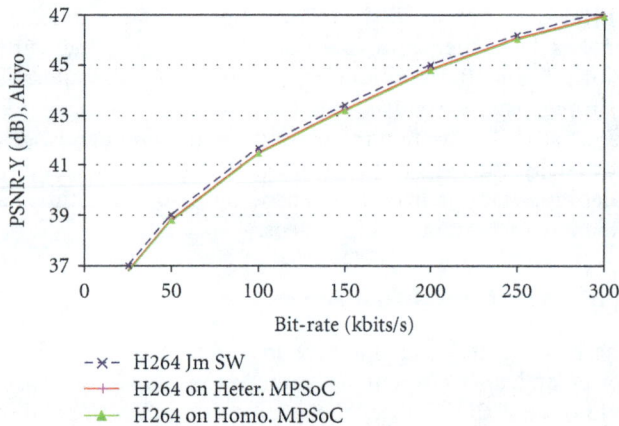

-×- H264 Jm SW
-+- H264 on Heter. MPSoC
-▲- H264 on Homo. MPSoC

FIGURE 13: PSNR versus bit-rate for Akiyo, a fast ME is used for homogeneous and heterogeneous MPSoC.

TABLE 2: THD results for the reproduction of 1 kHz and 10 kHz audio signals (16-bit 44.1 KHz CD quality).

MPSoC	% THD at 1 kHz	% THD at 10 kHz
Homogeneous	0.09	0.12
Heterogeneous	0.09	0.13

design and verification time and is easier to scale. In case of the heterogeneous architecture, all tiles are different and have to be designed and tested independently before the integration and test in the whole MPSoC.

To assess the functional performances of the proposed MPSoCs, we show the results achieved when exploiting their instruction set and computing/memory capabilities for two applications (selected as case studies among the large set of algorithms that are supported). The first selected case study is the real-time compression of a video with an H.264 AVC compliant hybrid coder scheme configured as follows: fast search adaptive ME with ±16 pixel search displacement, maximum 5 previous reference frames for best block matching, SAD error cost function, CABAC entropy coding, 2D Integer DCT transform as well as, in-loop de-blocking filter. This is a very challenging computing intensive task. With reference to scenes with different degrees of dynamism, such as the sport scene Stefan at 30 Hz VGA video format, and a videoconferencing scene such as Akiyo 30 Hz CIF, a low-dynamic scene, Figures 12 and 13 present the obtained rate-distortion curves (PSNR in dB versus bit-rate). The achieved performances are compared to the ones of a reference MPEG AVC encoder, using JM software implementation, configured using CABAC and FS ME with similar parameter set as described above. From Figures 12 and 13, it is clear that both MPSoCs offer optimal performances for video processing with a PSNR degradation for a fixed bit-rate of 0.1-0.2 dB versus the reference JM software implementation. The same coding efficiency with respect to the JM reference software

has been obtained with other test scenes and 30 Hz video formats including HD frames.

As proved in [11], the original JM software does not allow for real-time processing, even considering small image formats. On the contrary, the proposed MPSoCs have a computational capability enough to ensure real-time processing of 30 Hz HD videos.

Comparing in terms of maximum computational power the two platforms is not easy since the heterogeneous platform is dedicated to specific tasks while the homogenous one has more general purpose DSP capabilities. As example considering the specific H.264/AVC encoder the heterogeneous platform, offering 128 G AD/s, allows for a frame rate up to 200 frames/s while the homogenous platform is 5 times slower. Considering a more general purpose algorithm, such as the FFT, the heterogeneous platform allows for a throughput of 250 Msamples/s while the homogenous one can sustain a throughput 10 times faster.

The proposed MPSoCs can sustain in real-time lots of image, video, or audio processing algorithms. As example of audio processing both MPSoCs have been used to reproduce audio tracks from a CD-quality source. The incoming audio signal is stored in the on-chip memories, processed following the chain in Figure 8, and then sent to an off-chip output power stage accepting PWM signals and driving 8 Ω loudspeaker. Table 2 reports the achieved performance measured in terms of THD (total harmonic distortion) at 1 kHz, and 10 KHz, respectively. The results are compliant with high-fidelity audio applications.

7. CMOS Implementation Results

The proposed MPSoC architectures have been designed using VHDL language and then synthesized in STMicroelectronics 45 nm CMOS technology with 1.1 V supply. The RTL to gate-level synthesis has been accomplished within Synopsys CAD environment while the back-end phases have been conducted within Cadence environment. The back-end of both the homogeneous MPSoC (within the SHAPES

Homogeneous and Heterogeneous MPSoC Architectures with Network-On-Chip Connectivity for Low-Power and
Real-Time Multimedia Signal Processing

183

FIGURE 14: Layout of the SHAPES homogeneous MPSoC architecture with NoC.

project) and the heterogeneous MPSoC architectures have been successfully realized in 45 nm STMicroelectronics CMOS technology. As far as speed performance is concerned, the maximum achievable frequency depends on the used standard-cell library version. Indeed, the used technology provides 3 types of standard-cells: beside the SVT (standard voltage threshold) version there are also an HVT (high voltage threshold) version optimized in terms of leakage power consumption, and a LVT (Low voltage threshold) optimized in terms of speed but with a much higher power cost. By using a mix of HVT and SVT cells (HVT for all paths with noncritical time performances, while SVT only for time-critical paths limiting maximum. clock speed), we were able to run the NoC at 500 MHz while keeping the cost figure power consumption/(MHz·gate) near the minimum value permitted by the technology library. A faster clock, about 800 MHz, could be achieved using LVT cells; however, the power cost with LVT would be much higher and the performances achieved at 500 MHz with the HVT/SVT cells are enough for the target mobile and/or handheld applications.

To reduce the power consumption by a factor of 2 or 4 for applications not requiring the full-bandwidth capability the NoC clock frequency can be reduced at 250 MHz or 125 MHz. The clock frequency of the IP tiles can be reduced from 250 MHz to 125 MHz.

7.1. Homogeneous MPSoC Results. With reference to the scheme in Figure 2, Figure 14 reports the layout of the 8-tile NoC-based MPSoC configuration of SHAPES. The area of each tile is $2 \times 3.5 \, \text{mm}^2$ and the number of logic gates is 4.46 Millions. The density of the tile chip, that is, the ratio between the area occupied by devices (logic gates, memory, I/O,...) and the total occupied silicon area after place and route, is roughly 70%. Considering the whole platform in Figures 2 and 14, the area is $8 \times 7.1 \, \text{mm}^2$, that is, about $56 \, \text{mm}^2$, and the number of gates is about 36 Millions. The number of transistors is roughly 150 millions. The density of the whole MPSoC is 68%. The typical value for chip density is about 75%. In the 8-tile SHAPES chip,

density has been kept a little bit lower in order to minimize crosstalk effects and improve timing performance. The power consumption depends on the workload. Considering for the homogeneous MPSoC platform a target frequency of 250 MHz and a computing intensive application such as the H.264/AVC video encoder applied to the Stefan video sequence in Section 6, the total dynamic core power consumption amounts to roughly 2.8 Watts (350 mW per each tile). The leakage power consumption is 65 mW.

As far as the NoC communication infrastructure is concerned, from Figure 14, it is clear that its overhead in terms of area is minimal. The occupied area of the synthesized NoC interconnect, after place and route, is $0.123 \, \text{mm}^2$ in 45 nm CMOS. The area occupancy is approximately due for 2/3 to the 8 4-port routers and for the remaining 1/3 to the 8 NIs. Power consumption analysis has been performed for the NoC at a target frequency of 250 MHz and using a 1.1 V supply voltage. The results show that the overall power consumption (dynamic power plus leakage power) is less than 4 mW, 60% due to the 8 4-port routers and 40% due to the 8 NIs. These data confirm that the overhead due to the NoC infrastructure, $0.123 \, \text{mm}^2$ area and few mW of power, is negligible versus the power (2.8 W) and area ($56 \, \text{mm}^2$) costs of the interconnected tiles in the MPSoC. The above NoC results refer to the following configuration: NI with data bus size and flit size of 32 bits; tiles and NoC running at 250 MHz and 500 MHz, respectively; header and payload FIFOs in the Kernel have 2 locations of 32 bits; advanced NI services (security, order handling, error management, frequency, and data size conversion,etc.) disabled.

7.2. Heterogeneous MPSoC Results. A similar characterization has been done for the heterogeneous MPSoC architecture of Figure 4 in the same CMOS technology. The computational and memory complexity of all tiles in Figure 4 amounts to 550 k logic gates plus 22.5 Mbits of on-chip SRAM, hierarchically organized in 21 Mbits level 2 frame memories (*memory tiles* in Figure 4) accessible via NoC, and around 1.5 Mbits level 1 local memories distributed in the tiles. The overall circuit complexity of the heterogeneous MPSoC is determined by both the tiles discussed in Section 4 and the resources required by the NoC infrastructure presented in Section 5. To meet the functional requirements of the proposed MPSoC each NI is configured to support 32-bit IP AHB bus, 128-bit flits on the NoC side, conversion of protocol, data size and frequency up to a maximum of 500 MHz, FIFOs sized to store 2 locations (data or headers). The complexity of this NI configuration is 14 k logic gates. As far as the router is concerned, it has 5 ports, 128-bit flits data size, an input buffer for each port of 2 locations, and supports the features described in Section 5 with a clock frequency of 500 MHz and a complexity of 32 k logic gates. The FIFOs in the NoC building blocks are not memory-based, since their small size makes more convenient a register-based implementation. The overall NoC complexity (16 NIs plus 8 5-port routers) for the heterogeneous MPSoC amounts to 480 k logic gates. Hence, the total heterogeneous MPSoC complexity, due to computing and memory tiles plus NoC infrastructure, amounts to 1030 k logic gates and

TABLE 3: Comparison of the proposed MPSoCs versus the state of the art (all in 45 nm technology node except Tile64 in 90 nm CNOS and AsAP2 in 65 nm CMOS).

Processor	Area	Tr. count	Power	CMOS Tech.	Computational Power	Encoder supported/format
Het. MPSoC	49 mm^2	120 M	1 W	45 nm	1.5 GOPS + 128 G AD/s	H264/AVC video codec, HD
Hom. MPSoC	56 mm^2	150 M	2.8 W	45 nm	2.2 GOPS + 20 GFLOPS (extended-precision)	H264/AVC video codec, HD
CELL-BE [20]	115 mm^2	320 M	40 W	45 nm SOI	200 GFLOPS (single-precision)	H264/AVC video codec, HD
Core2 [34]	107 mm^2	290 M	44 W	45 nm	22 GOPS at 2.4 GHz	3D DCT codec, VGA
AsAP2 [42]	40 mm^2	55 M	10.5 W	65 nm	196.8 GMAC/s + 15 G AD/s + 680 M FFT-sample/s	H264/AVC video codec, HD
ATOM 330 [58]	52 mm^2	94 M	<8 W	45 nm	3.8 GOPS at 1.6 GHz	H264/AVC video codec, CIF
ATOM 230	26 mm^2	47 M	<4 W	45 nm	7.6 GOPS at 1.6 GHz	H264/AVC video codec, QCIF
Tile64 [44]	430 mm^2	615 M	11 W	90 nm	144 GOPS at 750 MHz	H264/AVC video codec, HD

22.5 Mbits of on-chip SRAM. The number of transistors of the whole platform is about 130 millions while the area occupied by the heterogeneous MPSoC is about 49 mm^2. The power consumption depends on the workload and on the configuration status of the different tiles. For the applications described in Section 6, the estimated power consumption amounts to 1 W in case of the H.264/AVC encoding task of the Stefan sequence, the same test used for power analysis of the homogeneous MPSoC. The main contribution to the above power consumption figures is due to the computing and memory tiles.

7.3. Comparison. Table 3 summarizes the area, power consumption results, and computational capabilities of the proposed MPSoCs and some state-of-art multimedia processors. Also, the target encoder that can be supported in real-time by the platform is reported.

To be noted that computational power is expressed, depending on the target platform, as sum of operations on integer data (GOPS), on floating-point operations (GFLOPS, 32-bit single or 40-bit extended precision) or custom instructions such as multiply and accumulate (MAC) or absolute difference (AD). From Tables 1 and 3, it emerges that the heterogeneous MPSoC, made up of application specific tiles, has a smaller area and power cost versus the homogeneous MPSoC, which instead ensures a higher flexibility, being composed by general purpose tiles with DSP capabilities, and easier scalability. The heterogeneous MPSoC is particularly suited for multimedia processing on mobile and handheld devices.

Compared to state-of-the-art solutions reviewed in Section 2, the proposed MPSoCs offer the possibility of achieving real-time processing for complex multimedia tasks with a power consumption much lower than GPU. The power cost is even lower than low-power general purpose CPUs, such as the Intel ATOM 330, or application-specific multicore architecture as AsAP2 [42]. With respect to such low-power solutions the proposed MPSoCs offer a higher computational capability: for example, real-time H264/AVC encoding up to 30 Hz HD frames is supported by our MPSoCs but not by AsAP, limited to JPEG encoding of still images [41], or Intel Atom 330, limited to H264/AVC video encoding of CIF frame in surveillance applications [58]. With

respect to ASICs achieving real-time and very low-power performance but for a specific algorithm, the proposed MPSoCs offer a higher flexibility and the support, even in case of the heterogeneous approach, of a large number of audio and image/video algorithmic classes.

8. Conclusions

The paper has presented the design and characterization in submicron CMOS technology of homogeneous and heterogeneous MPSoC architectures for real-time multimedia processing. In both architectures, the multiple tiles are interconnected by a network-on-chip infrastructure with Spidergon topologies and minimum crossing-latency NIs and Routers. The proposed packet-switched data transfer scheme avoids communication bottlenecks when more tiles are working concurrently. The heterogeneous architecture ensures a higher power efficiency and a smaller area occupation while the homogeneous scheme allows for higher flexibility since the basic tile offers more general purpose computation resources. With respect to the state-of-the-art, a better trade-off between cost (area, power) and performance (throughput, latency, flexibility) is achieved. The functional performances of the NoC-based MPSoC architectures are assessed with reference to H264/MPEG AVC video coding and digital input audio reproduction case studies. The heterogeneous MPSoC is particularly suited for multimedia processing in mobile and handheld terminals.

Acknowledgments

This work has been partially supported by the FP6 EU project SHAPES in collaboration with STMicroelectronics (M. Coppola, R. Locatelli, G. Maruccia, V. Catalano, AST, Grenoble), University of Cagliari (L. Raffo), ATMEL Roma (P. S. Paolucci).

References

[1] J. U. Garbas, B. Pesquet-Popescu, and A. Kaup, "Methods and tools for wavelet-based scalable multiview video coding," *IEEE Transactions on Circuits and Systems for Video Technology*, vol. 21, no. 2, pp. 113–126, 2011.

Homogeneous and Heterogeneous MPSoC Architectures with Network-On-Chip Connectivity for Low-Power and
Real-Time Multimedia Signal Processing

185

[2] G. D. Hines, Z. U. Rahman, D. J. Jobson, G. A. Woodell, and S. D. Harrah, "Real-time enhanced vision system," in *Enhanced and Synthetic Vision*, vol. 5802 of *Proceedings of SPIE*, pp. 127–134, March 2005.

[3] D. J. Jobson, Z. U. Rahman, and G. A. Woodell, "Properties and performance of a center/surround retinex," *IEEE Transactions on Image Processing*, vol. 6, no. 3, pp. 451–462, 1997.

[4] L. Shao, H. Hu, and G. De Haan, "Coding artifacts robust resolution up-conversion," in *Proceedings of the 14th IEEE International Conference on Image Processing (ICIP '07)*, pp. V409–V412, September 2007.

[5] N. E. L'insalata, S. Saponara, L. Fanucci, and P. Terreni, "Automatic synthesis of cost effective FFT/FFT cores for VLSI OFDM systems," *IEICE Transactions on Electronics*, vol. E91-C, no. 4, pp. 487–496, 2008.

[6] F. Luisier, T. Blu, and M. Unser, "Image denoising in mixed poissongaussian noise," *IEEE Transactions on Image Processing*, vol. 20, no. 3, pp. 696–708, 2011.

[7] S. Marshall and G. L. Sicuranza, *Advances in Nonlinear Signal and Image Processing*, Hindawi Publishing Corporation, New York, NY, USA, 2006.

[8] S. Marsi, G. Impoco, A. Ukovich, G. Ramponi, and S. Carrato, "Using a recursive rational filter to enhance color images," *IEEE Transactions on Instrumentation and Measurement*, vol. 57, no. 6, pp. 1230–1236, 2008.

[9] J. Ostermann, J. Bormans, P. List et al., "Video coding with H.264/AVC: tools, performance, and complexity," *IEEE Circuits and Systems Magazine*, vol. 4, no. 1, pp. 7–28, 2004.

[10] C. Pascual, Z. Song, P. T. Krein, D. V. Sarwate, P. Midya, and W. B. J. Roeckner, "High-fidelity PWM inverter for digital audio amplification: spectral analysis, real-time DSP implementation, and results," *IEEE Transactions on Power Electronics*, vol. 18, no. 1, pp. 473–485, 2003.

[11] S. Saponara, K. Denolf, G. Lafruit, C. Blanch, and J. Bormans, "Performance and complexity co-evaluation of the Advanced Video Coding standard for cost-effective multimedia communications," *EURASIP Journal on Applied Signal Processing*, vol. 2004, no. 2, pp. 220–235, 2004.

[12] S. Saponara, P. Nuzzo, C. Nani, G. Van Der Plas, and L. Fanucci, "Architectural exploration and design of Time-interleaved SAR arrays for low-power and high speed A/D converters," *IEICE Transactions on Electronics*, vol. E92-C, no. 6, pp. 843–851, 2009.

[13] A. Yoneya, "Pulse width and position modulation for fully digital audio amplifier," in *Proceedings of the IEEE International Symposium on Circuits and Systems (ISCAS '08)*, pp. 1692–1695, May 2008.

[14] A. Chonka, W. Zhou, L. Ngo, and Y. Xiang, "Ubiquitous multicore (UM) methodology for multimedia," in *Proceedings of the International Symposium on Computer Science and Its Applications (CSA '08)*, pp. 131–136, October 2008.

[15] K. Popovici, X. Guerin, F. Rousseau, P. S. Paolucci, and A. A. Jerraya, "Platform-based software design flow for heterogeneous MPSoC," *Transactions on Embedded Computing Systems*, vol. 7, no. 4, article 39, 2008.

[16] J. Park and S. Ha, "Performance analysis of parallel execution of H.264 encoder on the cell processor," in *Proceedings of the 5th Workshop on Embedded Systems for Real-Time Multimedia (ESTIMedia '07)*, pp. 27–32, October 2007.

[17] K. Kim, J. Lee, H. W. Park, and S. Ha, "Automatic H.264 encoder synthesis for the cell processor from a target independent specification," in *Proceedings of the IEEE/ACM/IFIP Workshop on Embedded Systems for Real-Time Multimedia (ESTIMedia '08)*, pp. 95–100, October 2008.

[18] J. Nickolls and W. J. Dally, "The GPU computing era," *IEEE Micro*, vol. 30, no. 2, pp. 56–69, 2010.

[19] D. C. Pham, T. Aipperspach, D. Boerstler et al., "Overview of the architecture, circuit design, and physical implementation of a first-generation cell processor," *IEEE Journal of Solid-State Circuits*, vol. 41, no. 1, pp. 179–196, 2006.

[20] J. Pille, C. Adams, T. Christensen et al., "Implementation of the CELL broadband engine in a 65nm SOI technology featuring dual-supply SRAM arrays supporting 6GHz at 1.3V," *IEEE Journal of Solid State Circuits*, vol. 43, no. 1, pp. 163–171, 2008.

[21] X. Ma, M. Dong, L. Zhong, and Z. Deng, "Statistical power consumption analysis and modeling for GPU-based computing," in *Proceedings of the Workshop on Power Aware Computing and Systems*, Big Sky, Mont, USA, October 2009.

[22] B. G. Nam, J. Lee, K. Kim, S. Lee, and H. J. Yoo, "Cost-effective low-power graphics processing unit for handheld devices," *IEEE Communications Magazine*, vol. 46, no. 4, pp. 152–159, 2008.

[23] B. G. Nam and H. J. Yoo, "An embedded stream processor core based on logarithmic arithmetic for a low-power 3-D graphics SoC," *IEEE Journal of Solid-State Circuits*, vol. 44, no. 5, pp. 1554–1570, 2009.

[24] C. M. Chang, S. Y. Chien, Y. M. Tsao, C. H. Sun, K. H. Lok, and Y. J. Cheng, "Energy-saving techniques for low-power graphics procrssing unit," in *Proceedings of the International SoC Design Conference (ISOCC '08)*, pp. I242–I245, November 2008.

[25] S. Y. Chien, Y. W. Huang, C. Y. Chen, H. H. Chen, and L. G. Chen, "Hardware architecture design of video compression for multimedia communication systems," *IEEE Communications Magazine*, vol. 43, no. 8, pp. 123–131, 2005.

[26] M. Murphy, K. Keutzer, and H. Wang, "Image feature extraction for mobile processors," in *Proceedings of the IEEE International Symposium on Workload Characterization (IISWC '09)*, pp. 138–147, October 2009.

[27] Quick Logic's Visual Enhancement Engine (VEE) Brings iridix to Mobile Devices, 2010.

[28] S. Saponara, M. Martina, M. Casula, L. Fanucci, and G. Masera, "Motion estimation and CABAC VLSI co-processors for real-time high-quality H.264/AVC video coding," *Microprocessors and Microsystems*, vol. 34, no. 7-8, pp. 316–328, 2010.

[29] S. Saponara, L. Fanucci, and P. Terreni, "Design of a low-power VLSI macrocell for nonlinear adaptive video noise reduction," *EURASIP Journal on Applied Signal Processing*, vol. 2004, no. 12, pp. 1921–1930, 2004.

[30] NXP, UDA1355H: audio stereo codec with SPIDIF interface, 2003.

[31] K. Masselos, F. Catthoor, C. E. Goutis, and H. Deman, "A systematic methodology for the application of data transfer and storage optimizing code transformations for power consumption and execution time reduction in realizations of multimedia algorithms on programmable processors," *IEEE Transactions on Very Large Scale Integration (VLSI) Systems*, vol. 10, no. 4, pp. 515–518, 2002.

[32] L. Fanucci, R. Saletti, and S. Saponara, "Parametrized and reusable VLSI macro cells for the low-power realization of 2-D discrete-cosine-transform," *Microelectronics Journal*, vol. 32, no. 12, pp. 1035–1045, 2001.

[33] R. Hameed, W. Qadeer, M. Wachs et al., "Understanding sources of inefficiency in general-purpose chips," in *Proceedings of the 37th International Symposium on Computer Architecture (ISCA '10)*, pp. 37–47, June 2010.

[34] T. Fryza, "Introduction to implementation of real time video compression method," in *Proceedings of the 15th International*

Conference on Systems, Signals and Image Processing (IWSSIP '08), pp. 217–219, June 2008.

[35] E. S. Chung, P. A. Milder, J. C. Hoe, and K. Mai, "Single-chip heterogeneous computing: does the future include custom logic, FPGAs, and GPGPUs?" in *Proceedings of the 43rd Annual IEEE/ACM International Symposium on Microarchitecture (MICRO '10)*, pp. 225–236, December 2010.

[36] Apple, 2011, http://www.apple.com/iphone/specs.html.

[37] Mali Graphics Hardware, http://www.arm.com/products/multimedia/mali-graphics-hardware/index.php.

[38] O. Takahashi, C. Adams, D. Ault et al., "Migration of Cell Broadband Engine from 65nm SOI to 45nm SOI," in *Proceedings of the IEEE International Solid State Circuits Conference (ISSCC '08)*, pp. 81–87, February 2008.

[39] L. Fanucci, S. Saponara, and L. Bertini, "A parametric VLSI architecture for video motion estimation," *Integration, the VLSI Journal*, vol. 31, no. 1, pp. 79–100, 2001.

[40] P. Pirsch, N. Demassieux, and W. Gehrke, "VLSI architectures for video compression—a survey," *Proceedings of the IEEE*, vol. 83, no. 2, pp. 220–246, 1995.

[41] Z. Yu, M. J. Meeuwsen, R. W. Apperson et al., "AsAP: an asynchronous array of simple processors," *IEEE Journal of Solid-State Circuits*, vol. 43, no. 3, pp. 695–705, 2008.

[42] D. N. Truong, W. H. Cheng, T. Mohsenin et al., "A 167-processor computational platform in 65 nm CMOS," *IEEE Journal of Solid-State Circuits*, vol. 44, no. 4, pp. 1130–1144, 2009.

[43] S. Saponara, L. Fanucci, and E. Petri, "A multi-processor NoC-based architecture forreal-time image/video enhancement," *Journal of Real-Time Image processing*. In press.

[44] S. Bell, B. Edwards, J. Amann et al., "TILE64 processor: a 64-core SoC with mesh interconnect," in *Proceedings of the IEEE International Solid State Circuits Conference (ISSCC '08)*, vol. 51, pp. 88–89, 2008.

[45] P. S. Paolucci, A. A. Jerraya, R. Leupers, L. Thiele, and P. Vicini, "SHAPES: a tiled scalable software hardware architecture platform for embedded systems," in *Proceedings of the 4th International Conference on Hardware Software Codesign and System Synthesis (CODES+ISSS '06)*, pp. 167–172, October 2006.

[46] P. S. Paolucci, "Four levels of parallelism to be managed in the DIOPSIS based SHAPES multi-tiled architecture," in *Proceedings of the 8th International Forum on Application-Specific Multi-Processor SoC (MPSOC '08)*, pp. 23–27, Aachen, Germany, 2006.

[47] Z. Zhou, W. Wu, M. He, and L. Hou, "A SoPC design based on LEON3 SoC platform," in *Proceedings of the 1st Asia Pacific Conference on Postgraduate Research in Microelectronics and Electronics (PrimeAsia '09)*, pp. 400–403, November 2009.

[48] L. Fanucci, S. Saponara, and A. Morello, "Power optimization of an 8051-compliant IP microcontroller," *IEICE Transactions on Electronics*, vol. E88-C, no. 4, pp. 597–600, 2005.

[49] A. Chimienti, C. Ferraris, and D. Pau, "A complexity-bounded motion estimation algorithm," *IEEE Transactions on Image Processing*, vol. 11, no. 4, pp. 387–392, 2002.

[50] Infineon, "Data Sheet BSO200N03S: Optimos 2 Power-Transistor. rev 1. 6," 2008.

[51] L. Benini and G. De Micheli, "Networks on chips: a new SoC paradigm," *Computer*, vol. 35, no. 1, pp. 70–78, 2002.

[52] M. Coppola, M. D. Grammatikakis, R. Locatelli, G. Maruccia, and L. Pieralisi, *Design of Cost-Efficient Interconnect Processing Units: Spidergon STNoC*, CRC Press, Boca Raton, Fla, USA, 2008.

[53] H. G. Lee, N. Chang, U. Y. Ogras, and R. Marculescu, "On-chip communication architecture exploration: a quantitative evaluation of point-to-point, bus, and network-on-chip approaches," *ACM Transactions on Design Automation of Electronic Systems*, vol. 12, no. 3, Article ID 1255460, 2007.

[54] F. Vitullo, N. E. L'Insalata, E. Petri et al., "Low-complexity link microarchitecture for mesochronous communication in networks-on-chip," *IEEE Transactions on Computers*, vol. 57, no. 9, pp. 1196–1201, 2008.

[55] S. Saponara, T. Bacchillone, E. Petri, and L. Fanucci, "Design of a NoC interface Macrocell with hardware support of advanced networking functionalities," *IEEE Transactions on Computers*. In press.

[56] M. Palesi, G. Ascia, F. Fazzino, and V. Catania, "Data encoding schemes in networks on chip," *IEEE Transactions on Computer-Aided Design of Integrated Circuits and Systems*, vol. 30, no. 5, pp. 774–786, 2011.

[57] M. Coppola, R. Locatelli, S. Saponara, E. Petri, T. Bacchillone, and L. Fanucci, "Network on Chip Router," Patent Number 11-GR2CO-0356EP01, France, 2011.

[58] Intel, "Optimizing H. 264 software codec on Intel Atom and Intel Core2 processors, targeting Intel digital security surveillance applications," Tech. Rep. 323064, 2010.

17

Fast and Near-Optimal Timing-Driven Cell Sizing under Cell Area and Leakage Power Constraints Using a Simplified Discrete Network Flow Algorithm

Huan Ren and Shantanu Dutt

Department of ECE, University of Illinois at Chicago, Chicago, IL 60607, USA

Correspondence should be addressed to Shantanu Dutt; dutt@ece.uic.edu

Academic Editor: Gi-Joon Nam

We propose a timing-driven discrete cell-sizing algorithm that can address total cell size and/or leakage power constraints. We model cell sizing as a "discretized" mincost network flow problem, wherein available sizes of each cell are modeled as nodes. Flow passing through a node indicates the choice of the corresponding cell size, and the total flow cost reflects the timing objective function value corresponding to these choices. Compared to other discrete optimization methods for cell sizing, our method can obtain near-optimal solutions in a time-efficient manner. We tested our algorithm on ISCAS'85 benchmarks, and compared our results to those produced by an optimal dynamic programming- (DP-) based method. The results show that compared to the optimal method, the improvements to an initial sizing solution obtained by our method is only 1% (3%) worse when using a 180 nm (90 nm) library, while being 40–60 times faster. We also obtained results for ISPD'12 cell-sizing benchmarks, under leakage power constraint, and compared them to those of a state-of-the-art approximate DP method (optimal DP runs out of memory for the smallest of these circuits). Our results show that we are only 0.9% worse than the approximate DP method, while being more than twice as fast.

1. Introduction

In order to achieve a balance between design quality and time-to-market, cell library-based design is becoming the dominant design methodology over the custom design method even for high-performance ICs. Usually in a cell library, several different cell implementations are available for the same function with different sizes, intrinsic delays, driving resistances, and input capacitances. Choosing the cell with an appropriate size, that is, *cell sizing*, is a very effective approach to improve timing.

The cell-sizing problem has been studied for a long time. Many methods [1, 2] assume the availability of a continuous range of cell sizes; that is, the size of a cell can take any value in a range. Then, the obtained gate size is rounded to the nearest available size in the library. However, a large number of realistic cell libraries are "sparse," for example, geometrically spaced instead of uniformly spaced [3]. Geometrically spaced gate sizes are desired in order to cover a large size range

with a relatively small number of cell instances. Also it has been proved in [4] that, under certain conditions, the set of optimal gate sizes must satisfy the geometric progression. With a sparse library, the simple rounding scheme can introduce huge deterioration from the continuous solution, which often causes the sizing results to fail to meet given timing requirements [3].

On the other hand, few time-efficient methods are known that they can directly handle timing optimization with discrete cell sizes since this problem is NP complete. The technique in [5] uses multidimensional descent optimization that iteratively improves a current solution by changing the size choices of a set of cells that produces the largest improvement. It is not clear how well this method can avoid being trapped in a local optimum. In [3], a new rounding method is developed based on an initial continuous solution. Instead of only rounding to the nearest available size, the method visits cells in topological order in the circuit, and tries several discrete sizes around the continuous solution for each cell. In order to reduce the search space, after a new cell is visited, it performs a pruning step that discards obviously inferior solutions and a merging step that keeps only several representative solutions within a certain quality region. The run time of this method is significantly larger than the method in [5].

More recently, [6] has proposed a method that uses a combination of continuous and discrete optimization techniques for the power-driven timing-constrained cell-sizing problem. The problem is first simplified to an unconstrained problem using Lagrange relaxation. Then, the resulting unconstrained discrete sizing problem is solved using dynamic programming. Through Lagrange relaxation-based simplification, it is able to handle complex delay constraints of current industry designs. However, as a continuous convex optimization technique, the quality and convergence of the Lagrange relaxation method is not guaranteed when applied to discrete problems. On the other hand, our proposed method directly handles constraint satisfaction simultaneously with objective optimization in a unified and specially designed mincost network flow model.

In this paper, we propose a network-flow-based method for the discrete gate-sizing problem. In our method, the different size options of a cell are modeled by nodes in the network graph. The flow cost represents the change in the timing objective function value when the cell sizes corresponding to the nodes in the graph which have flows through them are chosen. Hence, by solving a mincost flow in the network graph, near-optimal cell sizing can be determined by choosing cell sizes whose corresponding nodes have the mincost flow through them—the near optimality comes from having to constraint the flow to adhere to certain discrete requirements like going through exactly one size-option node per cell.

By modeling the gate-sizing problem as a mincost network flow problem, we can solve it using standard network flow algorithms, which are very time efficient. Also problem constraints, like the maximum allowable total cell area, can be handled efficiently by making the flow amount proportional to the chosen cell area and using an arc with an appropriate capacity to limit the total flow amount.

However, network flow is a continuous optimization method. Thus when applying it to solve the discrete option selection problem, invalid solutions may be produced. For example, the mincost flow may pass through two sizing options for the same cell. We solve such problems by using *min lookahead-cost* or *max flow* selection heuristics. Network flow has been used to solve various EDA problems including placement [7–9] and placement legalization [10]. In the recent technique of [10], only the network graph modeling the legalization problem, and which nodes (representing bins of cells) are overfull (these are then supply nodes) and under full (these are demand nodes) have been determined a priori to solve the mincost flow problem. Costs of arcs, which represent shipment of cells, between adjacent bins are not known in advance since this depends on the exact set of cells being shipped from one bin to an adjacent one. This set of cells and the cost of arcs out of the "current" bin are determined dynamically within Dijstra's shortest path

computation that is used iteratively to solve an approximate mincost flow problem. Our problem in using network flow to solve the cell-sizing problem is different: while we know the exact cost and capacities of arcs in the network graph modeling the problem, the issue we need to tackle is that of preventing splitting of the flow among each subset of arcs that represent the selection of the sizes of each cell, so that flow can go through only one of these arcs, thus providing a unique selection of a size for each cell. We solve this problem in an outer loop enveloping the mincost flow computation, as opposed to the technique in [10], which solves it within an inner loop of the mincost flow computation.

This paper is an extended version of the workshop paper [11] that appeared in a internal workshop compendium of papers, which is not considered a published proceedings. Thus it is not strictly necessary to discuss the issue of extensions of that paper. However, there are extensions, which include an updated discussion of previous work, results for benchmark circuits using Synopsys's 90 nm library (in addition to the 180 nm library used in [11]), and an analysis of the differences in the two sets of results for the two libraries.

The rest of the paper is organized as follows. Section 3 provides an overview of our method. A general view of our size selection network graph (SSG) is presented in Section 4. In Sections 5.1–5.4, we discuss various issues of the SSG. In Section 6, we show how to obtain a valid mincost flow in the SSG. Section 7 briefly describes an optimal exhaustive search method to which we compare our network-flow-based technique. Section 8 presents experimental results and we conclude in Section 9.

2. New Algorithmic Approach Used

In this paper we use a simplified version of *discretized network flow* (DNF) that has been recently introduced as a versatile optimization technique for CAD problems over the last four years [7, 9, 12–14]. The DNF technique imposes certain discrete requirements on the flow through the network graph G like a mutual exclusivity constraint on certain arc sets, called *mutually exclusive arc* (MEA) sets, in which at most one arc can have a nonzero flow in each MEA set. In the above cited works, the DNF problems is modeled as a fixed-charge network flow problem [15] in order to solve complex physical synthesis problems using multiple transforms and constraints with significant efficacy. However, it is possible to solve the discrete cell-sizing problem near optimally with a simple version of DNF in which max-flow or min-lookahead-cost (subsequently termed "mincost" for brevity) heuristics for determining which arc in each MEA set should have nonzero flow. We present this simpler version of DNF as the new algorithmic technique in this paper.

3. Overview of Our Method

Our cell-sizing method starts from an initial sizing solution that may be far from the optimal. The objective is to improve the critical path delay of the circuit by resizing cells. We define $\mathscr{P}_\alpha$ to be the set of paths with a delay greater than $1 - \alpha$

Fast and Near-Optimal Timing-Driven Cell Sizing under Cell Area and Leakage Power Constraints Using a Simplified
Discrete Network Flow Algorithm

189

fraction ($\alpha < 1$) of the most critical path delay. In order to reduce the complexity of the problem we only consider changing the sizes of cells in $\mathscr{P}_\alpha$. In our experiments, α is set to be 0.1. This simplification does not limit the optimization potential of our method, since our method is incremental in its nature. Thus we can iterate it several times to take more paths into consideration.

We define $CS(n_j)$ to be the set of critical sinks of a net n_j, which are (1) all sinks in $\mathscr{P}_\alpha$ if the net is in $\mathscr{P}_\alpha$, or (2) the sink with the minimum slack otherwise. Our method tries to improve the critical path delay by minimizing the objective function proposed in [9]. A timing cost $t_c(n_i)$ of a net n_i is defined as [9]:

$$t_c(n_i) = \sum_{u_j \in CS(n_i)} \frac{D(u_j, n_i)}{S_a^\beta(n_i)}, \qquad (1)$$

where u_j is a sink cell of net n_i, $D(u_j, n_i)$ is the net delay of n_i to u_j, and $S_a(n_i)$ is the *allocated slack* of a net in the initial sizing solution; the allocated slack is defined as the path slack divided by the number of nets in the path. β is the exponent of the allocated slack used to adjust the weight difference between costs of nets on critical and noncritical paths. Based on experimental results, $\beta = 1$ works best in this scenario, since only nets in $\mathscr{P}_\alpha$ and those connected to it are considered in the optimization function (see below). Let $\widehat{\mathscr{P}}_\alpha$ be the set of nets that are either in $\mathscr{P}_\alpha$ or connected to nodes in it. The timing objective function F_t is the summation of $t_c(n_i)$ of all nets in $\widehat{\mathscr{P}}_\alpha$ given as

$$F_t = \sum_{n_i \in \widehat{\mathscr{P}}_\alpha} t_c(n_i). \qquad (2)$$

Here we only consider nets in $\widehat{\mathscr{P}}_\alpha$, since the delays of only these nets will change with our selection of resizable cells. Note that in this objective function the nets on more critical paths have a higher contribution since the net cost is inversely proportional to the allocated slack and thus will be optimized more—a desirable outcome, especially in a scenario where there is a quota on the resource (e.g., total area) available for optimization.

Usually, the total area is given as a constraint for timing driven cell sizing. Our algorithm can also handle this constrained optimization problem.

4. Overview of the Size Selection Graph (SSG)

We model the timing-minimization cell-size selection problem as a mincost network flow problem. A network flow graph called *the size selection graph* (SSG) is constructed in which the set of sizing options of each cell is modeled as a set of *sizing option nodes*, and each sizing option node corresponds to one sizing option. We use flows in the SSG to model cell-size selection; that is, a size option of a cell is chosen by a flow in the SSG if its corresponding option node in the SSG is on the path of the flow. Hence, each flow through the SSG that passes through one option node in every set of sizing options in the SSG corresponds to a sizing

selection for the cells in $\mathscr{P}_\alpha$. Furthermore, if we can set the costs of arcs between option nodes in the SSG in such a way that the total cost incurred by a flow through the SSG is equal to the change in the timing objective function F_t (2) corresponding to the sizing scheme selected by the flow, then the mincost flow in the SSG selects the optimal cell-sizing scheme for F_t. Hence the problem of finding the optimal cell sizing is converted to the problem of finding the mincost flow in a graph for which several efficient algorithms such as the network simplex algorithm and the enhanced scaling algorithm [16] are available. The general structure of our SSG is described below.

4.1. The SSG Structure. Since F_t is the summation of the timing costs t_c of nets, we employ a divide-and-conquer approach in constructing the SSG; that is, first a mininetwork flow graph called a *net structure* is constructed for each net in $\mathscr{P}_\alpha$, and then net structures of connected nets are connected by *net spanning structures* to form the complete SSG as shown in Figure 1. Thus, the SSG has a similar topology as the paths in $\mathscr{P}_\alpha$. Note that the source node S is the only supply node with total flow of $F(S)$ (whose determination is discussed in Section 5.4), and the sink T is the only demand node of flow amount $F(S)$ in the SSG. A net structure N_j is a *child net structure* of N_i if they are connected, and N_j follows N_i in the flow direction in the SSG (i.e., the signal direction in the circuit is from net n_i to net n_j); correspondingly N_i is also a *parent net structure* of N_j. Each net structure contains the sets of option nodes for cells in the corresponding net.

Let us denote the set of sizing options of a cell u as S_u, and the lth sizing option in it as S_u^l. For a net n_i with a driving cell u_d, the net structure is constructed by connecting each sizing option $S_{u_d}^l$ of u_d to all sizing option nodes in S_{u_k} of every sink cell u_k to form a complete bipartite subgraph between S_{u_d} and S_{u_k}. An example is shown in Figure 2. The net has one driving cell u_d and two critical sink cells u_j and u_k; see Figure 2(a). The corresponding net structure is shown in Figure 2(b). Here, we only show two sizing options in each option set. A flow f through the net structure is also shown, which selects size option $S_{u_d}^1$ for u_d, $S_{u_k}^1$ for u_k, and $S_{u_j}^2$ for u_j. With the complete bipartite subgraphs between the sizing option sets of the drive cell and sink cells in a net structure, every possible combination of size choices of cells in the net has a corresponding flow through the net structure and hence is considered in our size selection process.

There are two major issues that need to be tackled in constructing the SSG in order to correctly map the mincost flow in the SSG to an optimal sizing choice. They are as follow.

(1) In each net structure, the cost of each arc needs to be determined so the total cost of a flow through the net structure can accurately capture the change value of the timing cost t_c for the net corresponding to the sizing options chosen by the flow. A detailed description of this issue is given in Section 5.1.

(2) The flow that is determined must be a valid flow, that is, can be converted to a valid sizing option selection. A valid sizing option selection requires the satisfaction of two types of *consistencies*. First, sizing option nodes of a particular cell

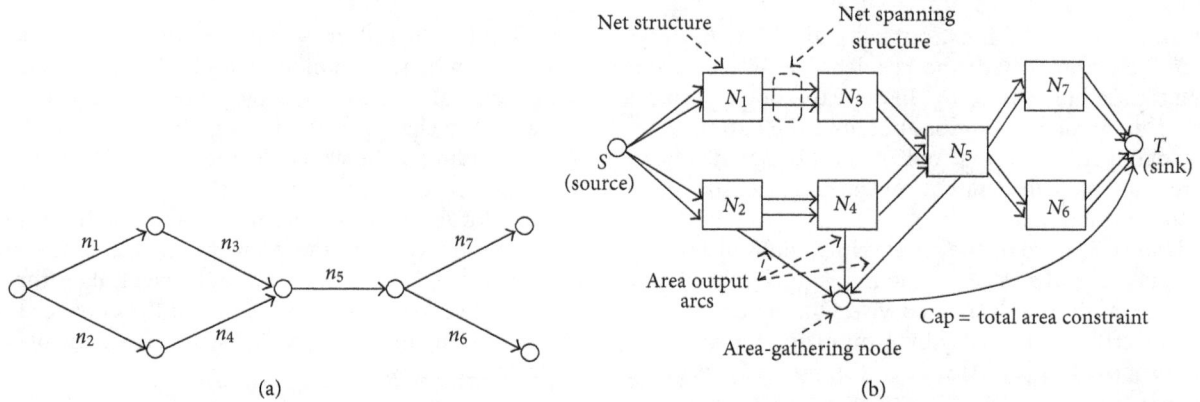

FIGURE 1: (a) Nets in $\mathscr{P}_\alpha$ of a circuit. (b) The corresponding SSG, which includes net structures and net spanning structures. N_i is the net structure corresponding to net n_i. Each net structure will send a flow of amount equal to the total cell size it chooses through the area output arc to the area-gathering node. All these flows are routed to the sink through an arc from the area-gathering node with a capacity equal to the given total cell area constraint.

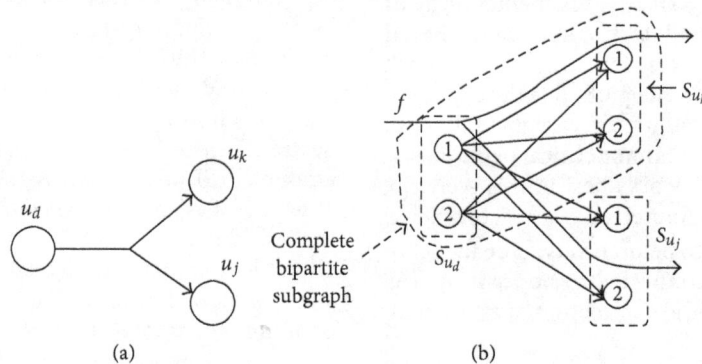

FIGURE 2: (a) A net in critical paths. (b) The net structure corresponding to the net; f is a flow through the net structure.

have to be chosen in a mutually exclusive manner (consistency in a sizing option set). Second, if a cell is connected to multiple nets, its sizing options will also be included in each of the corresponding net structures. In such cases, the selected sizing options for the cell must be consistent across all these net structures (consistency across net structures). As described in Section 5.2, the net spanning structure is designed to guide flows between net structures in the SSG to satisfy the second type of consistency. To guarantee the first type of consistency, two heuristic methods are proposed that prune flows corresponding to invalid option selections when determining the mincost flow; these are discussed in Section 6.

4.2. Handling Cumulative Constraints.

We define a *cumulative* metric as one, that is, the sum over all relevant circuit components (cells in our case) of a function $f(s_{i,j})$ of the chosen option (cell size in our case) $s_{i,j}$ of component/cell c_j. A cumulative constraint is then an upper-bound or lower-bound constraint on a cumulative metric. We tackle upper-bound cumulative metrics, specifically, total cell area and total leakage power in this paper.

To handle any given total cell area constraint, each net structure has an outgoing arc called the *area output arc*.

A *shunting structure* as explained in Section 5.3 is present in each net structure and connects the area output arc with the option nodes in the net structure. The function of the shunting structure is that when an option node is chosen, the shunting structure diverts a flow of amount that equals the chosen size to the area output arc from the incoming flow through the option node. All these flows are gathered at the *area-gathering node* as shown in Figure 1(b) which is connected to the sink. By setting the capacity of the arc between the area gathering node and the sink to be the given total cell area limit, we make sure the total selected cell area, which is equal to the total incoming flow amount to the area-gathering node, is smaller than or equal to the given limit. The shunting structure is discussed in detail in Section 5.3. Note that, as mentioned before, the sizing option of a cell can be contained in more than one net structure; in this case, the flow amount equal to its selected area will be sent to the area output arc in only one of the net structures that contain the cell.

A leakage power constraint is handled similarly, by having a flow equal to the leakage power of a cell corresponding to its chosen size option go through the shunting structure into a *power-gathering node*, and having an arc from this node to the sink with capacity equal to the leakage power upper-bound constraint.

Fast and Near-Optimal Timing-Driven Cell Sizing under Cell Area and Leakage Power Constraints Using a Simplified
Discrete Network Flow Algorithm

191

Algorithm FlowSize

(1) Construct a net structure as depicted in Figure 2 for each net in $\mathscr{P}_\alpha$.

(2) Determine the cost of each arc in the net structures, so that the change in timing cost is accurately incurred by flows through them.

(3) Connect net structures with net spanning structures that maintain consistency of size selection of common cells across multiple net structures; see Section 5.2.

(4) Add the shunting structure and an area output arc (Section 5.3) to each net structure to divert flow of amount equal to the selected size options (nodes) to the area output arc.

(5) Connect area output arcs to the area gathering node that has only one outgoing arc with capacity equal to the area-constraint to limit total selected cell area (= flow amount into the node).

(6) Determine a valid min-cost flow in the resulting SSG by applying the standard min-cost flow algorithm and the min-cost/max-flow heuristics (Section 6) iteratively.

(7) Select the size options chosen by the valid min-cost flow as cell sizes to obtain a near-optimal critical path delay for the circuit.

ALGORITHM 1: Network flow algorithm for the cell sizing problem for optimizing timing under a given area constraint.

The high-level pseudocode of our cell-sizing method FlowSize is given in Algorithm 1.

5. Further Details of the SSG

In this section, we discuss important details of the SSG, including determining arc costs and capacities, net spanning structures, and further details of the structures for constraint satisfaction.

5.1. Arc Cost Determination. To explain our arc cost formulation that accurately captures the timing cost change, we assume a lumped capacitance and resistance net delay model, which is widely used in cell sizing [2, 4, 17]. For net n_i, the delay $D(u_j, n_i)$ to a sink cell u_j of n_i is

$$D\left(u_j, n_i\right) = R_d \left(c \cdot L_{n_i} + \sum_{u_k \in \text{SC}(n_i)} C_{u_k} \right), \quad (3)$$

where R_d is the driving resistance of n_i, c is the unit WL capacitance, L_{n_i} is the WL of n_i, $\text{SC}(n_i)$ is the set of sink cells of n_i, and C_{u_k} is the input capacitance of cell u_k. In the pre-placement sizing stage, the WL of a net is usually estimated according to the fan-out number of the net. In the postplacement resizing stage, it can be estimated using one of several well-know models, for example, HPBB. With this delay model, for a critical net n_i with m critical sinks, the timing cost $t_c(n_i)$ of n_i is

$$t_c\left(n_i\right) = \frac{m}{S_a^\beta\left(n_i\right)} \cdot R_d \left(c \cdot L_{n_i} + \sum_{u_k \in \text{SC}(n_i)} C_{u_k} \right). \quad (4)$$

In the above expression, let us denote the coefficient $m/S_a^\beta(n_i)$ as α. The parameters affected by cell-sizing options are R_d and C_{u_k}. Hence, if a term in the formula includes R_d, its value is determined by the size of u_d, and if a term includes C_{u_k}, its value is determined by the size of u_k. For example, the term $\alpha R_d \cdot C_{u_k}$ is determined by choices in sizes of both u_d and u_k, and the value change of this term $\Delta \alpha R_d C_{u_k}(S_{u_d}^l, S_{u_k}^m)$

when the two sizing options $S_{u_d}^l$ and $S_{u_k}^m$ are chosen can be written as

$$\Delta \alpha R_d C_{u_k}\left(S_{u_d}^l, S_{u_k}^m\right)$$
$$= \alpha \left[R_d\left(S_{u_d}^l\right) \cdot C_{u_k}\left(S_{u_k}^m\right) - R_d\left(S_{u_d}'\right) \cdot C_{u_k}\left(S_{u_k}'\right) \right], \quad (5)$$

where $R_d(S_{u_d}^l)$ is the driving resistant corresponding to the driver cell u_d with size option $S_{u_d}^l$, $C_{u_k}(S_{u_k}^m)$ is the input capacitance of u_k with size option $S_{u_k}^m$, and S_{u_d}' and S_{u_k}' are the original sizes of u_d and u_k, respectively. The term $\alpha R_d \cdot c \cdot L_{n_i}$ is determined only by the size of u_d, and its value change $\Delta \alpha R_d c L_{n_i}(S_{u_d}^l)$ when the option $S_{u_d}^l$ is selected is

$$\Delta \alpha R_d c L_{n_i}\left(S_{u_d}^l\right) = \alpha \cdot c \cdot L_{n_i}\left(R_d\left(S_{u_d}^l\right) - R_d\left(S_{u_d}'\right)\right). \quad (6)$$

We denote the set of arcs between S_{u_d} and S_{u_k} as $E(S_{u_d}, S_{u_k})$, where u_d is the driving cell, and u_k is a sink cell. A valid flow will pass through only one arc in each such arc set, since only one option in S_{u_d} and S_{u_k} can be meaningfully chosen. Hence, in order to make the valid flow cost equal to the change of the timing cost, the cost of an arc in an arc set is set as the sum of the changes of all terms in the timing cost function that is functions of the cell-size options represented by the arc. Furthermore, if a term in t_c is a function of only the size of one cell v, then we arbitrarily choose an arc set $E(S_v, S_w)$ among all those connected to S_v, and the term value change determined by each S_v^l is added to all arcs starting from option node S_v^l in the arc set.

Thus, the value change of term $\alpha R_d C_{u_k}$ is included in the cost of the arc set $E(S_{u_d}, S_{u_k})$. Specifically, the cost of an arc $(S_{u_d}^l, S_{u_k}^m)$ in the set includes $\Delta \alpha R_d C_{u_k}(S_{u_d}^l, S_{u_k}^m)$. The value change of term $\alpha R_d \cdot L(n_i)$ is a function only of the driving cell size and thus is included in the cost of arcs in only one arc set $E(S_{u_d}, S_{u_j})$, where u_j is an arbitrary chosen sink cell of n_i. Specifically, the cost of an arc $(S_{u_d}^l, S_{u_j}^m)$ in the set includes $\Delta \alpha R_d c L_{n_i}(S_{u_d}^l)$.

Finally, the size change of a cell in a critical net also affects the timing cost of noncritical nets that are connected to the cell. Instead of also constructing net structures for these affected noncritical nets, we use a much simpler method, which is including the timing cost changes of noncritical nets in the net structures of their connected critical nets. Let u be a cell in $\mathcal{P}_\alpha$. We have two cases with respect to u and any noncritical net (a net not in $\mathcal{P}_\alpha$) it may be connected to: (1) if u is the driver of a noncritical net n_j, the timing cost change of n_j is $(R_d(S_u^l) - R_d(S_u^l))(c \cdot L_{n_j} + C_{\text{load}}(n_j))/S_a^\beta(n_j)$, where S_u^l is the chosen option of u, S_u^l is the initial size of u, and $C_{\text{load}}(n_j)$ is the total load capacitance of n_j. (2) Otherwise, if u is a sink cell of a noncritical net n_j, the timing cost change of n_j is $R_d(C_u(S_u^l) - C_u(S_u^l))/S_a^\beta(n_j)$. Let $\Delta t_c^u(S_u^l)$ denote the total timing cost change of all noncritical nets connected to u. If $u = u_d$ is the driving cell of the critical net n_i, $\Delta t_c^u(S_u^l)$ is included in arcs in one arc set $E(S_{u_d}, S_{u_j})$, where, again, u_j is the arbitrarily chosen sink cell of n_i. Otherwise, if $u = u_k$ is a sink cell, $\Delta t_c^u(S_u^l)$ is included in the arc set $E(S_{u_d}, S_{u_k})$. Specifically, the cost of an arc $(S_{u_d}^m, S_{u_k}^l)$ includes $\Delta t_c^u(S_u^l)$.

To sum up, for an arc $(S_{u_d}^l, S_{u_k}^m)$ in a net structure, if $u_k = u_j$ (u_j is the chosen sink in whose arc set $E(S_{u_d}, S_{u_j})$ cost, that is, costs of the arcs in this set, and we include the change in the terms in F_t that are only dependent on the cell size of driver u_d), its cost $(S_{u_d}^l, S_{u_k}^m)$ is

$$\text{cost}\left(S_{u_d}^l, S_{u_k}^m\right) = \Delta\alpha R_d C_{u_k}\left(S_{u_d}^l, S_{u_k}^m\right) + \Delta\alpha R_d c L_{n_i}\left(S_{u_d}^l\right)$$
$$+ \Delta t_c^{u_d}\left(S_{u_d}^l\right) + \Delta t_c^{u_k}\left(S_{u_k}^m\right). \tag{7}$$

Otherwise if $u_k \neq u_j$, its cost is

$$\text{cost}\left(S_{u_d}^l, S_{u_k}^m\right) = \Delta\alpha R_d C_{u_k}\left(S_{u_d}^l, S_{u_k}^m\right) + \Delta t_c^{u_k}\left(S_{u_k}^m\right). \tag{8}$$

We should note that our cost formulation is accurate under the assumption that the delay is linearly proportional to the load capacitance change. Only under such assumption, for a multiple fanout net, we can sum up the cost for the size change of each fanout to obtain the total delay change of the net. Unfortunately, in modern libraries with small feature sizes, the delay of a cell usually shows a nonlinear relationship with load capacitance. To handle this issue of a nonlinear delay function (as a function of capacitive load) in the ISPD'12 library (where the non-linearity is very pronounced), we determine in each iteration of the mincost flow computation (see Section 6), a min-square error linear approximation around the current design point.

5.2. Maintaining Consistency across Net Spanning Structures.
As we mentioned before, the net spanning structure is designed to maintain consistency across net structures. Note that if multiple nets have a common cell, their net structures must be connected in the SSG by net spanning structures.

We first consider the situation in which a cell u is a sink cell of a net n_i, and the driving cell of k nets $n_{j_1}, \ldots, n_{j_k}$ ($k \geq 1$); see Figure 3(a). The size option set S_u is present in all the net structures corresponding to these connected nets. We connect these net structures by adding arcs from

each option node in S_u of N_i to the equivalent option node (indicating the same size choice) in the S_u's of $N_{j_1}, \ldots, N_{j_k}$, where N_i is the net structure corresponding to net n_i. The resulting spanning structure is shown in Figure 3(b). With this structure, it is easy to see that if one size option of u is chosen by the flow through N_i, then this flow will also pass through the equivalent option nodes in $N_{j_1}, \ldots, N_{j_k}$. Thus, the required consistency is maintained.

The other situation is that the common cell u is the sink cell of more than one net $n_{i_1}, \ldots, n_{i_l}$ ($l > 1$) and the driver of at least one net n_j, as shown in Figure 3(c). In this situation, we will treat each N_{i_m} ($1 \leq m \leq l$) individually and make the connections between each N_{i_r} and N_j as stated in the first situation. However, in this case the spanning structure cannot guarantee the consistency of the option selected for u in $N_{i_1}, \ldots, N_{i_l}$. We use a *mincost* heuristic to tackle this problem; this is described in Section 6.

The costs of all arcs in the spanning structure are 0.

5.3. Structures for Cumulative Constraint Satisfaction.
In this subsection we discuss further details on the structures for satisfying the cell area constraint. As described earlier in Section 4.2, a leakage power constraint is handled by a similar structure.

As we mentioned before, for each net structure, there is an *area output arc*. Once a sizing option node is chosen, a flow with an amount equal to the chosen size needs to be sent to this arc—this flow is then sent to the sink via a common *area-gathering node* and its capacity-constrained arc in order to satisfy the total cell area constraint. The simplest way to achieve this is adding to each option node an arc leaving the net structure to the area output arc with a capacity equal to the option size. Then if enough flow comes into the option node, the desired amount will be sent to the area output arc.

An example is shown in Figure 4. A flow f selects two sizing options $S_{u_d}^l$ and $S_{u_k}^m$ and incurs the corresponding cost of arc $(S_{u_d}^l, S_{u_k}^m)$ in a net structure. $A(S_u^l)$ is the size of an option S_u^l. At each of the two option nodes, a branch of flow diverges a part of f to the area output arc with amount equal to the corresponding option sizes.

However, with this structure, the amount of flow that leaves a net structure is dependent on the option selected and is thus a variable. This is not desirable for determining the capacities of arcs in the spanning structures—the capacities of arcs in a spanning structure from net structure N_i to N_j need to be a constant, that is, independent of the size choices made in N_i by the flow entering N_i. We thus need a structure to diverge a constant amount of the flow that enters N_i; this amount is $\sum_{u \in n_i} A_{\max}(S_u)$, where $A_{\max}(S_u)$ is the maximum size of options in S_u.

Our complete structure for diverting flows to the area output arc is shown in Figure 5. In the structure, the capacity of the arc leaving towards the sink (called the *leaving arc*) from each option node in an option set S_u is set to be $A_{\max}(S_u)$. Therefore, the amount of flow leaving the net structure from any option node $\in S_u$ is always $A_{\max}(S_u)$. However, not all this amount is sent to the area output arc. A *shunting node* is connected to each of these leaving arcs

Fast and Near-Optimal Timing-Driven Cell Sizing under Cell Area and Leakage Power Constraints Using a Simplified
Discrete Network Flow Algorithm

193

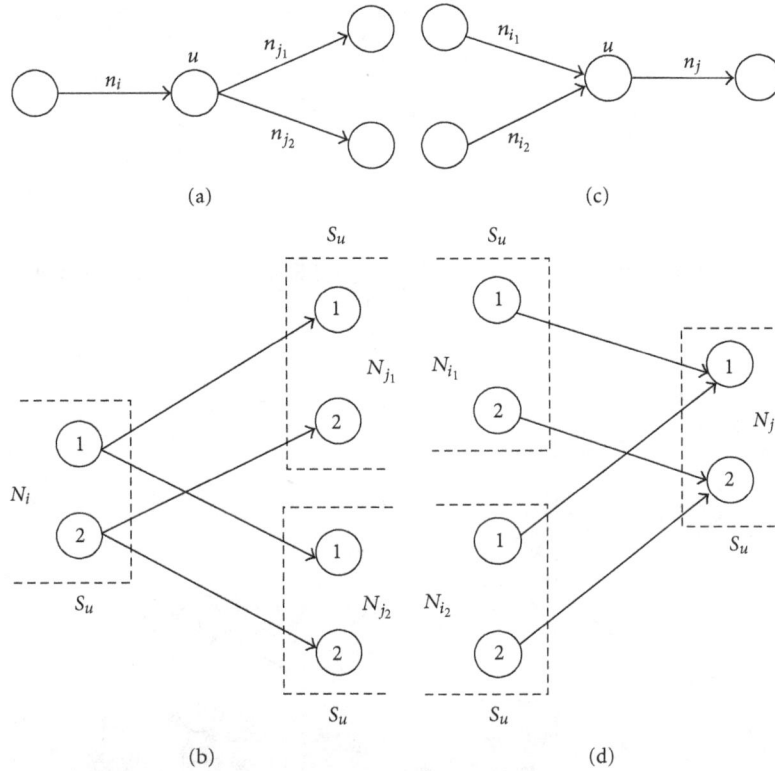

FIGURE 3: The net spanning structure. (a) The common cell u is a sink cell of one net n_i, and the driver of multiple nets n_{j_1} and n_{j_2}. (b) The corresponding spanning structure of (a), where N_k is the net structure corresponding to net n_k. (c) u is a sink cell of multiple nets n_{i_1} and n_{i_2} and the driver of n_j. (d) The corresponding spanning structure of (c).

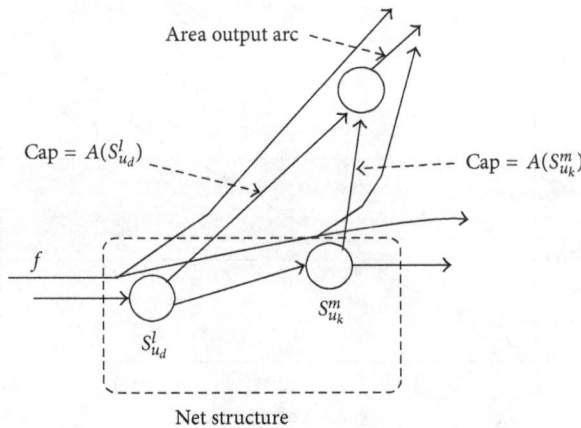

FIGURE 4: Flows to the area output arc of a net structure.

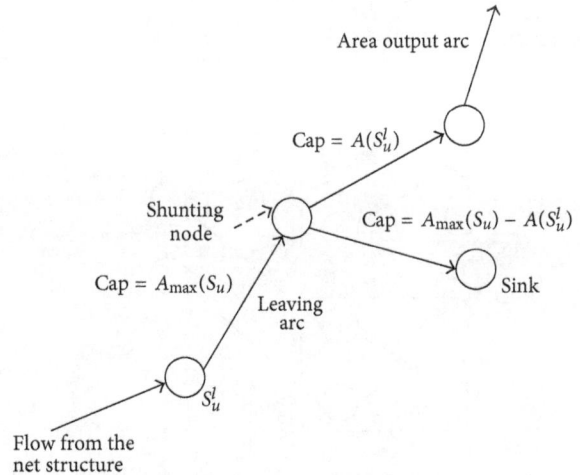

FIGURE 5: The shunting structure for outgoing flows from a net structure.

and divides the flow into two parts, one to the area output arc and the other one *shunted* (i.e., sent directly) to the sink. If the leaving arc is from an option node S_u^l, the capacity of the arc between the shunting node and the area output arc is then $A(S_u^l)$, and the capacity of the arc to the sink is $A_{\max}(S_u) - A(S_u^l)$. Therefore, if S_u^l is chosen, the amount of flow sent to the area output arc is exactly $A(S_u^l)$, and the rest of the amount $A_{\max}(S_u) - A(S_u^l)$ is shunted to the sink. In this way, we send the correct amount of flow into the area output arc of each net structure N_i and also make the total amount

of flow leaving N_i to the sink be $\sum_{u \in n_i} A_{\max}(S_u)$. The costs of all arcs in this structure between an option node and the area output arc are 0.

5.4. *Arc Capacity Determination.* Setting proper capacities for arcs is very important for the correct functioning of the SSG. The capacities should be set such that (1) sufficient flow can be sent to each net structure in the SSG to meet

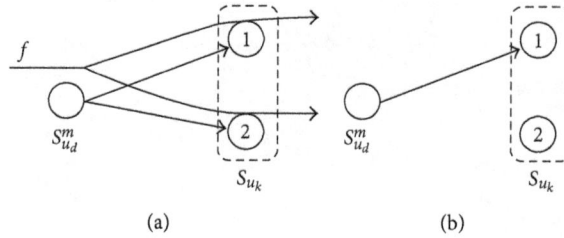

FIGURE 6: (a) Invalid flow f through two sizing options of the same cell u_k. (b) Selecting only one option of S_{u_k} for the next iteration based on some criteria of the previous "split flow."

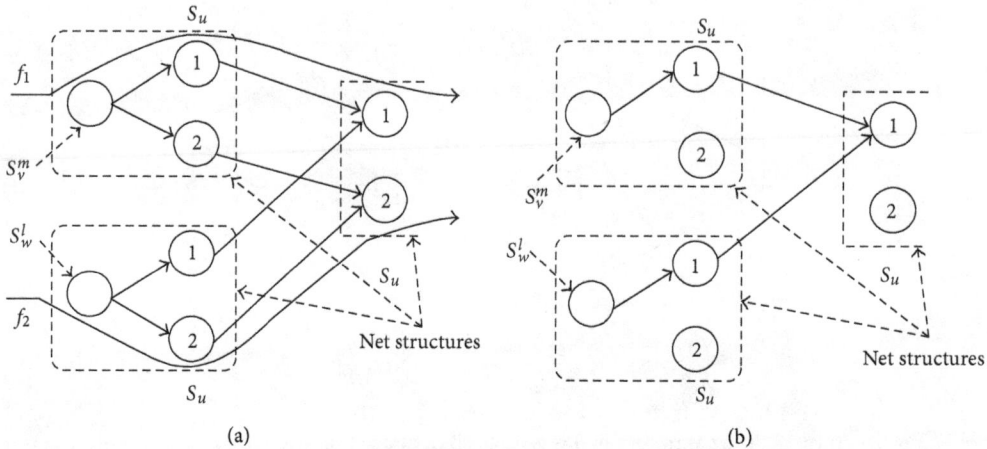

FIGURE 7: (a) Inconsistent option selections of the same cell u in different net structures. f_1 chooses option S_u^1; f_2 chooses option S_u^2. (b) The same option S_u^1 of u is selected in the next iteration for all net structures contains u as a sink cell based on some criteria of the previous flow of (a).

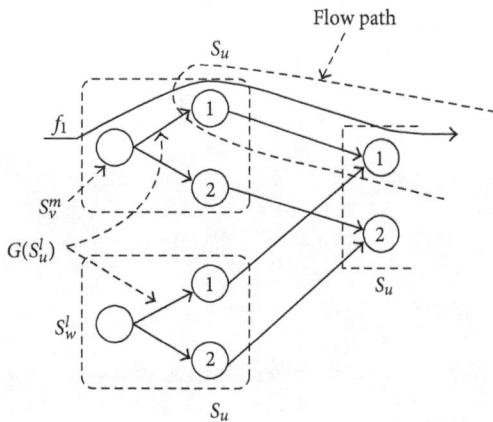

FIGURE 8: The flow path for determining the path cost of S_u^1, and the arcs for determining the arc cost of S_u^1.

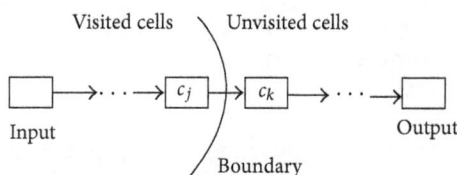

FIGURE 9: Visited cells, unvisited cells, and the boundary between them.

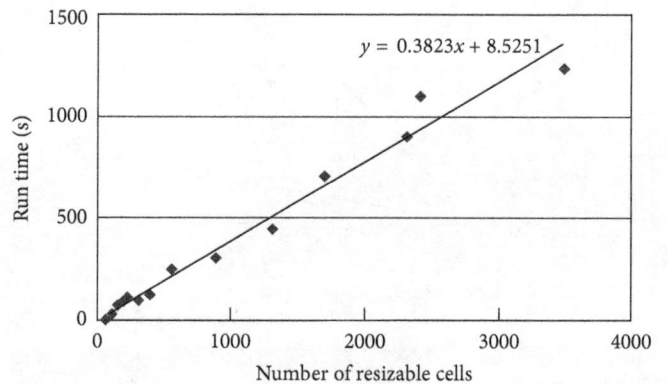

FIGURE 10: Run time versus number of resizable cells.

the flow demand on its area/power output arc; (2) within a net structure, the total incoming flow can be distributed to all sink and the driver cell option sets.

In order to determine the arc capacity, we first determine how much incoming flow is needed for each net structure. A net structure has two types of outgoing flows: (1) into the area output arc for area constraint satisfaction and (2) supply flow to its child net structures. The first type of outgoing flow has a fixed amount $\sum_{u \in n_i} A_{\max}(S_u)$ for a net structure N_i, irrespective of the chosen sizing options, as discussed in Section 5.3. The incoming flow amount must be sufficient to

Fast and Near-Optimal Timing-Driven Cell Sizing under Cell Area and Leakage Power Constraints Using a Simplified Discrete Network Flow Algorithm

195

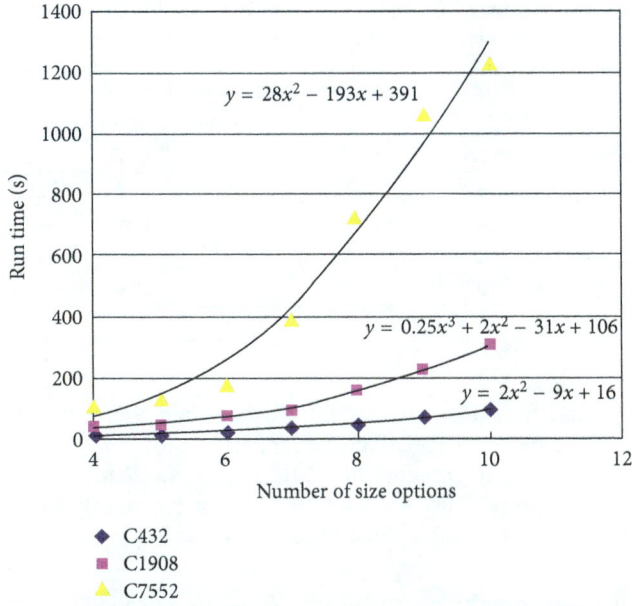

FIGURE 11: Run time versus number of available size options for each cell for circuits C432, C1908, and C7552. The approximation functions for the empirical time complexity of our method on these circuits are shown.

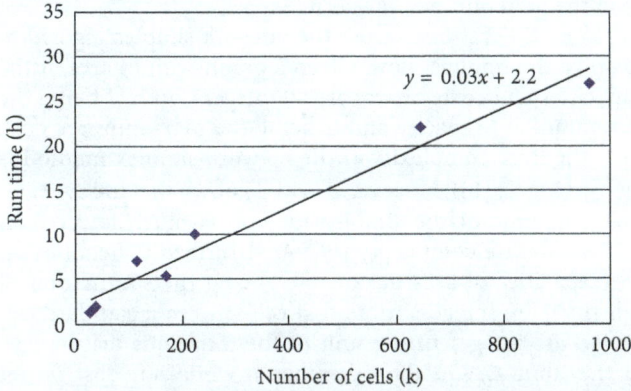

FIGURE 12: Run time versus number of cells for ISPD 2012 benchmark circuits.

cover the two outgoing flows, and thus the required incoming amount $f_{\text{in}}(N_i)$ of a net structure N_i is recursively given as

$$f_{\text{in}}(N_i) = \sum_{u \in n_i} A_{\max}(S_u) + \sum_{N_j \in \text{child}(N_i)} \frac{f_{\text{in}}(N_j)}{d_{\text{in}}(N_j)}$$

if N_i has any child net structure (9)

$$f_{\text{in}}(N_i) = \sum_{u \in n_i} A_{\max}(S_u),$$

otherwise (N_i) is a "leaf" net structure,

where child (N_i) is the set of child net structures of N_i, and $d_{\text{in}}(N_j)$ is the incoming degree of N_j, that is, the number of parent net structures of N_j. In (9), we assume that the

required flow amount for a net structure is sent uniformly from all its parent net structures. The determination of the flow needed in each structure starts from the boundary condition of "leaf" net structures given in (9) that are directly connected to the sink. The incoming flow amount needed for these net structures is the total of their first type of outgoing flows. Starting from the leaf net structures, we visit other net structures in a reverse topological order and determine their required incoming flow amount according to the formulation in (9). The total flow $F(S)$ to be supplied from the source node S is then

$$F(S) = \sum_{N_i \text{ is a "root" net structure}} f_{\text{in}}(N_i), \qquad (10)$$

where a root net structure is a net structure of a net that is driven by an I/O cell in the circuit and thus has no "parent" net in the circuit; in the SSG, the parent of all root net structures is S.

After obtaining f_{in} for each net structure, we can determine its arc capacities as follows.

(i) For an arc in the net spanning structure from N_i to N_j, its capacity is $f_{\text{in}}(N_j)/d_{\text{in}}(N_j)$, which equals the flow amount sent from N_i to N_j according to (9). This makes the total incoming flow amount to a net structure N_i exactly $f_{\text{in}}(N_i)$. As mentioned before, the cost of this arc is 0.

(ii) Within a net structure, the capacities of arcs in each arc set $E(S_{u_d}, S_{u_k})$ are the same as is derived below; u_d is the driving cell and u_k is any sink cell of the corresponding net.

For an arc set $E(S_{u_d}, S_{u_k})$, if S_{u_k} is not connected to any arc in the outgoing spanning structure of N_i, the capacity of each arc in it is set as $A_{\max}(S_{u_k})$, so that sufficient flow can be sent to the leaving arc for constraint satisfaction. Otherwise, let $N_{j_1}, \ldots, N_{j_t}$ be the child net structures of N_i that S_{u_k} is connected to (via spanning structures). Note that this means that u_k is a common cell in nets n_i and $n_{j_1}, \ldots, n_{j_t}$. Then, the capacity of each arc in the set is set to be $A_{\max}(S_{u_k}) + \sum_{r=1}^{t} f_{\text{in}}(N_{j_r})/d_{\text{in}}(N_{j_r})$.

(a) *Unit Flow Arc Cost.* When we gave the costs of arcs in a net structure in Section 5.1, we assumed that any flow on the arc will incur the cost. However, in a standard network flow graph, the cost of a flow on an arc is determined as the flow amount multiplied by the unit flow cost of the arc. With the above capacity assignment, a valid flow will always be a full flow on each arc it passes through in a net structure, since the summation of the arc capacity of a single arc in each arc set is equal to the incoming flow amount, and a valid flow uses exactly one arc in each arc set. In order to incur the same cost for a valid flow as determined in Section 5.1, the unit flow cost $C(S_{u_d}^l, S_{u_k}^m)$ of an arc $(S_{u_d}^l, S_{u_k}^m)$ is set to be the corresponding arc cost given in Section 5.1 divided by its capacity as determined above, that is,

$$C\left(S_{u_d}^l, S_{u_k}^m\right) = \frac{\text{cost}\left(S_{u_d}^l, S_{u_k}^m\right)}{\text{cap}\left(S_{u_d}^l, S_{u_k}^m\right)}. \qquad (11)$$

6. Finding a Valid (Discretized) Mincost Flow

The standard mincost network flow solves a linear programming problem (a continuous optimization method). Hence, it cannot automatically handle the consistency (mutual exclusiveness) requirement in a size option set for a particular cell, which may result in an invalid mincost flow for size selection. Therefore, after one iteration of a mincost network flow process, in a net structure, the obtained mincost flow may pass through an option node $S_{u_d}^m$ for the driving cell u_d and then to two or more sizing options in S_{u_k} for a sink cell u_k as shown in Figure 6(a). Furthermore, as explained in Section 5.2, the resulting flow may also violate the consistency requirement for the size option selection across net structures that have a common sink cell.

In the above two cases, we will start a new iteration of the network flow process by pruning out some options that lead to an invalid flow based on certain criteria of the flow so that a near-optimal size selection is obtained. In the new iteration, for the first case, $S_{u_d}^m$ will only connect to one of the option nodes in S_{u_k} that had flow through them in the first iteration as shown in Figure 6(b); the selection criterion is discussed shortly. Similarly for the second case, in all net structures whose corresponding nets have u as a sink cell, the chosen driving cell options in the first iteration, for example, S_v^m and S_w^l as shown in Figure 7(b), will only connect to the same sizing option of u selected from those that had flow through them in the first iteration. In this way, the same invalid flow will not occur in the second iteration.

We have used two alternative selection heuristics to choose a good option node from an illegal selection in each option set S_u to be part of the new iteration.

(i) Max-flow heuristic: always choose the option node with the largest flow amount through it.

(ii) Mincost heuristic: for the first situation, starting from $S_{u_d}^m$, follow the path of each branch of the mincost flow up to a length of l, and choose the option node that is on the branch that has the mincost path.

For the second situation, for each currently selected option of u, the cost that we use to determine whether it is a good option consists of two parts, output path cost and incoming arc cost. As shown in Figure 8, similar to the first situation, the outgoing path cost of an option S_u^l of u is the cost of the flow path starting from the option node. The incoming arc cost of an option S_u^l of u is the total cost of the set of incoming arcs $G(S_u^l)$ to all S_u^l's across all net structures that contain the option set S_u; see Figure 8. The summation of these two costs is a good estimation of the cost of a valid flow that chooses only option S_u^l for cell u. Hence, we choose the option with smallest summation of the path cost and the arc cost. Note that due to run time consideration we cannot always follow each branch flow to the sink. We thus set a limit l (in number of net structures) on the length of paths we follow.

The percentage timing improvement of four representative circuits in Table 1 for the ISCAS85 benchmarks reveals that the mincost heuristics with path length limits of 2,

TABLE 1: Percentage timing improvements of max-flow heuristic and min-cost heuristic with different path length limits.

Ckt	Max flow	Min cost of length			
		2	3	4	5
	%ΔT	%ΔT	%ΔT	%ΔT	%ΔT
C432	8.4	10.0	11.9	12.2	12.2
C1355	4.2	6.0	6.8	6.9	7.2
C3540	12.9	16.1	16.8	17.0	17.1
C7552	8.3	9.3	9.9	10.3	10.5
Average	8.4	10.4	11.4	11.6	11.7

3, 4, and 5 perform consistently better than the max-flow heuristic and have a relatively better performance in the range of 24–39%. The mincost heuristic is thus implemented in our algorithms, and we set $l = 3$ for a balance between computational complexity and accuracy.

6.1. Time Complexity of FlowSize. It is easy to see that, with the two pruning heuristics, if the option selection for a cell is inconsistent according to the mincost flow obtained in one iteration, in the following iterations the mincost flow will select a valid size option for the cell. Thus, the number of iterations required to reach a valid mincost flow is no more than the total number of cells in $\mathcal{P}_\alpha$.

In each iteration, we use the network simplex algorithm to solve the mincost flow. Given a graph with m arcs, if the capacities and costs of arcs are all integers, with U being the maximum arc capacity and C being the maximum arc cost, then the time complexity of the network simplex method is $O(Um^2 \log C)$ [18]; however, it is well known that the average-case run time of the simplex method is much lower than this worst-case complexity [19]. As described in Section 5.4, the capacities of arcs in our SSG, being the summation of cell sizes, are integers (note that cell sizes in a standard cell design are integer in the unit of the technique feature size of the library). On the other hand, while our cost is not integer, it can be converted to integer values by proper scaling. Hence though we do not actually do the scaling, we use this assumption here in order to derive an upper bound on the time complexity of our algorithm.

Let us first consider the total number of arcs in our SSG. It is dependent on the number of cells N, the number of nets M in the circuit, and the number of available sizes for each cell S. There are three types of arcs in our SSG: arcs between size option sets, arcs in net spanning structures, and arcs in shunting structures. The number of arcs between two size option sets is S^2, and in each net structure there are $d - 1$ sets of such arcs, where d is the degree of the corresponding net. Thus, the total number of these arcs in the SSG is $S^2(d_{avg} - 1)M$, where d_{avg} is the average degree of nets in a circuit. Since d_{avg} is usually no more than 4, the total number of arcs between size option sets is $O(S^2 M)$. The number of arcs in the shunting structure for each option node is three, and hence the total number is $3NS$. The net spanning structure only connects size option sets for the same cell in multiple net structures, and the number of arcs between two of such size option sets is S. Since the total number of size option sets is $O(N)$, the total number of arcs in the net spanning structure is thus $O(NS)$. To sum up, the total number of arcs in the SSG is $O(S^2 M + NS)$.

The maximum arc capacity is equal to the total cell size when all cells are chosen to be at their maximum width and thus is $O(N)$. The maximum arc cost is less than or equal to the maximum net delay. The delay of a net is dependent on the driving resistance of the driving cell, input capacitances of the sink cells, and the net fanout. Since the driving resistances and input capacitances of cells are constants specified by the library and the average fanout is usually a small constant in a VLSI circuit, C can be viewed as a constant.

Typically in a real circuit, N and M are about the same. Hence, the number of arcs in our SSG can be rewritten as $O(S^2 N)$. Therefore, the time complexity of each iteration is $O(S^4 N^3)$, and the total time complexity is then $O(S^4 N^4)$. The polynomially bounded time complexity is a highlight of our algorithm FlowSize, since other discrete cell sizing methods such as [3, 5] are not polynomially bounded in run time. Also, as we show in Section 8 (Figure 10), its actual run time reveals a much smaller complexity, that is, in keeping with the much smaller average-case run time of the network simplex algorithm compared to its worst-case complexity. Furthermore, our experiments show that the actual number of iterations needed in FlowSize is much less than the number of cells in $\mathscr{P}_\alpha$, for example, eight iterations for a circuit with 300 cells in $\mathscr{P}_\alpha$.

7. An Optimal Dynamic Programming Algorithm

Since we do not have the library or library parameters used in [3], we cannot directly compare our results for the benchmarks they use (ISCAS'85) to their published results for those benchmarks. Thus, we implemented an optimal dynamic programming (DP) method that can produce optimal solutions for the sizing problem of cells in $\mathscr{P}_\alpha$ and compared our solution quality to the optimal one.

In the DP algorithm we propose three pruning methods to reduce the number of partial solutions generated in the DP process that can maintain the optimality of the solution but greatly reduce the run time. We process cells in circuit topological order (from driver cells to sink cells). Each time we process a cell, a new set of partial solutions that involve the cell is generated by combining all partial solutions we have for previously processed cells with possible size choices of the cell. The pruning happens after new partial solutions are generated.

We propose three pruning conditions. A partial solution is pruned when (1) it fails to meet the area constraint; (2) it gives longer delay than the critical path delay produced by our method; (3) there is another better partial solution (generated in the search process or extracted from the complete solution of our method) that gives smaller total area, and better arrival time at the outputs of cells on the boundary of the visited region (connected to unvisited cells as shown in Figure 9) in both of the following cases: (a) the unvisited cells are all at their maximum sizes or (b) the unvisited cells are all at their minimum sizes.

The first two pruning methods obviously do not change the optimality of the exhaustive search method. For the third condition, let us first denote the arrival time at the output of a cell c_i as $A_o(c_i)$. Figure 9 shows a single path situation. c_j is the visited cell at the boundary of the visited region, and c_k is the unvisited cell connected to it. We have two partial solutions τ and τ', and τ' is a better solution according to our third pruning condition. Then for any complete solution τ_{comp} expanded from τ, we can also expand τ' to τ'_{comp} by choosing exactly the same sizes for unvisited cells in τ' as in τ_{comp}. Since τ' has smaller area than τ, if τ_{comp} meets the constraints, so does τ'_{comp}. Then the total delay at the output of c_j will be the same for both complete solutions. Since τ' is better than τ, we have $A_o^\tau(c_j) > A_o^{\tau'}(c_j)$, where $A_o^\tau(c_j)$ is the $A_o(c_j)$ value according to partial solution τ, and $A_o^{\tau'}(c_j)$ is the $A_o(c_j)$ value according to partial solution τ'. Hence, the total delay of the path for τ_{comp} > the total delay for τ'_{comp}. Therefore, τ cannot be expanded to an optimal solution. Thus, our third pruning method also does not negatively affect optimality of the method.

8. Experimental Results

We tested our algorithm on two sets of benchmarks, the ISCAS'85 suite, and the ISPD 2012 suite [21]. For the ISCAS'85 benchmark suite, we used two different libraries, a $0.18\,\mu m$ (180 nm) library and Synopsys's 90 nm library. We use the same industrial $0.18\,\mu m$ standard cell library as in [22], which provides four cell implementations for each function with different areas, driving resistances, input capacities, and intrinsic delays. The interval between the four available sizes $w_1 < w_2 < w_3 < w_4$ for each cell is increased about exponentially, that is, $(w_4 - w_3) \approx 2(w_3 - w - 2) \approx 4(w_2 - w_1)$. Other electrical parameters we use are unit length interconnect resistance $r = 7.6 \times 10^{-2}\,ohm/\mu m$ and unit length interconnect capacitance $c = 118 \times 10^{-18}\,f/\mu m$. For ISPD 2012 benchmark, it has its own artificial library, which has high nonlinear dependency between delay and load capacitance. Results were obtained on Pentium IV machines with 1 GB of main memory for ISCAS'85 benchmark and Xeon machine with 72 G memory for ISPD 2012 benchmark. Competing methods (the optimal DP method of Section 7 for the ISCAS'85 benchmarks and an approximate DP method [3] for the ISPD'12 benchmarks) were also run on the respective machines.

The ISCAS'85 benchmarks come with initial sizing solutions. We ran our algorithm with a 10% total cell area increase constraint, which means that the total cell area after cell sizing cannot increase more than 10% from the initial solution. The improvements compared to the initial solution obtained by our net work flow method and the optimal dynamic programming- (DP-) based exhaustive search method are listed in Table 2. Compared to the optimal solution from DP, the improvement obtained by our method is only 1% worse

TABLE 2: Results of our method and the optimal DP method. Four sizes are available for each cell. %ΔT is the percentage timing improvement, %ΔA is the percentage change of total cell area (negative value means deterioration).

Ckt	Number of cells	Number of crit. cells	DNF				Opt. DP			
			Delay (ns)	%ΔT	%ΔA	Runtime (sec)	Delay (ns)	%ΔT	%ΔA	Run time (sec)
C432	160	50	3.2	11.9	−9.9	9	3.1	13.2	−10.0	103
C499	202	51	3.5	11.9	−9.4	14	3.5	12.4	−9.3	119
C880	383	77	3.5	14.8	−9.9	19	3.4	16.2	−10.0	195
C1355	544	85	4.5	6.8	−8.1	22	4.4	7.0	−8.8	489
C1908	880	88	5.5	16.1	−9.8	39	5.4	17.0	−10.0	556
C2670	1.3 K	91	3.4	9.6	−9.5	38	3.4	11.1	−9.7	908
C3540	1.7 K	124	5.0	16.8	−9.0	69	4.8	19.0	−9.4	1724
C5315	2.3 K	138	5.3	10.0	−6.9	70	5.3	10.2	−7.9	3228
C6288	2.4 K	299	14.1	11.5	−8.1	97	14.1	11.7	−9.1	14998
C7552	3.5 K	199	7.3	9.9	−8.2	112	7.2	11.5	−7.5	6847
Average	1336	120		11.9	−8.9	48		12.9	−9.2	2916

TABLE 3: Results of our method and the optimal DP method with Synopsys 90 nm library [20].

Ckt	DNF				Opt. DP			
	Delay (ns)	%ΔT	%ΔA	Run time (secs)	Delay (ns)	%ΔT	%ΔA	Run time (secs)
C432	1.7	8.5	−9.1	10	1.7	11.7	−9.9	92
C499	1.0	12.8	−9.9	15	1.0	13.2	−9.9	144
C880	1.5	10.5	−9.8	17	1.4	14.7	−10.0	148
C1355	1.0	7.0	−8.8	40	1.0	8.1	−9.0	705
C1908	1.6	12.1	−10.0	72	1.5	16.2	−10.0	1024
C2670	1.3	9.4	−10.0	44	1.3	13.1	−10.0	884
C3540	2.2	13.5	−10.0	105	2.1	17.9	−10.0	1650
C5315	1.6	9.9	−9.4	280	1.5	11.8	−9.5	8410
C6288	6.0	14.2	−10.0	312	5.8	18.2	−10.0	25689
C7552	1.7	10.2	−9.8	364	1.7	13.8	−9.8	13204
Average		10.9	−9.7	126		13.9	−9.8	5195

(11.9% versus 12.9%); note that the solutions of both methods satisfy the 10% area increase constraint. Furthermore, our run time is 60X less than that of the exhaustive search method even with all three pruning conditions. Note again that we cannot compare our technique directly to that of [3], since we do not have their cell library or parameters.

We have also tested our algorithm using the more advanced and industry-like 90 nm library from Synopsys [20]. The results are given in Table 3. A similar trend to that of Table 2 is obtained is with the same initial sizing and 10% area relaxation, our method is only 3% worse (10.9% versus 13.9%) than the optimal solution. The run time we use is 40X less than the optimal one on average. We observe that there is a slightly increase in the optimality gap (from 1% to 3%) for 90 nm library compared to the 180 nm library. Our conjecture is that this increase is mainly due to the increased sensitivity in the 90 nm library; that is, for the same amount of cell size change, the relative delay change in the 90 nm library can be about 40% more than that of 180 nm library. Thus small errors in the optimal choice of cell sizes in near-optimal methods such as ours can lead to somewhat larger errors in the circuit delay results for libraries with larger sensitivity. However, we

should note that, even with such large sensitivity increase as mentioned above, our optimality gap increases by only 2%.

To show the scalability of our algorithm, two additional experiments were performed with the 180 nm library. In the first one, the sizes of all cells were considered for resizing rather than only cells in $\mathscr{P}_\alpha$. The results are listed in Table 4(a). Compared to only focusing on $\mathscr{P}_\alpha$, the average number of cells that are resizable is increased by 10 times from 120 to 1336, the run time is also increased by about 10 times from 48 secs to 525 secs, while the timing improvement % is an absolute of 2% better. The run time plot with respect to the number of resizable cells is shown in Figure 10, which best fits a linear function. This is much lower than the worst-case complexity we derived in Section 6, and in keeping with the well-known much smaller empirical time complexity of the network simplex method compared to its worst-case complexity [19].

In the second experiment, we expanded the cell library by adding six artificial size options with proportional driving resistances and input capacitances for each cell (three size options are added between w_3 and w_4 with uniform spacing between them, and the other three added options are made

Fast and Near-Optimal Timing-Driven Cell Sizing under Cell Area and Leakage Power Constraints Using a Simplified
Discrete Network Flow Algorithm

199

TABLE 4: (a) Results of our method when all circuit cells are considered for resizing with four sizes for each cell. (b) Results of our method when ten size options available for each cell; only cells in $\mathscr{P}_\alpha$ are resizable.

(a)

ckt	%ΔT	%ΔA	Run time (secs)
C432	12.9	−9.9	95
C499	13.5	−9.7	97
C880	14.9	−9.8	128
C1355	9.4	−9.5	249
C1908	19.9	−9.7	309
C2670	9.7	−9.4	449
C3540	22.6	−9.9	698
C5315	10.0	−6.9	901
C6288	13.2	−8.9	1097
C7552	12.1	−8.5	1229
Average	13.9	−9.2	525

(b)

ckt	%ΔT	%ΔA	Run time (secs)
C432	14.8	−9.7	72
C499	16.6	−9.8	140
C880	14.9	−9.2	144
C1355	13.3	−9.9	208
C1908	19.9	−9.1	344
C2670	12.7	−9.8	315
C3540	23.1	−9.7	527
C5315	10.8	−8.5	476
C6288	15.0	−9.5	698
C7552	14.9	−8.9	1087
Average	15.6	−9.4	401

larger than w_4. The intervals between the last three newly added size options are the same as between the first three newly added size options. We use linear approximation determined from the four options provided in the original library to calculate the driving resistances and input capacitances of the added options), so that each cell has ten different size options. The results with this larger library and cells in $\mathscr{P}_\alpha$ being resizable are shown in Table 4(b). With the larger library, we can only obtain the optimal results for the two smallest circuits using the exhaustive search method. Our results are only 2.5% worse (14.8% versus 17.3%) for circuit C432, and 2.0% worse (16.8% versus 18.8%) for C499 compared to the optimal solutions. The run times of our method are about 80X less than the exhaustive search method for C432 (72 secs versus 5343 secs) and over 135X less for C499 (140 secs versus 18993 secs). Compared to the four-option results in Table 2, our DNF method's run time is increased by about 7 times, while the timing improvement % is increased by an absolute of 3.7%. We also plot in Figure 11 the run time with respect to the number S of available size options for each cell for three representative circuits C432, C1908, and C7552. The plot for C1908 best matches a cubic function of S, and the plots for C432 and C7552 best match quadratic

functions, which are all consistent with the upper bound time complexity derivation given in Section 6 (that the run time is proportional to S^4). However, since, in current VLSI circuits, S and even S^4 are generally much smaller than N, the number of cells being re-sized, the dominant run time function is the one shown in Figure 10 that is linear in N.

Finally, we ran our DNF method on the complex ISPD 2012 contest benchmarks [21]. The sizes of the benchmarks range from 25 K to 959 K cells. For each type of cell, there are 30 different "sizes" (the sizes are combinations of actual sizes and different threshold voltages) in the library. For each circuit, the power constraint is determined by power optimizing the fast version of each ISPD benchmark using an internal tool (this tool uses the more complex DNF formulation modeled as a fixed-charge network flow problem that was alluded to in Section 2) that was part of the ISPD'12 competition and was in the top 6 (out of 17 teams) for 6 of the circuits; see [23]. Then, we perform our timing-driven sizing under this leakage power constraint. We run three rounds of our DNF-based sizing. In the first round, all cells in the circuit are sizable; this is needed as the ISPD'12 circuits also come with an initial sizing, and these correspond to very high delays and low power designs. In the second round, we perform further improvement by focusing on sizing only cells on critical and near critical paths (with delay ≥90% of the max path delay); note that this is the only round we do for the ISCAS'85 circuits. In the last round, we perform final adjustment, again using the DNF method, to only sub-paths in the critical paths that show delay reduction potential. The delay reduction potential is measured by the differences in the delay of the size selection s chosen in the second round, and the adjacent sizes of s, and the ratio of the corresponding power increases to the current positive leakage power slack of the circuit.

Both the circuit sizes of the ISPD'12 benchmark and the number of sizing options per cell are far beyond the capabilities of the optimal DP method (it runs out of memory for the smallest circuit). We thus compared our method to a state-of-the-art cell-sizing method in [3] (implemented by us), which also uses an approximate dynamic programming (DP)—it has a nonoptimal similarity-based partial solution pruning that can significantly reduce the number of partial solution generated. The results are shown in Table 5. As we can see, we achieve almost the same delay quality as the DP-based method (only 0.9% above it on the average) but use less than half of its run time. This highlights the efficiency of our method in achieving high quality results. We also show the run time plot with respect to the number of cells in Figure 12. Again, a linear scalability with respect to the number of cells is seen for our DNF method.

9. Conclusions

We presented a novel and efficient timing-driven network flow-based cell-sizing algorithm. We developed a size option selection graph, in which cell size options are modeled as nodes, and the cost of flows passing through various nodes is equal to the change in the timing objective function when the cell sizes corresponding to these nodes are chosen. Thus,

TABLE 5: Results of our DNF method and the approximate DP method of [3] for the ISPD 2012 cell-sizing benchmark suite. The "% Delay Impr." column shows the percentage difference of the delay between ours and the Approx. DP method (positive value indicates our method is better, negative indicates that the Approx. DP method is better).

Ckt	Number of of cells	Power Con-str. (w)	DNF				Aprrox. DP		
			Delay (ps)	Power (w)	Run time (h)	% Delay Impr.	Delay (ps)	Power (w)	Run time (h)
DMA	25 k	1.2	770	1.2	1.3	−0.7	765	1.2	2.8
pci_br32	33 k	0.9	676	0.8	2.0	−0.9	670	0.9	3.5
des_perf	111 k	7	795	7	7.1	−0.9	788	7	12.4
vga_lcd	165 k	0.7	673	0.7	5.6	0.6	677	0.7	14
b19	219 k	2.3	2126	2.3	10	−2	2084	2.3	19
leon	649 k	2.1	1632	2.1	22.2	−0.9	1616	2.1	48.4
Netcard	959 k	2.5	2218	2.5	27.2	−1.4	2186	2.5	58
Average					10.7	−0.9			22.6

by solving for a mincost flow, we can determine the cell sizes that can optimize the circuit delay. Various techniques are proposed to ensure that we can obtain, from the continuous optimization of standard mincost flow, a valid "discrete" mincost flow that meets the discrete mutual exclusiveness condition of cell-size selection. Area and leakage power constraint satisfaction is also taken care of by special network flow structures. The results show that the timing improvement obtained using our method is near optimal for ISCAS'85 benchmarks and similar to a state-of-the-art method for the ISPD'12 benchmarks (the near optimality could not be determined for these circuits as the optimal DP method ran out of memory for the smallest circuit) while being more than twice as fast as that technique. Furthermore, our technique scales well with problem size since its worst-case time complexity is polynomially bounded, and the empirical time complexity is linear.

Acknowledgment

This work was supported by NSF Grants CCF-0811855 and CCR-0204097.

References

[1] C. P. Chen, C. C. N. Chu, and D. F. Wong, "Fast and exact simultaneous gate and wire sizing by Lagrangian relaxation," *IEEE Transactions on Computer-Aided Design of Integrated Circuits and Systems*, vol. 18, no. 7, pp. 1014–1025, 1999.

[2] J. Fishburn and A. Dunlop, "Tilos: a posynomial programming approach to transistor sizing," in *Proceedings of International Conference on Computer-Aided Design*, pp. 326–328, 1985.

[3] S. Hu, M. Ketkar, and J. Hu, "Gate sizing for cell library-based designs," in *Proceedings of the 44th ACM/IEEE Design Automation Conference (DAC '07)*, pp. 847–852, June 2007.

[4] F. Beeftink, P. Kudva, D. Kung, and L. Stok, "Gate-size selection for standard cell libraries," in *Proceedings of the IEEE/ACM International Conference on Computer-Aided Design (ICCAD '98)*, pp. 545–550, November 1998.

[5] O. Coudert, "Gate sizing for constrained delay/power/area optimization," *IEEE Transactions on Very Large Scale Integration (VLSI) Systems*, vol. 5, no. 4, pp. 465–472, 1997.

[6] M. M. Ozdal, S. Burns, and J. Hu, "Gate sizing and device technology selection algorithms for high-performance industrial designs," in *Proceedings of the IEEE/ACM International Conference on Computer-Aided Design (ICCAD '11)*, pp. 724–731, November 2011.

[7] S. Dutt and H. Ren, "Discretized network flow techniques for timing and wire-length driven incremental placement with white-space satisfaction," *IEEE Transactions on Very Large Scale Integration (VLSI) Systems*, vol. 19, no. 7, pp. 1277–1290, 2011.

[8] H. Ren and S. Dutt, "A provably high-probability white-space satisfaction algorithm with good performance for standard-cell detailed placement," *IEEE Transactions on Very Large Scale Integration (VLSI) Systems*, vol. 19, no. 7, pp. 1291–1304, 2011.

[9] S. Dutt, H. Ren, F. Yuan, and V. Suthar, "A network-flow approach to timing-driven incremental placement for ASICs," in *Proceedings of the International Conference on Computer-Aided Design (ICCAD '06)*, pp. 375–382, November 2006.

[10] U. Brenner, "VLSI legalization with minimum perturbation by iterative augmentation," in *Proceedings of Design, Automation & Test in Europe Conference & Exhibition (DATE '12)*, pp. 1385–1390, March 2012.

[11] H. Ren and S. Dutt, "A network-flow based cell sizing algorithm," in *Proceedings of the 17th International Workshop on Logic & Synthesis*, pp. 7–14, 2008.

[12] S. Dutt and H. Ren, "Timing yield optimization via discrete gate sizing using globally-informed delay PDFs," in *Proceedings of the IEEE/ACM International Conference on Computer-Aided Design (ICCAD '10)*, pp. 570–577, November 2010.

[13] H. Ren and S. Dutt, "Effective power optimization under timing and voltage-island constraints via simultaneous VDD, Vth assignments, gate sizing, and placement," *IEEE Transactions on Computer-Aided Design of Integrated Circuits and Systems*, vol. 30, no. 5, pp. 746–759, 2011.

[14] H. Ren and S. Dutt, "Algorithms for simultaneous consideration of multiple physical synthesis transforms for timing closure," in *Proceedings of IEEE/ACM International Conference on Computer-Aided Design (ICCAD '08)*, November 2008.

[15] A. Nahapetyan and P. M. Pardalos, "A bilinear relaxation based algorithm for concave piecewise linear network flow problems," *Journal of Industrial and Management Optimization*, vol. 3, no. 1, pp. 71–85, 2007.

[16] R. K. Ahuja, T. L. Magnanti, and J. B. Orlin, *Network Flows: Theory, Algorithms, and Applications*, chapter 10-11, Prentice-Hall, Upper Saddle River, NJ, USA, 1993.

[17] D. Kim and P. Pardalos, "Gate sizing in MOS digital circuits with linear programming," in *Proceedings of the Conference on European Design Automation (EURO-DAC '90)*, pp. 217–221, 1990.

[18] R. K. Ahuja and J. B. Orlin, "Scaling network simplex algorithm," *Operations Research*, vol. 40, supplement 1, pp. S5–S13, 1992.

[19] I. Adler and N. Megiddo, "A simplex algorithm whose average number of steps is bounded between two quadratic functions of the smaller dimension," *Journal of the ACM*, vol. 32, no. 4, pp. 871–895, 1985.

[20] Synopsys 90 *nm* Library, http://www.synopsys.com/community /universityprogram/pages/library.aspx.

[21] ISPD 2012 cell sizing contest, http://www.ispd.cc/contests/12/ ispd2012_contest.html.

[22] X. Yang, B. K. Choi, and M. Sarrafzadeh, "Timing-driven placement using design hierarchy guided constraint generation," in *Proceedings of the IEEE/ACM International Conference on Computer Aided Design (ICCAD '02)*, pp. 177–180, November 2002.

[23] http://www.ispd.cc/contests/12/ISPD_2012_Contest_Results .pdf.

Permissions

The contributors of this book come from diverse backgrounds, making this book a truly international effort. This book will bring forth new frontiers with its revolutionizing research information and detailed analysis of the nascent developments around the world.

We would like to thank all the contributing authors for lending their expertise to make the book truly unique. They have played a crucial role in the development of this book. Without their invaluable contributions this book wouldn't have been possible. They have made vital efforts to compile up to date information on the varied aspects of this subject to make this book a valuable addition to the collection of many professionals and students.

This book was conceptualized with the vision of imparting up-to-date information and advanced data in this field. To ensure the same, a matchless editorial board was set up. Every individual on the board went through rigorous rounds of assessment to prove their worth. After which they invested a large part of their time researching and compiling the most relevant data for our readers. Conferences and sessions were held from time to time between the editorial board and the contributing authors to present the data in the most comprehensible form. The editorial team has worked tirelessly to provide valuable and valid information to help people across the globe.

Every chapter published in this book has been scrutinized by our experts. Their significance has been extensively debated. The topics covered herein carry significant findings which will fuel the growth of the discipline. They may even be implemented as practical applications or may be referred to as a beginning point for another development. Chapters in this book were first published by Hindawi Publishing Corporation; hereby published with permission under the Creative Commons Attribution License or equivalent.

The editorial board has been involved in producing this book since its inception. They have spent rigorous hours researching and exploring the diverse topics which have resulted in the successful publishing of this book. They have passed on their knowledge of decades through this book. To expedite this challenging task, the publisher supported the team at every step. A small team of assistant editors was also appointed to further simplify the editing procedure and attain best results for the readers.

Our editorial team has been hand-picked from every corner of the world. Their multi-ethnicity adds dynamic inputs to the discussions which result in innovative outcomes. These outcomes are then further discussed with the researchers and contributors who give their valuable feedback and opinion regarding the same. The feedback is then collaborated with the researches and they are edited in a comprehensive manner to aid the understanding of the subject.

Apart from the editorial board, the designing team has also invested a significant amount of their time in understanding the subject and creating the most relevant covers. They scrutinized every image to scout for the most suitable representation of the subject and create an appropriate cover for the book.

The publishing team has been involved in this book since its early stages. They were actively engaged in every process, be it collecting the data, connecting with the contributors or procuring relevant information. The team has been an ardent support to the editorial, designing and production team. Their endless efforts to recruit the best for this project, has resulted in the accomplishment of this book. They are a veteran in the field of academics and their pool of knowledge is as vast as their experience in printing. Their expertise and guidance has proved useful at every step. Their uncompromising quality standards have made this book an exceptional effort. Their encouragement from time to time has been an inspiration for everyone.

The publisher and the editorial board hope that this book will prove to be a valuable piece of knowledge for researchers, students, practitioners and scholars across the globe.

List of Contributors

A. Kishore Kumar
ECE Department, Hindustan College of Engineering & Technology, Coimbatore 641 032, India

D. Somasundareswari
Department of Electrical Sciences, Adithya Institute of Technology, Coimbatore 641 107, India

V. Duraisamy
Maharaja Institute of Technology, Coimbatore 641 407, India

T. Shunbaga Pradeepa
ECE Department, Coimbatore Institute of Technology, Coimbatore 641 014, India

Subodh Wairya
Department of Electronics Engineering, Institute of Engineering & Technology (IET), Lucknow 226021, India

Rajendra Kumar Nagaria and Sudarshan Tiwari
Department of Electronics and Communication Engineering, Motilal Nehru National Institute of Technology (MNNIT), Allahabad 211004, India

Roberta Piscitelli and Andy D. Pimentel
Computer Systems Architecture Group, Informatics Institute, University of Amsterdam, 1098 XH Amsterdam, The Netherlands

Diego Javier Reinoso Chisaguano and Minoru Okada
Graduate School of Information Science, Nara Institute of Science and Technology, 8916-5 Takayama, Ikoma-shi, Nara 630-0192, Japan

Maher Jridi
Vision Department, L@bIsen, ISEN–Brest, CS 42807, 29228 Brest Cedex 2, France

Ayman Alfalou
Vision Department, L@bIsen, ISEN–Brest, CS 42807, 29228 Brest Cedex2, France

Pramod Kumar Meher
Department of Embedded Systems, Institute for Infocomm Research, Singapore 138632

Yu-Ming Hsiao, Miin-Shyue Shiau, Kuen-Han Li, Jing-Jhong Hou, Heng-Shou Hsu, Hong-Chong Wu and Don-Gey Liu
Department of Electronic Engineering, Feng Chia University, Taichung 40724, Taiwan

Carna Radojicic and Christoph Grimm
Design of Cyber-Physical Systems, Kaiserslautern University of Technology, Postfach 3049, 67663 Kaiserslautern, Germany

Florian Schupfer and Michael Rathmair
Institute of Computer Technology, Vienna University of Technology, Gushausstraße 27-29, 1040 Vienna, Austria

Muhammad Martuza and Khan A. Wahid
Department of Electrical and Computer Engineering, University of Saskatchewan, Saskatoon, SK, Canada S7N 5A9

Shiwani Singh and B. P. Singh
Faculty of Engineering & Technology, MITS (Deemed University), Lakshmangarh 332311, India

Tripti Sharma and K. G. Sharma
Department of Electronics & Communication, Suresh Gyan Vihar University, Jaipur, India

Khaled Jerbi
IETR/INSA, UMR CNRS 6164, 35043 Rennes, France
CES Laboratory, National Engineering School of Sfax, 3038 Sfax, Tunisia

Olivier Deforges and Mickael Raulet
IETR/INSA, UMR CNRS 6164, 35043 Rennes, France

Mohamed Abid
CES Laboratory, National Engineering School of Sfax, 3038 Sfax, Tunisia

Lilia Zaourar
SOC Department, LIP6 Laboratory, University Pierre and Marie Curie, 4 Place Jussieu, 75252 Paris Cedex 05, France

Yann Kieffer
LCIS, Grenoble Institute of Technology and University of Grenoble, 50 Rue Barthelemy de Laffemas, 26000 Valence Cedex, France

Chouki Aktouf
DeFacTo Technologies, 167 Rue de Mayoussard, 38430 Moirans, France

Liyuan Liu
Tsinghua University, Beijing 100084, China
Institute of Semiconductors, Chinese Academy of Sciences, Beijing 100083, China

Dongmei Li and Zhihua Wang
Tsinghua University, Beijing 100084, China

L. Rakai, A. Farshidi, L. Behjat and D. Westwick
Department of Electrical and Computer Engineering, University of Calgary, 2500 University Drive NW Calgary, AB, Canada T2N 1N4

M. Bariani, P. Lambruschini and M. Raggio
Department of Biophysical and Electronic Engineering, University of Genova, Via Opera Pia 11 A, 16145 Genova, Italy

Ching-Hwa Cheng
Department of Electronic Engineering, Feng-Chia University, 100Wen-Hwa Road, Taichung, Taiwan

Sergio Saponara and Luca Fanucci
Department of Information Engineering, University of Pisa, Via G. Caruso 16, 56122 Pisa, Italy

Huan Ren and Shantanu Dutt
Department of ECE, University of Illinois at Chicago, Chicago, IL 60607, USA